NUREG-1924

# Electric Raceway Fire Barrier Systems in U.S. Nuclear Power Plants

Final Report

Manuscript Completed: May 2010
Date Published: May 2010

Prepared by:
G. Taylor and M.H. Salley

Office of Nuclear Regulatory Research

## ABSTRACT

In response to the 1975 Browns Ferry Nuclear Power Plant fire, the U.S. Nuclear Regulatory Commission (NRC) issued Appendix R to Title 10 of the *Code of Federal Regulations* Part 50. To support fire protection defense-in-depth 1- or 3-hour electric raceway fire barrier systems (ERFBS) were permitted for use as an acceptable method to protect electrical cables essential to fire protection safe shutdown capability. However, ERFBS were a new approach to fire barrier applications and as the initial installation of the ERFBS began, there was uncertainty regarding the ERFBS performance and definitive test standards for ERFBS qualification. Following review and research efforts, NRC resolved many concerns with ERFBS, including the fire resistance, ampacity derating, and seismic position retention. This report documents the history of various ERFBS and how U.S. commercial nuclear power plants use ERFBS for compliance. This report also documents the current state of the use of ERFBS and evaluates the effectiveness of these barriers in achieving adequate protection for nuclear power plants.

# TABLE OF CONTENTS

| Section | Page |
|---|---|
| ABSTRACT | iii |
| TABLE OF CONTENTS | v |
| LIST OF FIGURES | vii |
| LIST OF TABLES | viii |
| EXECUTIVE SUMMARY | xiii |
| ACKNOWLEDGMENTS | xv |
| ACRONYMS AND ABBREVIATIONS | xvii |

1. INTRODUCTION ............................................................................................................. 1-1

2. DEFENSE IN DEPTH AND THE ROLE OF ELECTRIC RACEWAY FIRE BARRIER SYSTEMS ........................................................................................................................ 2-1

3. ERFBS AND FIRE PROTECTION REGULATIONS ....................................................... 3-1

4. ERFBS FIRE RESISTANCE TESTING CRITERIA ........................................................ 4-1
   4.1 History of Testing Criteria .......................................................................................... 4-1
   4.2 Fire Endurance Rating ............................................................................................... 4-2
   4.3 Acceptance Criteria & Test Standards ...................................................................... 4-3

5. ELECTRICAL RACEWAY FIRE BARRIER SYSTEMS (ERFBSS) ............................... 5-1
   5.1 3M Interam™ E-50 Series & Rigid Panel System .................................................... 5-3
   5.2 Thermo-Lag 330-1   (Ablative Material) ................................................................... 5-21
   5.3 DARMATT KM-1   (Hydrate) ERFBS ...................................................................... 5-44
   5.4 Hemyc and MT   (Insulative and Insulative/Hydrate ERFBS) ................................. 5-50
   5.5 Mecatiss   (Insulative ERFBS) ................................................................................. 5-68
   5.6 Kaowool and FP-60   (Insulative ERFBS) ............................................................... 5-76
   5.7 Promat   (Hydrate) ................................................................................................... 5-88
   5.8 Pyrocrete   (Hydrate) ............................................................................................... 5-92
   5.9 Versawrap   (Hydrate/Insulative/Intumescent) ........................................................ 5-95
   5.10 Pabco   (Intumescent ERFBS) ............................................................................... 5-100
   5.11 Concrete   (Insulative ERFBS) ............................................................................... 5-101

6. PLANT SPECIFIC USAGE AND RESOLUTION OF ERFBS ISSUES ......................... 6-1

7. SUMMARY OF FINDINGS ............................................................................................. 7-1

8. CONCLUSION ................................................................................................................. 8-1

9. DEFINITIONS .................................................................................................................. 9-1

10. REFERENCES ............................................................................................................... 10-1

# TABLE OF CONTENTS
## (Continued)

**Section**         **Page**

Appendix A    The Browns Ferry Fire ................................................................................................ A-1

Appendix B    Ampacity Derating ....................................................................................................... B-1

Appendix C    Summaries of NRC Generic Communications on ERFBS ............................... C-1

Appendix D    Supplemental Test Result Summaries ............................................................... D-1

Appendix E    Fire Protection Regulations cited from 10 CFR 50 ........................................... E-1

Appendix F    Summary of GL 06-03 Responses ........................................................................ F-1

Appendix G    Additional Information on ERFBS Acceptance Criteria ................................... G-1

# LIST OF FIGURES

| Figure | | Page |
|---|---|---|
| 4-1. | ERFBS Fire Resistance Testing Acceptance Criteria Logic Diagram | 4-4 |
| 5-1. | 3M Interam E-54 Series ERFBS | 5-3 |
| 5-2. | Comparison of Butt and Finger Barrier Jointing Methods | 5-9 |
| 5-3. | Thermo-Lag 330-1 ERFBS Conduit Application 1- or 3-hour | 5-23 |
| 5-4. | Thermo-Lag 330-1 ERFBS Cable Tray Application 1- or 3-hour | 5-23 |
| 5-5. | SNL Full-Scale Thermo-Lag Test Article Shown in Various States | 5-34 |
| 5-6. | Photo of Two Layer Darmatt KM-1 System | 5-45 |
| 5-7. | (a) Sketch and (b) Photo of Hemyc Mat | 5-50 |
| 5-8. | Hemyc 1-hour ERFBS Conduit Construction | 5-51 |
| 5-9. | Hemyc 1-hour ERFBS Banding | 5-51 |
| 5-10. | Sectional View and Photos of Hemyc Joint Techniques | 5-52 |
| 5-11. | MT ERFBS Construction | 5-53 |
| 5-12. | Post-test Photo of Hemyc ERFBS Showing Shrinkage at Junction | 5-57 |
| 5-13. | MT Installations Process for NRC Testing | 5-65 |
| 5-14. | Photo of Mecatiss ERFBS | 5-68 |
| 5-15. | Test Assembly of 1- and 3-hour Thermo-Lag/Mecatiss Test | 5-70 |
| 5-16. | Picture of (a) Kaowool and (b) FP-60 Material | 5-76 |
| 5-17. | Sketch of Kaowool / FP-50 Installation | 5-77 |
| 5-18. | Fire Engulfing Cable Trays Clad with Kaowool. | 5-80 |
| 5-19. | Promat-H Cable Tray Protection | 5-88 |
| 5-20. | Promat-H Conduit Protection | 5-89 |
| 5-21. | Sketch of Layers Used in Versawarp ERFBS | 5-95 |
| 5-22. | Cut away of Versawrap ERFBS showing individual Layers | 5-95 |
| A-1. | Photograph of Conduit Damaged from Fire (NUREG/BR-0361) | A-1 |
| G-1. | Effects of Cable Mass on Cable Tray ERFBS Performance | G-6 |
| G-2. | Cable Tray System Weight vs. Endpoint Temperatures – Test 1 | G-7 |
| G-3. | Cable Tray System Weight vs. Endpoint Temperatures – Test 2 | G-8 |
| G-4. | Conduit System Weight vs. Endpoint Temperatures Measured on the Conduit | G-9 |
| G-5. | Conduit System Weight vs. Endpoint Temperatures Measured on the Bare Copper Conductor Inside the Raceway | G-10 |
| G-6. | Post-Fire Exposure – Hose Stream Test | G-12 |
| G-7. | Thermo-Lag 90 Degree Test Assembly | G-18 |
| G-8. | Isometric View of Typical Base Horizontal Test Assembly Prior to ERFBS Installation | G-19 |
| G-9. | 12' x 18' Horizontal Furnace (Top View) | G-19 |

# LIST OF TABLES

| Table | | Page |
|---|---|---|
| 1-1. | ERFBS Currently Used in U.S. NPPs | 1-2 |
| 4-1. | NFPA 251 and ASTM E-119 Temperature Time Curve Values | 4-6 |
| 5-1. | 3M Interam Design Comparison Old-to-New | 5-4 |
| 5-2. | 3M Interam E-50 Series Generic Installation Requirements | 5-7 |
| 5-3. | Summary of April 20, 1995 Fire Endurance Test | 5-11 |
| 5-4. | Summary of May 17, 1995 Fire Endurance Test | 5-12 |
| 5-5. | Summary of July 7, 1995 Fire Endurance Test | 5-12 |
| 5-6. | Peak Seals Test Results | 5-14 |
| 5-7. | Peak Seals Test Results with One Additional Layer | 5-14 |
| 5-8. | Peak Seals Test Results 1-Hour | 5-14 |
| 5-9. | 3M Combustibiltiy Test Results | 5-15 |
| 5-10. | Components of 3M Interam™ E54A ERFBS | 5-17 |
| 5-11. | 3M Interam™ E54A Fire Endurance Test Results | 5-17 |
| 5-12. | Summary of Peak Seals Upgrade Testing of 3-hr 3M Interam Design to GL 86-10 Supplement 1 Criteria | 5-18 |
| 5-13. | 3M Interam Combustibility Test Results Conducted at OPL | 5-18 |
| 5-14. | 3M Interam™ E54A Ampacity Results | 5-19 |
| 5-15. | Thermo-Lag 330-1 Confirmatory Order Documentation | 5-27 |
| 5-16. | NIST Results of ASTM E-1354 Thermo-Lag 330-1 Testing | 5-36 |
| 5-17. | Ampacity Derating Test Results - Thermo-Lag 330-1 | 5-42 |
| 5-18. | Darmatt KM-1 Specifications | 5-45 |
| 5-19. | KM-1 1-hr Fire Endurance Results (FTCR/94/0060) | 5-48 |
| 5-20. | Ampacity Results Faverdale (Test Report FTCR/96/0077) | 5-48 |
| 5-21. | One & four-inch RSC Ampacity Results 3-hour KM-1 | 5-49 |
| 5-22. | 600mm Cable Tray Ampacity Results (Test Report FTCR/96/0108) | 5-49 |
| 5-23. | Hemyc and MT Test Matrix (NRC) | 5-62 |
| 5-24. | Summary of NRC 1-hour Hemyc ERFBS Tests | 5-63 |
| 5-25. | Summary of NRC 3-hour MT ERFBS Tests | 5-66 |
| 5-26. | 1-hour FPC Mecatiss Testing in France | 5-72 |
| 5-27. | 3-hour Mecatiss Testing in France | 5-72 |
| 5-28. | Results of UL Mecatiss Testing | 5-74 |
| 5-29. | Kaowool ERFBS Test Results (10/24/1978) | 5-82 |
| 5-30. | Summary of UL FP-60 Fire Resistance Test Results | 5-84 |
| 5-31. | VCSNS Kaowool Testing Results | 5-85 |
| 5-32. | PROMAT Properties | 5-88 |
| 5-33. | List of Promat Test Reports | 5-90 |
| 5-34. | PROMAT-H Ampacity Derating | 5-90 |
| 5-35. | Results of Pyrocrete 241 Thermal Transmission | 5-94 |
| 5-36. | Summary of UL Testing of Versawrap | 5-97 |
| 5-37. | Summary of Omega Point Testing of Versawrap | 5-98 |

# LIST OF TABLES
## (Continued)

**Table** | **Page**

| Table | | Page |
|---|---|---|
| D-1. | 3M Interam E54A Ampacity Results 1" & 4" Conduits | D-3 |
| D-2. | 3M Interam™ E54A Ampacity Results 12" Wide Cable Tray | D-3 |
| D-3. | 3M Interam E54A Ampacity Results | D-4 |
| D-4. | 3M Interam™ E54A Ampacity Results 12" Wide Cable Tray | D-5 |
| D-5. | 3M Interam™ E54A Fire Endurance Testing 1", 3", & 5" RSC (OPL) | D-6 |
| D-6. | 3M Interam™ E54A Fire Endurance Testing Steel Cable Trays (OPL) | D-7 |
| D-7. | 3M Interam E54A Fire Endurance Test Conduits & Junction Boxes (OPL) | D-8 |
| D-8. | 3M Interam E54A Test Results (OPL Project No. 8610-102570) | D-9 |
| D-9. | 3M Interam E54A Fire Endurance Test Tray & Conduits (OPL) | D-11 |
| D-10. | E-50A 1-hr (SwRI 01-7912) | D-12 |
| D-11. | E-50A 1-hr (SwRI 01-7912a(1)) | D-13 |
| D-12. | UL Test of 3M E-50D 3-hr (R10125, 86NK2919) | D-14 |
| D-13. | E-50A 1-hr (SwRI 01-7912(2)) | D-15 |
| D-14. | UL Subject 1724 Test Results of E-50A 3-hr (R10125-3, 84NK23288) | D-16 |
| D-15. | UL Subject 1724 Test Results for E-50D 3-hr (R10125-3, 84NK2919) | D-17 |
| D-16. | UL Subject 1724 Test Results of E-50A 1-hr (R10125, 82NK21937) | D-17 |
| D-17. | UL Testing Results for E-50A 1-hr (R10125, 82NK21937) | D-18 |
| D-18. | E-50A UL Test Results (R10125, 82NK21937) | D-18 |
| D-19. | E-50A UL Test Results (R10125, 84NK23299) | D-19 |
| D-20. | E-54A 3M Test Results #86-78 | D-20 |
| D-21. | E-54A 3M Test Results #92-115 | D-20 |
| D-22. | E-54A 3M Test Results #87-82 | D-21 |
| D-23. | E-54A 3M Test Results (July 1992) | D-21 |
| D-24. | E-53A 3M Test Results #92-167 | D-22 |
| D-25. | E-54A 3M Test Results #92-141 | D-22 |
| D-26. | E-54A 3M Test Results #87-40 | D-23 |
| D-27. | E-53A & E-54A 3M Test Results #87-57 | D-23 |
| D-28. | 3M E-53A Test Results (June 1987) | D-24 |
| D-29. | Twin Cities Testing 3M E-50D Test Results | D-25 |
| D-30. | Twin Cities Testing 3M E-50 Series 1-hr Test Results | D-26 |
| D-31. | SwRI Ampacity Testing Results at 20°C | D-27 |
| D-32. | Ampacity Results SwRI Conduits | D-28 |
| D-33. | Ampacity Results SwRI Tray/Conduit | D-28 |
| D-34. | Ampacity Results SwRI without Baseline | D-29 |
| D-35. | 3M Ampacity Results Conduit | D-29 |
| D-36. | UL Ampacity Test Results for E-50A Cable Tray Configurations | D-30 |
| D-37. | UL Ampacity Test Results for E-54A 1-hour Cable Tray Configurations | D-30 |
| D-38. | UL Interpretations of Previously Completed Test Results | D-31 |
| D-39. | UL Test Report on M20-A 1-hr Cable Tray | D-32 |
| D-40. | UL Results for 3M M20-A/CS-195 1-hr Junction Box | D-32 |
| D-41. | UL Report on 3M Testing of 3/4, 2, 3-inch Steel Conduits | D-33 |

# LIST OF TABLES
## (Continued)

| Table | | Page |
|---|---|---|
| D-42. | UL M20-A 1-hour Test Results | D-33 |
| D-43. | 3M Test Results of M20-A Conduit 1-hr | D-34 |
| D-44. | UL 723 Test Results for FS-195 | D-35 |
| D-45. | 3M Results of Thermo-Lag upgraded with E-54A 3-hr 2" conduit | D-35 |
| D-46. | 3M Results of Thermo-Lag upgraded with E-54A 3-hr 1.5" conduit | D-36 |
| D-47. | TVA Ampacity Derating of 3M M20A | D-36 |
| D-48. | SwRI Ampacity Test Results for M20-A and CS-195 | D-37 |
| D-49. | NUMARC Thermo-Lag Test 1-1 | D-38 |
| D-50. | NUMARC Thermo-Lag Test 1-3 | D-39 |
| D-51. | NUMARC Thermo-Lag Test 1-4 | D-40 |
| D-52. | NUMARC Thermo-Lag Test 1-5 | D-41 |
| D-53. | NUMARC Thermo-Lag Test 1-6 | D-41 |
| D-54. | NUMARC Thermo-Lag Test 1-7 | D-42 |
| D-55. | NUMARC Thermo-Lag Test 2-1 | D-43 |
| D-56. | NUMARC Thermo-Lag Test 2-2 | D-44 |
| D-57. | NUMARC Thermo-Lag Test 2-3 | D-45 |
| D-58. | NUMARC Thermo-Lag Test 2-7 | D-46 |
| D-59. | NUMARC Thermo-Lag Test 2-8 | D-47 |
| D-60. | NUMARC Thermo-Lag Test 2-9 | D-48 |
| D-61. | NUMARC Thermo-Lag Test 2-10 | D-49 |
| D-62. | TU Electric Thermo-Lag Test 9-1 | D-50 |
| D-63. | TU Electric Thermo-Lag Test 9-3 | D-51 |
| D-64. | TU Electric Thermo-Lag Test 10-1 | D-52 |
| D-65. | TU Electric Thermo-Lag Test 10-2 | D-53 |
| D-66. | TU Electric Thermo-Lag Test 11-1 | D-54 |
| D-67. | TU Electric Thermo-Lag Test 11-2 | D-55 |
| D-68. | TU Electric Thermo-Lag Test 11-4 | D-56 |
| D-69. | TU Electric Thermo-Lag Test 11-5 | D-57 |
| D-70. | TU Electric Thermo-Lag Test 12-1 | D-58 |
| D-71. | TU Electric Thermo-Lag Test 12-2 | D-59 |
| D-72. | TU Electric Thermo-Lag Test 13-1 | D-60 |
| D-73. | TU Electric Thermo-Lag Test 13-2 | D-60 |
| D-74. | TU Electric Thermo-Lag Test 13-3 | D-62 |
| D-75. | TU Electric Thermo-Lag Test 14-1 | D-63 |
| D-76. | TU Electric Thermo-Lag Test 15-1 | D-64 |
| D-77. | TU Electric Thermo-Lag Test 15-2 | D-65 |
| D-78. | TVA Thermo-Lag Test 6.1.1 | D-66 |
| D-79. | TVA Thermo-Lag Test 6.1.2 | D-67 |
| D-80. | TVA Thermo-Lag Test 6.1.3 | D-68 |
| D-81. | TVA Thermo-Lag Test 6.1.4 | D-69 |
| D-82. | TVA Thermo-Lag Test 6.1.5 | D-70 |

# LIST OF TABLES
## (Continued)

| Table | | Page |
|---|---|---|
| D-83. | TVA Thermo-Lag Test 6.1.6 | D-71 |
| D-84. | TVA Thermo-Lag Test 6.1.7 | D-72 |
| D-85. | TVA Thermo-Lag Test 6.1.8 | D-73 |
| D-86. | TVA Thermo-Lag Test 6.1.9 | D-74 |
| D-87. | TVA Thermo-Lag Test 6.1.10 | D-75 |
| D-88. | TVA Thermo-Lag Test 6.1.11 | D-76 |
| D-89. | TVA Thermo-Lag Test 6.1.12 | D-77 |
| D-90. | TVA Thermo-Lag Test 6.1.13 | D-78 |
| D-91. | TVA Thermo-Lag Test 6.1.14 | D-79 |
| D-92. | TVA Thermo-Lag Test 6.1.15 | D-80 |
| D-93. | TVA Thermo-Lag Test 6.2.2 | D-82 |
| F-1. | Summary of GL 06-03 Responses | F-1 |
| G-1. | Effects of Cable Mass on ERFBS Thermal Performance | G-10 |
| G-2. | ERFBS Protected Conduit – External Raceway Surface vs. Internal Area Temperature Differential as a Function of Cable Mass | G-11 |
| G-3. | Summary of Acceptable Cable Insulation Testing Approach | G-14 |

# EXECUTIVE SUMMARY

In the United States, commercial nuclear power plants (NPPs) are designed with robust redundant safety systems, but the 1975 Brown Ferry Nuclear Plant (BFN) fire demonstrated the vulnerability of these redundant systems to fire damage from a single fire. At BFN, the fire damaged over 1600 cables, rendered numerous systems unavailable, and caused several systems to operate inadvertently. This event resulted in additional regulatory attention to fire protection aspects of NPP design. In response to this review, the U.S. Nuclear Regulatory Commission (NRC) issued new regulatory requirements backfit onto the licensee and developed to reduce the likelihood of a single fire causing damage to reactor safety systems required to safely shutdown and maintain the plant in a safe condition. As part of these new regulatory requirements, Section III.G.2 of Appendix R to 10 CFR Part 50 provided three prescriptive means to ensure redundant trains within a single fire area were protected from the effects of a fire. Two of the three approaches required the use of a 1- or 3-hour rated fire barrier, to protect equipment required for post-fire safe-shutdown.

Electrical cables are often the primary component requiring protection within a single fire area and the barriers used to protect cables became known as electric raceway fire barrier systems (ERFBS). ERFBS are widely used in U.S. commercial NPP applications to protect critical components (i.e., electrical cables) from a fire not promptly extinguished by the fire suppression activities. Employing ERFBS properly will ensure the safe shutdown of the reactor. Although ERFBS are not required where other fire protection features can provide adequate protection, there is a large fraction of NPPs that use ERFBS to meet the fire protection regulations as well as to provide the third level of protection in the fire protection defense-in-depth philosophy, with the first level being fire prevention, and the second level being rapid detection and suppression.

Numerous ERFBS vendors began developing systems that would provide the protection required by NRC fire protection regulations in the late 1970s and early 1980s. However, unclear guidance on the acceptable level of protection resulted in barriers from different vendors being qualified to different acceptance criteria making it difficult to evaluate the performance of the various barrier systems available. Clear acceptance criteria were eventually published and it soon became clear to NRC that several designs were not providing the required protection. This resulted in extensive time and effort by NRC staff and utilities to evaluate and confirm that the barriers used were capable of providing the required protection.

Beginning with the questionable test reports related to the initial design and testing of Thermo-Lag 330-1 in the early 1980's up until the most recent Generic Letter (GL) 2006-03 regarding Hemyc in 2006, the NRC has had a large role in the review, development and ultimate acceptance of the adequate use of ERFBS in NPPs to assure public health and safety. NRC staff review identified numerous deficiencies with several ERFBS designs and communicated these findings to the U.S. nuclear operating fleet and stakeholders via generic communications in the form of GLs, bulletins, and information notices (INs). However, review of ERFBS is not the only role of fire protection staff at NRC and in some instances, closure of ERFBS issues may have taken longer than should be expected from both NRC and the publics' perspective.

In the mid-2000's the U.S. Government Accountability Office (GAO) was asked to examine NRC oversight of fire protection at U.S. commercial NPPs. GAO documented its conclusion in a GAO report issued in June 2008 titled, "Nuclear Safety – NRC's Oversight of Fire Protection at U.S. Commercial Nuclear Reactor Units Could Be Strengthened, GAO-08-0747." The conclusions of the GAO report found it critical, in the opinion of the GAO, for the need of NRC to test and resolve the effectiveness of ERFBS installed at nuclear units. This NUREG presents a review of the effectiveness of the various ERFBS used in NPPs, including problems identified and methods used to resolve these deficiencies.

Today ERFBS are used in almost every NPP in the U.S. At this writing ERFBS are found in all but 12 of the 104 operational U.S. NPPs, adequately performing their passive fire protection function. In many NPPs more than one type of ERFBS is used. The rapid development and use of ERFBS resulted in regulatory attention to the proper testing, design, installation, maintenance, age management, and ability of the barrier to perform the desired function without affecting the operability and reliability of other structure, systems, and components important to safety.

This report attempts to provide the complete history of ERFBS use at the U.S. commercial NPPs. The history includes discussion of NRC fire protection defense-in-depth philosophy, development of NRC fire protection regulations, qualification testing criteria, information on the individual ERFBS products, and a review of ERFBS used at each NPP currently operating in the U.S. Its purpose is to provide a single document compiling information on the various ERFBS used in the United States. This report provides a description of regulatory requirements and ERFBS testing acceptance criteria, a detailed evaluation of each type of ERFBS, and a review of individual plants use of ERFBS. It also provides the regulatory footprint as to how the NRC achieved closure for various ERFBS issues. The report presents a history of the problems and benefits of using ERFBS to protect critical components. In addition to providing a historical perspective, each ERFBS is evaluated and an attempt has been made to identify the use and acceptance of the ERFBS. An electronic media (DVD) is included in the back cover of this NUREG to provide an understanding for both the construction and testing of ERFBS as well as informative video recordings to allow the reader to better understand the design construction, testing and operation of ERFBS.

The information presented has been collected from hundreds of publically available documents and interviews with staff involved with review of these issues. The vast amount of information available makes it evident that NRC and nuclear industry have undertaken a substantial amount of effort to ensure that ERFBS are performing their design function to ensure public health and safety. This report provides additional verification that there are no outstanding generic safety issues related to ERFBS known to NRC and past ERFBS deficiencies have been addressed or are in the process of resolution via the risk-informed performance-based approach outlined in 10 CFR 50.48(c), *National Fire Protection Association Standard NFPA 805*.

This report also shows that there is reasonable assurance that the ERFBS currently used in NPPs to provide protection for safe shutdown capability are sufficient for adequate protection of the public health and safety. In addition, the report shows that there are sufficient controls in place for future installations of new materials and industry inspections of the existing materials, to provide for safety of NPPs.

# ACKNOWLEDGEMENTS

The authors offer their thanks and appreciation to the many individuals who provided support and comments during the development of this report. First we acknowledge the contributions of Daniel Breedlove. Dan's hard work and diligence ensured that this report was published on time and include all available information, including past test report and various aspects of plant ERFBS activities.

We also thank H.W. 'Roy' Woods who provided invaluable guidance throughout this project as well as sharing his experiences with testing ERFBS. Mr. Stephen P. Nowlen of Sandia National Laboratories also provided supplemental discussions on the history of ERFBS testing that Sandia had conducted in the past. We must also acknowledge Mr. Patrick Madden of the Nuclear Regulatory Commission, although Pat is no longer as heavily involved in the Fire Protection arena as he once was, he was undoubtedly the most involved individual at NRC during the resolution of the Thermo-Lag issues in the mid- to late 1990's. Pats down to the point discussions provided great insights in translating the written history of ERFBS.

Finally, we would like to thank the internal and external stakeholders who took the time to provide comments and suggestions on the draft of this report when it was published in the Federal Register (74 FR 51621) on October 7, 2009. Those stakeholders who commented are listed and acknowledged below.

Daniel Frumkin, Nuclear Regulatory Commission, NRR

Ray Gallucci, Nuclear Regulatory Commission, NRR

Steven Nowlen, Sandia National Laboratories

Frank Wyant, Sandia National Laboratories

Wayne Harper, Energy Northwest

Patricia Campbell, GE Hitachi Nuclear Energy

Fred Madden, Luminant Power

John Butler, Nuclear Energy Institute

David Helker, Exelon Nuclear

Alan Holder, Progress Energy

Randal Brown, PCI Promatec

# ACRONYMS AND ABBREVIATIONS

| | |
|---|---|
| 3M | Minnesota Mining and Manufacturing |
| | |
| ADAMS | Agencywide Document Access and Management System |
| ADF | ampacity derating factor |
| ANI | American Nuclear Insurers |
| ANO | Arkansas Nuclear One |
| APCSB | Auxiliary and Power Conversion Systems Branch |
| ASTM | American Society of Testing and Materials |
| AWG | American Wire Gauge |
| | |
| BFN | Browns Ferry Nuclear Plant |
| BL | Bulletin |
| BSEP | Brunswick Steam Electric Plant |
| BTP | Branch Technical Position |
| BVPS | Beaver Valley Power Station |
| | |
| CPL | Carolina Power and Light Company |
| CFR | *Code of Federal Regulations* |
| CMEB | Chemical and Mechanical Engineering Branch |
| $CO_2$ | carbon dioxide |
| CPSES | Comanche Peak Steam Electric Station |
| CRD | control rod drive |
| CR-3 | Crystal River Unit 3 Nuclear Generating Station |
| CSPE | chlorosulphonated polyethylene |
| CSR | control spreading room |
| | |
| DAEC | Duane Arnold Energy Center |
| DBNPP | Davis Besse Nuclear Power Plant |
| DC Cook | Donald C. Cook Nuclear Power Plant |
| DCNPP | Diablo Canyon Nuclear Power Plant |
| | |
| EQ | environmental qualification |
| ERFBS | electric raceway fire barrier system |
| | |
| FCS | Fort Calhoun Station |
| FHA | fire hazards analysis |
| FNP | Joseph M. Farley Nuclear Plant |
| FPC | Florida Power Corporation |
| FSC | flame spread classification |
| FSSD | fire safe shutdown |
| FR | *Federal Register* |

# ACRONYMS AND ABBREVIATIONS
## (Continued)

| | |
|---|---|
| GAO | Government Accountability Office |
| GDC | generic design criteria |
| GGNS | Grand Gulf Nuclear Station |
| GL | Generic Letter |
| gpm | gallons per minute |
| GSU | Gulf States Utilities |
| | |
| HNP | Shearon Harris Nuclear Plant |
| HRR | heat release rate |
| | |
| IEEE | Institute of Electrical and Electronic Engineers |
| IN | Information Notice |
| IP-2, IP-3 | Indian Point Nuclear Generating Units 2 and 3 |
| ITL | Industrial Testing Laboratories |
| | |
| JAFNPP | James A. Fitzpatrick Nuclear Power Plant |
| | |
| LCS | LaSalle County Station |
| LER | licensee event report |
| | |
| MAERP | Mutual Atomic Energy Reinsurance Pool |
| | |
| NAPS | North Anna Power Station |
| NEI | Nuclear Energy Institute |
| NFPA | National Fire Protection Association |
| NIST | National Institute of Standards and Technology |
| NMS | Nine Mile Point Nuclear Station |
| NPP | nuclear power plant |
| NRC | Nuclear Regulatory Commission |
| NRR | Office of Nuclear Reactor Regulation |
| NUMARC | Nuclear Management and Resources Council |
| | |
| OBE | operating basis earthquake |
| OCGS | Oyster Creek Nuclear Generating Station |
| OIG | Office of Inspector General |
| OPL | Omega Point Laboratories |
| | |
| PE | polyethylene |
| PINGP | Prairie Island Nuclear Generating Plant |
| PNP | Palisades Nuclear Plant |
| PRA | probabilistic risk assessment |
| PVC | polyvinyl chloride |
| PVNGS | Palo Verde Nuclear Generating Station |
| | |
| QCNPS | Quad Cities Nuclear Power Station |

# ACRONYMS AND ABBREVIATIONS
## (Continued)

| | |
|---|---|
| RBS | River Bend Station |
| RES | radiant energy shield |
| RG | Regulatory Guide |
| RRS | required response spectra |
| RSC | rigid steel conduit |
| | |
| SCE&G | South Carolina Electric and Gas Company |
| SER | safety evaluation report |
| SFPE | Society of Fire Protection Engineers |
| SNL | Sandia National Laboratories |
| SONGS | San Onofre Nuclear Generating Station |
| SQN | Sequoyah Nuclear Plant |
| SRP | standard review plan |
| SSE | safe shutdown earthquake |
| STP | South Texas Project |
| SwRI | Southwest Research Institute |
| | |
| TET | thermal exposure threshold |
| TIA | Task Interface Agreement |
| TMI-1 | Three Mile Island Unit 1 |
| TSI | Thermal Science, Inc. |
| TU | Texas Utilities |
| TVA | Tennessee Valley Authority |
| | |
| UL | Underwriters Laboratories, Inc. |
| URI | unresolved item |
| U.S. | United States |
| | |
| VCSNS | Virgil C. Summer Nuclear Station |
| | |
| WBN | Watts Bar Nuclear Plant |
| | |
| XLPE | cross-linked polyethylene |

# 1. INTRODUCTION

On March 22, 1975, the Tennessee Valley Authority (TVA) Browns Ferry Nuclear Plant (BFN) experienced a serious fire in its cable spreading room (CSR) and Unit 1 reactor building. The fire burnt for over 7 hours and damaged over 1,600 electrical cables, rendering all of Unit 1 and many of Unit 2 Emergency Core Cooling Systems inoperable. This near-miss accident illustrated the vulnerability of essential electric cables to fire damage. In response to this fire, the U.S. Nuclear Regulatory Commission (NRC) issued Appendix R to Title 10 of the *Code of Federal Regulations* Part 50 (10 CFR 50) as a backfit to operating reactors and similar guidance was implemented on reactors under construction.

For compliance[1] via Appendix R, Section III.G.2, two of the three options for lack of separation of cables within a single fire area involve protecting cables and/or equipment that are needed for post fire safe shutdown or could cause maloperation of post-fire safe shutdown equipment. The authors of Appendix R envisioned classical fire-rated walls being installed to separate or protect these cables. In actual application of the regulation, often times, classical fire walls could not be installed and the need was to protect just a train / division of equipment located in the electrical raceway. This is the origin of the ERFBS.

An ERFBS is defined as a non-load-bearing partition type envelope system installed around electrical components and cabling that are rated by test laboratories in hours of fire resistance and are used to maintain safe-shutdown functions free of fire damage (RG 1.189). ERFBSs are used in NPPs to provide separation between redundant safety-related components and safe shutdown functions. ERFBS come in numerous designs and configurations, however, they all encase the component they are protecting to reduce the thermal exposure to the protected component during elevated fire conditions. Although the majority of ERFBS are used to protect electrical cables, there are some applications that use ERFBS to protect piping and other components important to safety. They provide fire resistance protection to one safe shutdown train in those fire areas that contain both trains. The objective of the safe-shutdown fire barrier is to ensure that a safe-shutdown train is conservatively protected from fire-related thermal insult. The necessity for ERFBS has been verified by probabilistic risk assessments (PRAs). These PRAs indicated that, even with fire barriers installed, fires are still major contributors to core melt probabilities.

In June 2008, the GAO issued its report titled, "NRC's Oversight of Fire Protection at U.S. Commercial Nuclear Reactor Units Could Be Strengthened, GAO-08-747." One conclusion identified the need for NRC to test and resolve the effectiveness of fire wraps[2] at NPPs. This report provides the history, effectiveness, and plant resolution of ERFBS (i.e., fire wraps).

Fire barriers are one level of protection used in fire protection programs to ensure the safety of the public and to protect the environment. Fire barriers are often employed to ensure that the plant can safely shut down in the event of a fire. ERFBSs are non-structural fire-rated assemblies that protect the electrical cables they enclose. In NPP applications, ERFBS are required to have a fire-resistance rating of either 1- or 3-hours, based on the specific application. One-hour ERFBSs require detection and automatic suppression to be installed within the same fire area. For some areas, licensees have requested exemptions to these

---

[1] Compliance can also be achieved through III.G.1, III.G.3, or through the exemption process (10 CFR 50.12)
[2] Fire wrap is synonymous with ERFBS

requirements based on the specific area configuration and low combustible loading. Exemptions are reviewed by the NRC staff under the agency's normal exemption process and the staff approves or disapproves the exemptions, as appropriate. ERFBSs in use at NPPs include Thermo-Lag 330-1, Darmatt, Hemyc, MT, Versa Wrap, Mecatiss, Pyrocrete, FP-60, Pabco, Promat, Cerablanket, Kaowool, and Minnesota Mining and Manufacturing (3M) Interam.

Table 1-1, provides a summary of the ERFBS use at individual NPP sites. The table is ordered by plant name alphabetically in left column and by barrier popularity along the top header row with the most popular to least popular barriers arranged from left to right. As is shown in the table, many plants use more than one type of ERFBS. Although the choice to use multiple ERFBS is site specific, some factors that may have influence the use of multiple barriers are costs, ease of installation, new product, technical problems with other barriers, better performance, etc. Section 5 provides the specific details and history of each ERFBS.

Table 1-1. ERFBS Currently Used in U.S. NPPs

| Plant Name | 3M Interam [1] | Thermo-Lag | Darmatt | Hemyc & MT | Mecatiss | Kaowool & FP-60 | Promat | Pyrocrete | VersaWrap | Pabco | Concrete |
|---|---|---|---|---|---|---|---|---|---|---|---|
| Arkansas Nuclear One Units, 1 and 2 | | X | | X | | | | | X | | |
| Beaver Valley Power Station, Units 1 and 2 | X | X | X | | | | | | | | |
| Braidwood Station, Units, 1 and 2 | X | | | | | | | | | | |
| Browns Ferry Nuclear Plant Units, 1, 2, and 3 | | X | | | | | | | | | |
| Brunswick Steam Electric Plant Units, 1 and 2 | X | | | | | | | | | | |
| Byron Station, Units, 1 and 2 | | | X | | | | | | | | |
| Callaway Plant, Unit 1 | | | X | | | | | | | | |
| Calvert Cliffs Nuclear Power Plant Units, 1 and 2 | | | | | | | | | | | |
| Catawba Nuclear Station, Units, 1 and 2 | | | | | X | | | | | | |
| Clinton Power Station, Unit 1 | X | X | | | | | | | | | |
| Columbia Generating Station | X | | X | | | | | | | | |
| Comanche Peak Steam Electric Station, Units, 1 and 2 | | X | | X | | | | | | | |
| Cooper Nuclear Station | | | | | | | | | | | |
| Crystal River Unit 3 Nuclear Generating Plant | | X | | | | X | | | | | |
| Davis Besse Nuclear Power Station, Unit 1 | X | | | | | | | | | | |

[1] 3M Interam includes CS-195 and FS-195

Table 1-1. ERFBS Currently Used in U.S. NPPs (Continued)

| Plant Name | 3M Interam [1] | Thermo-Lag | Darmatt | Hemyc & MT | Mecatiss | Kaowool & FP-60 | Promat | Pyrocrete | VersaWrap | Pabco | Concrete |
|---|---|---|---|---|---|---|---|---|---|---|---|
| Diablo Canyon Nuclear Power Plant, Units, 1 and 2 | X | | | | | | | X | | | |
| Donald C. Cook Nuclear Power Plant Units, 1 and 2 | X | X | X | | X | | | | | | |
| Dresden Nuclear Power Station, Units, 2 and 3 | X | | | | | | | | | | |
| Duane Arnold Energy Center | | | X | | | | | | | | |
| Edwin I. Hatch Nuclear Plant | | | | | | | X | | | | |
| Fermi-2 | X | | | | | | | | | | |
| Fort Calhoun Station, Unit 1 | X | | | | | | | X | | X | |
| Grand Gulf Nuclear Station | X | X | | | | | | | | | |
| H. B. Robinson Steam Electric Plant, Unit 2 | X | | | | | | | | | | |
| Edwin I. Hatch Nuclear Plant | | | | | | | X | | | | |
| Hope Creek Generating Station | | | | | | | | | | | |
| Indian Point Nuclear Generating, Units, 2 and 3 | X | | | X | | | | | | | |
| James A. Fitzpatrick Nuclear Power Plant | X | | | X | | X | | | | | |
| Joseph M. Farley Nuclear Plant | X | | | | | | X | | | | |
| Kewaunee Power Station | X | | | | | | | | | | |
| LaSalle County Station, Units, 1 and 2 | | | X | | | X | | | | | |
| Limerick Generating Station, Units, 1 and 2 | | X | X | | | | | | | | |
| McGuire Nuclear Station, Units, 1 and 2 | | | | X | | | | | | | |
| Millstone Power Station, Units, 2 and 3 | | | | | | | | | | | |
| Monticello Nuclear Generating Plant, Unit 1 | | | | | | | | | | | |
| Nine Mile Point Nuclear Station, Units, 1 and 2 | X | | | | | | | | | | |
| North Anna Power Station, Units, 1 and 2 | X | | | | | | | | | | |
| Oconee Nuclear Station, Units, 1, 2, and 3 | | | | | | | | | | | |
| Oyster Creek Nuclear Generating Station | X | X | | | X | | | | | | |

Table 1-1. ERFBS Currently Used in U.S. NPPs (Continued)

| Plant Name | 3M Interam [1] | Thermo-Lag | Darmatt | Hemyc & MT | Mecatiss | Kaowool & FP-60 | Promat | Pyrocrete | VersaWrap | Pabco | Concrete |
|---|---|---|---|---|---|---|---|---|---|---|---|
| Palisades Nuclear Plant | | | | | | | | | | | X |
| Palo Verde Nuclear Generating Station, Units, 1, 2, and 3 | | X | | | | | | | | | |
| Peach Bottom Atomic Power Station, Units, 2 and 3 | | X | X | | | | | | | | |
| Perry Nuclear Power Plant, Unit 1 | X | | | | | | | | | | |
| Pilgrim Nuclear Power Station | X | | | | | X | | | | | |
| Point Beach Nuclear Plant, Units, 1 and 2 | X | | | | | | | | | | |
| Prairie Island Nuclear Generating Plant, Units, 1 and 2 | X | X | X | | | | | | | | |
| Quad Cities Nuclear Power Station, Units, 1 and 2 | X | | X | | | | | | X | | |
| River Bend Station, Unit 1 | X | X | | | | | | | | | |
| R. E Ginna Nuclear Power Plant | | | | X | | | | | | | |
| Salem Nuclear Generating Station, Units, 1 and 2 | X | | | | | | | | | | |
| San Onofre Nuclear Generating Station, Units, 2 and 3 | X | | | | | X[2] | | | | | |
| Seabrook Station, Unit 1 | X | | | | | | | | | | |
| Sequoyah Nuclear Plant, Units, 1 and 2 | | X | | | | | | | | | |
| Shearon Harris Nuclear Power Plant, Unit 1 | X | X | | X | | | | | | | |
| South Texas Project, Units, 1 and 2 | | X | | | | | | | | | |
| St. Lucie Plant, Units, 1 and 2 | | X | | X | X | | | | | | |
| Surry Power Station, Units, 1 and 2 | X | | | | | | | X | | | |
| Susquehanna Steam Electric Station, Units 1 and 2 | | X | X | | | | | | | | |
| Three Mile Island Nuclear Station, Unit 1 | | X | | | X | | | | | | |
| Turkey Point Nuclear Generating, Units, 3 and 4 | | X | | | | | | | | | |
| Vermont Yankee Nuclear Power Station | X | | | | | | | | | | |
| Virgil C. Summer Nuclear Station, Unit 1 | X | | | | | X | | | | | |

[2] San Onofre uses Certablanket which is a similar product to Kaowool

Table 1-1. ERFBS Currently Used in U.S. NPPs (Continued)

| Plant Name | 3M Interam [1] | Thermo-Lag | Darmatt | Hemyc & MT | Mecatiss | Kaowool & FP-60 | Promat | Pyrocrete | VersaWrap | Pabco | Concrete |
|---|---|---|---|---|---|---|---|---|---|---|---|
| Vogtle Electric Generating Plant, Units, 1 and 2 | X | | | | | | | | | | |
| Waterford Steam Electric Station, Unit 3 | X | | | X | | | | | | | |
| Watts Bar Nuclear Plant, Unit 1 | | X | | | | | | | | | |
| Wolf Creek Generating Station, Unit 1 | | X | X | | | | | | | | |

Some licensees have installed ERFBS identified in this report as non-combustible radiant energy shields. A non-combustible radiant energy shield is a heat shield designed to provide protection of redundant essential raceways or fire safe shutdown equipment against the radiant energy heat transfer from an exposure fire inside containment. GL 86-10, Question 3.7.1 provides additional information on non-combustible radiant energy shields, as does IN 92-82 and 95-27. Although ERFBS have been used as non-combustible radiant energy shields, there are numerous other non-combustible materials and systems that have been used to construct these non-combustible radiant energy shields, many of which are plant and licensee specific. Due to these variables, non-combustible radiant energy shields are beyond the scope of this report.

## 2. DEFENSE IN DEPTH AND THE ROLE OF ELECTRIC RACEWAY FIRE BARRIER SYSTEMS

NPPs licensed to operate by the U.S. NRC use the defense-in-depth concept of echelons of fire protection features to achieve a high degree of fire safety.

Section II, "General Requirements" of Appendix R to 10 CFR Part 50 states that the fire protection program shall extend the defense-in-depth concept to fire protection in fire areas important to safety, with the following objectives:

- prevent fires from starting;
- detect rapidly, control, and extinguish promptly those fires that do occur; and
- provide protection for structures, systems, and components important to safety so that a fire that is not promptly extinguished will not prevent the safe shutdown of the plant.

The multiple layers of fire protection provided by the defense-in-depth concept provide reasonable assurance that weakness or deficiencies in any echelon will not present an undue risk to public health and safety. To achieve defense-in-depth, each operating reactor has an NRC-approved fire protection program which, when properly designed, implemented, and maintained, will satisfy Section 50.48, "Fire protection," of Title 10 of the *Code of Federal Regulations*, Part 50 (10 CFR 50.48).

The licensees have designed the fire protection programs by analyses that (1) consider potential fire hazards, (2) determined the effects of fires in the plant on the ability to safely shutdown the reactor or on the ability to minimize and control the release of radioactivity to the environment, and (3) specified measures for fire prevention, fire confinement, fire detection, automatic and manual fire suppression, and post-fire safe-shutdown capability.

To confine a fire and limit fire damage, licensees divide NPP buildings into separate fire areas. These are generally rooms or plant areas that have fire-rated walls and fire-rated floor-ceiling assemblies (structural fire barriers) having sufficient fire resistance rating to withstand the fire hazards located in the fire area and, as necessary, to protect important equipment within the area from a fire outside the area. This passive fire protection concept, which is called "compartmentation," is a fundamental fire safety measure at U.S. commercial NPPs.

The fire barriers, which accomplish their intended design function simply by being in place during a fire, are important because they are the first and last lines of defense against a fire. That is, during the early stages of a fire, the barriers protect important equipment until the fire detection and automatic fire suppression systems operate. In addition, in the unlikely event that an automatic fire protection system fails to operate, the barriers continue to provide passive fire protection.

NPP operations rely on electrical cables to power, control, and provide indication of systems and components. Licensees must design structures, systems and components important to safety to minimize the probability and effects of fire and explosions. To protect cables from the adverse effects of fire, licensees have and continue to use ERFBS.

ERFBS are a component specific application of the passive fire protection discussed above. The purpose of an ERFBS is to provide thermal protection to cables or other equipment important to the safe operation of the NPP.

Following the design, review, testing (if applicable) and installation of an ERFBS, the licensees maintain their ERFBS to identify and prevent barrier degradation. Licensees perform this function through their maintenance and surveillance procedures. This continual review of ERFBS functionality maintains the defense-in-depth aspect of the ERFBS, thus providing the reasonable assurance that the ERFBS provides the protection for safe shutdown capability sufficient for adequate protection of the public health and safety. In addition, the NRC staff performs numerous fire protection inspections at each of the operating plants in the U.S. to confirm the licensees are using and maintaining their fire protection features as documented in their fire protection plan or other licensing basis documents.

# 3. ERFBS AND FIRE PROTECTION REGULATIONS

During the early stage of NPP construction and operation licensing, fire protection compliance was implemented based on the performance objective of General Design Criterion (GDC) 3 in Appendix A to Title 10 of the *Code of Federal Regulations* (10 CFR) Part 50.

GDC 3 states,
> Structures, systems, and components important to safety shall be designed and located to minimize, consistent with other safety requirements, the probability and effect of fires and explosions. Noncombustible and heat resistant materials shall be used whenever practical throughout the unit, particularly in locations such as the containment and control room. Fire detection and fighting systems of appropriate capacity and capability shall be provided and designed to minimize the adverse effects of fires on structures, systems, and components important to safety. Fire fighting systems shall be designed to assure that their rupture or inadvertent operation does not significantly impair the safety capability of these structures, systems, and components. (Appendix A to 10 CFR 50)

GDC 3 set high-level goals for the fire protection program but did not provide specific implementation guidance. At the time, fire protection was largely based on compliance with local fire codes and with the requirements of insurance underwriters, since there were no specific regulatory requirements. As a result, fire protection was based largely on best practices as established from other industrial facilities including, in particular, fossil fuel power plants.

Following the Browns Ferry Nuclear Plant Fire[3] in 1975 and the subsequent inspections, fundamental changes in the regulatory approach to NPP fire protection was made. The first change was new guidance published in Branch Technical Position, Auxiliary and Power Conversion Systems Branch 9.5-1 (BTP APCSB 9.5-1) that established the "defense-in-depth" concept for fire protection. This concept involved a layered approach to fire protection. As discussed previously, the fire protection defense-in-depth principles are aimed at achieving the following objectives:

- Preventing fires from starting,

- Promptly detecting, controlling, and extinguishing those fires that do occur,

- Providing protection of structures, systems, and components important to safety to ensure that a fire not promptly extinguished by the fire suppression activities will not prevent the safe shutdown of the plant or result in release of radioactive materials to the environment.

It should also be mentioned that this "defense-in-depth" philosophy for fire protection actually came out of the Browns Ferry Special Review Group recommendations. (NRC IN 92-46, Attachment 1)

In the years following the Browns Ferry Nuclear Plant Fire, the NRC performed numerous inspections and re-evaluated the fire risks at NPPs and, in November 1980, NRC published a

---

[3] The Browns Ferry fire was the root cause for developing NPP fire protection regulations (10 CFR 50.48, and Appendix R). A brief overview of the Browns Ferry Fire is provided in Appendix A of this document.

new set of fire protection requirements as 10 CFR 50.48 and Appendix R to 10 CFR Part 50.[4] The new regulations imposed a minimum set of fire protection program and post-fire safe shutdown requirements on plants operating prior to January 1, 1979. The primary focus of the requirements establishes fire protection criteria for systems needed to safely shutdown and maintain the reactor in a safe condition in the event of a fire. As described in more detail below, the new Appendix R requirements were imposed on NPPs operating before January 1, 1979, with several sections applicable to newer NPPs. However, Section 50.48 allowed for exceptions to meeting the requirements of these three sections of Appendix R, if the licensee had already received NRC acceptance of fire protection features meeting Appendix A to BTP APCSB 9.5-1 or by a comprehensive fire protection safety evaluation prepared by the NRC staff. It is therefore important to understand that a NPPs licensing basis documents the requirements for determining individual licensee compliance with regulations.

BTP APCSB 9.5-1 was applicable to plants that were issued a construction permit after July 1, 1976, while Appendix A to BTP APCSB 9.5-1 was applicable to plants for which application for construction permits were docketed prior to July 1, 1976, and plants that were operating or were issued construction permits prior to July 1, 1976.

All facilities operating prior to January 1, 1979, were backfit to 10 CFR 50.48 Appendix R, Sections III.G, III.J, and III.O. In addition, all plants to receive their operating license after January 1, 1979, have license condition that satisfy specific requirements of Appendix R, including Section III.G for redundant trains located in a fire area. Section III.G.2 of Appendix R, which states three prescriptive options for ensuring one redundant trains located in the same fire area remain free of fire damage[5], is reproduced here:

> Section III.G.2 Except as provided for in paragraph G.3 of this section, where cables or equipment, including associated non-safety circuits that could prevent operation or cause maloperation due to hot shorts, open circuits, or shorts to ground, of redundant trains of systems necessary to achieve and maintain hot shutdown conditions are located within the same fire area outside of primary containment, one of the following means of ensuring that one of the redundant trains is free of fire damage shall be provided:
>
> a. Separation of cables and equipment and associated non-safety circuits of redundant trains by a fire barrier having a 3-hour rating. Structural steel forming a part of or supporting such fire barriers shall be protected to provide fire resistance equivalent to that required of the barrier;
>
> b. Separation of cables and equipment and associated non-safety circuits of redundant trains by a horizontal distance of more than 20 feet with no intervening combustible or fire hazards. In addition, fire detectors and an automatic fire suppression system shall be installed in the fire area; or
>
> c. Enclosure of cable and equipment and associated non-safety circuits of one redundant train in a fire barrier having a 1-hour rating, In addition, fire detectors and an automatic fire suppression system shall be installed in the fire area;

---

[4] 10 CFR 50.48 and Appendix R to Part 50 are reproduced in full in Appendix F and became effective on February 19, 1981.

[5] The technical basis for Appendix R states that "(i)f specific plant conditions preclude the installation of a 3-hour fire barrier to separate the redundant trains, a 1-hour fire barrier and automatic fire suppression and detection system for each redundant train will be considered the equivalent of a 3-hour barrier."

The underlying purpose of Section III.G of Appendix R is to ensure that where redundant trains are located in the same fire area at least one means of achieving and maintaining safe shutdown conditions will remain available during and after any postulated fire in the plant. Section III.G specifies three options for limiting fire damage so that one train of systems necessary to achieve and maintain hot shutdown remains free of fire damage. Two of the options rely on fire-rated barriers.

**Fire-Rated Cables**

In lieu of ERFBS to protect post-fire safe-shutdown circuits, some licensees have elected to use fire-rated cables. Fire rated cables are capable of withstanding high temperature (e.g., 927 °C [1700°F]), however such cables do not represent rated fire barriers and do not meet the requirements of 10 CFR 50, Appendix R, Section III.G.2, for fire rated barriers. The reason is that the cables themselves have the rating, not an external barrier. Use of fire rated cables in lieu of fire rated barriers should be in accordance with the applicable rules and the plant's approved fire protection program. For instance, pre-1979 plants would require an exemption from the requirements of Appendix R Sections III.G.2 or III.G.3, while post-1979 plants would require a deviation from the guidelines to allow use of a fire rated cable to meet the safety requirements. GL 86-10 provides information on the use of these cables in Question 3.7.1, "Radiant Energy Shield Fire Rating" and Question 8.10, "ASTM E-119 Design Basis."

On January 19, 2006, the NRC issued IN 2006-02 to inform addresses of a generic safety issue related to the use of a fire rated cable namely, Meggitt Si 2400 stainless-steel-jacketed. This IN identified a failure mode of the Meggitt cable where the cables exterior jacket contacts galvanized support members. The degradation was attributed to liquid metal embrittlement of the stainless steel cable jacket directly contacting the galvanized support material at high temperatures. This failure mode was not observed when the cable samples were not in direct contact with the galvanized material.

Examples of the use of these fire rated cables in presented in Section 6.11 Columbia Generating Station, Section 6.20 Farley Nuclear Plant, and Section 6.33 McGuire Nuclear Station. Although, other plants may use these fire rated cables, these cables are not the focus of this report, however, due to their close association with ERFBS, this information is provided for completeness. Therefore, an extensive survey of fire-rated cable use has not been conducted in the development of this report.

**National Fire Protection Association (NFPA) 805**

10 CFR 50.48(c) allows licensees to use the NFPA 805, "Performance-Based Standard for Fire Protection for Light Water Reactor Electric Generating Plants," on a voluntary basis, as a risk informed performance-based approach to fire protection at NPPs. As is discussed in this report, several plants that have indicated their interest to transition to NFPA 805 have ERFBS that are not in compliance with the requirements.

In general, the majority of plants with Hemyc or MT ERFBS installed have notified NRC of their intent to transition to NFPA 805, which they believe will allow for resolution of the Hemyc or MT ERFBS issues. For those licensees transitioning to NFPA 805, NRC staff has confirmed that appropriate compensatory measures are in place for deficient Hemyc materials during the quarterly inspections conducted by the resident inspectors (onsite inspectors). These compensatory measures will remain in place pending the completion of the transition.

If any of the NPP units intending to transition to NFPA 805 change their mind and do not transition, then they will have to submit exemption requests for approval to the NRC providing justification for the specific ERFBS configurations, or modify these configurations to come into compliance.

# 4. ERFBS FIRE RESISTANCE TESTING CRITERIA

## 4.1 History of Testing Criteria

When the NRC developed Appendix R to 10 CFR Part 50 and other fire protection guidance, there were no rated ERFBS or standardized tests for ERFBS. At that time there was only the ANI testing standard, which was intended for insurance purposes only. The NRC would review and accept an applicant's use of the American Nuclear Insurers (ANI) standard for qualification on a case-by-case basis. During the implementation phase of Appendix R, licensees became unclear as to the acceptance criteria for ERFBS.

Following issuance of the fire protection rule in 1980, NRC began receiving questions related to the implementation of the rule. NRC developed responses to these questions and presented them in draft form in 1984 at NRC sponsored regional workshops on implementation of NRC fire protection requirements at NPPs. In 1986, NRC issued the final form of these responses in GL 86-10, "Implementation of Fire Protection Requirements." Enclosure 2 to GL 86-10 provided the staff position on several questions raised by licensees: specifically, Question 3.2.1 asked NRC staff to clarify the origin of the 163°C (325°F) temperature rise criterion. Enclosure 2 to GL 86-10 provided the staff position on fire endurance or resistance test acceptance criteria for fire barrier cable-tray wraps (ERFBS), as follows;

> The acceptance criteria contained in Chapter 7, "Tests of Nonbearing Walls and Partitions," of National Fire Protection Association (NFPA) Standard 251, "Standard Methods of Fire Tests of Building Construction," was applicable to cable-tray fire wraps. These criteria stipulate that transmission of heat through the barrier "shall not have been such as to raise the temperature on its unexposed surface more than 139°C (250°F) above its initial temperature. It is generally recognized that 24°C (75°F) represents an acceptable norm. The resulting 163°C (325°F) cold side temperature criterion is used for cable tray wraps because they perform the fire barrier function to preserve the cables free of fire damage. It is clear that cable that begins to degrade at 232°C (450°F) is free of fire damage at 163°C (325°F).

Therefore, the origin of the 163°C (325°F) single point acceptance criteria was based on NPFA 251 and the American Society for Testing and Materials (ASTM) E-119, "Standard Test Methods for Fire Tests of Building Construction and Materials," testing standards, along with the thermal damage threshold of cables found in use at NPPs.

It is important to understand that when ERFBS started showing up in NPPs the ANI standard was the only available method for testing ERFBS and was developed for insurance purposes only. NFPA 251 and ASTM E-119 testing standards are used for qualifying traditional building members (walls, floors, beams, columns, etc.) under fire exposure conditions, and licensees were unclear as to how to apply those standards to ERFBS. The lack of an understanding and guidance on acceptable testing standards resulted in uncertainty as to the method of qualifying ERFBS. Most nuclear utilities and ERFBS manufacturers originally tested their ERFBS to the ANI Criterion. The ANI standard, as discussed below, has its deficiencies and following issuance of GL 86-10 and Supplement 1 to GL 86-10, the NRC no longer considered the ANI standard to be an acceptable method to NRC staff for qualifying ERFBS. However, these GLs were not retroactive staff positions, so licensees with approved ERFBS did not have to re-evaluate their barriers.

Following issuance of IN 91-47 and IN 91-79, the Texas Utilities (TU) Electric Company conducted their own fire endurance test program in the summer of 1992 to qualify their ERFBS following the guidance and acceptance criteria of ANI as specified in ANI Information Bulletin #5 (79), "ANI/MAERP Standard Fire Endurance Test Method to Qualify a Protective Envelope for Class 1E Electrical Circuits," July 1979. This ANI standard had been developed for insurance purposes only and provided a method that was acceptable to ANI for demonstrating that an ERFBS was capable of protecting Redundant Class 1E cables in the same fire area for particular qualification fire resistance duration.

Subsequent to several interactions between NRC and TU staff, NRC concluded that the licensees were uncertain as to whether the ANI test method established a level of fire-barrier performance equivalent to that established by the GL 86-10 acceptance criteria. In recognizing that the 1-hour and 3-hour fire rated ERFBS are unique and additional guidance on the proper implementation of GL 86-10 would be helpful, NRC issued Supplement 1 to GL 86-10, "Fire Endurance Test Acceptance Criteria for Fire Barrier Systems Used to Separate Redundant Safe Shutdown Trains Within the Same Fire Area," in 1994. This supplement provided acceptance criteria that were satisfactory to NRC for qualifying an ERFBS fire rating. Supplement 1 to GL 86-10, also included performance criteria based on the type of cable, and other factors to achieve an acceptable barrier without meeting the prescriptive test limits. This guidance was not applied retroactively to already approved plant-specific ERFBS.

The general approach for licensees to qualify an ERFBS is to evaluate ERFBS testing results and related data to ensure it applies to the conditions under which they intend to install the barriers. If test results are not available for specific applications, the licensees are encouraged to perform independent qualification testing to provide adequate results. If all configurations cannot be tested, then an engineering analysis must be performed to demonstrate that cables would be protected adequately during and after exposure to fire. Enclosure 2 to GL 86-10 also provided guidance for instances where exact replication of plant configurations could not be tested. This guidance stated that an exemption would not be required if the following five criteria are met:

1. The continuity of the fire barrier material is maintained.
2. The thickness of the barrier is maintained.
3. The nature of the support assembly is unchanged from the tested configuration.
4. The application or "end use" of the fire barrier is unchanged from the tested configuration.
5. The configuration has been reviewed by a qualified fire protection engineer and found to provide an equivalent level of protection.

### 4.2  Fire Endurance Rating

The fire protection features required to satisfy GDC 3 include features to ensure that one train of those systems necessary to achieve and maintain shutdown conditions be maintained free of fire damage. One means of complying with this requirement is to separate one safe shutdown train from its redundant train in a fire area with a fire barrier having a 1- or 3-hour rating. But what exactly does this "rating" mean?

Fire rating is defined as the endurance period of a fire barrier or structure, which relates to the period of resistance to a standard fire exposure before the first critical point in behavior is observed (Regulatory Guide 1.189). The level of fire resistance required of the barrier—1 hour or 3 hours—depends on the other fire protection features in the fire area.

The statement of considerations for Appendix R (45 FR 76602), stipulated the following:

> "Fire Barriers are 'rated' for fire resistance by being exposed to a 'standard test fire.' This standard test fire is defined by the American Society for Testing and Materials in ASTM E-119, 'Standard for Fire Resistance of Building Materials.' Fire barriers are commonly rated as having a fire resistance of from 1 to 8 hours."

Fire endurance ratings of building construction and materials are demonstrated by testing fire barrier assemblies in accordance with the provisions of the applicable sections of NFPA 251 and ASTM E-119. Assemblies that pass specified acceptance criteria (e.g., standard time-temperature fire endurance exposure, unexposed side temperature rise, hose stream impingement) are considered to have a specific fire-resistance rating. The standard time-temperature curve, which is generally accepted for evaluating and rating the fire resistance of all types of building fire barriers, is considered to represent a severe fire exposure. However, the fire endurance tests are not intended to model any specific room fire or the conditions under which the ERFBS will be exposed during a fire, but rather provides a specific standard fire exposure against which similar fire-rated assemblies can be evaluated.

Documentation required to establish the fire rating of a fire barrier should include the design description of the barrier and the test reports that verify its fire rating.

### 4.3 Acceptance Criteria & Test Standards

NRC issued the following guidance on acceptable methods of satisfying the regulatory requirements of GDC 3:

- BTP APCSB 9.5-1, "Guidelines for Fire Protection for NPPs,"
- Appendix A to BTP APCSB 9.5-1,
- BTP CMEB 9.5-1, "Fire Protection for NPPs,"
- GL 86-10, which took precedence over previous staff guidance, and
- Supplement 1 to GL 86-10

In these guidance documents, NRC staff stated that as a minimum, the design of fire barriers for horizontal and vertical cable trays should meet the requirements of the American Standard ASTM E-119, "Fire Test of Building Construction and Materials," including the hose stream test. NRC also stated in GL 86-10 Supplement 1 that the acceptance criteria contained in NFPA 251, "Standard Methods of Fire Tests of Building Construction and Materials," pertaining to nonbearing fire barriers was applicable to cable-tray fire barrier wraps. Figure 4-1 on the next page, is a logic diagram for the qualification and acceptance criteria for ERFBS that was provided in Supplement 1 to GL 86-10. Please note the "NRC Review" box in Figure 4-1. NRC review of engineering evaluations of ERFBS functionality is conducted at the request of the licensee or during NRC staff inspections.

ASTM E-119 and NFPA 251 provided acceptance criteria for testing ERFBS. However, prior to NRC issuing Supplement 1 to GL 86-10, industry had no clear understanding of the specifics on how ERFBS testing was supposed to be conducted to ensure adequate testing to NRC. In developing Supplement 1, NRC staff relied on input from industry and public stakeholders concerning various methods of testing. In particular, the TVA had developed a detailed and sturdy engineering position on the proper way to test ERFBS, which is presented in Appendix G to this report. This position was presented to NRC in the early 1990s and serves as a basis to the guidance of GL 86-10 Supplement 1. Figure 4-1 provides the logic diagram presented in GL 86-10 Supplement 1 for qualifying ERFBS. Note that the "NRC Review" is done, either at the request of the licenses or during routine NRC staff inspections.

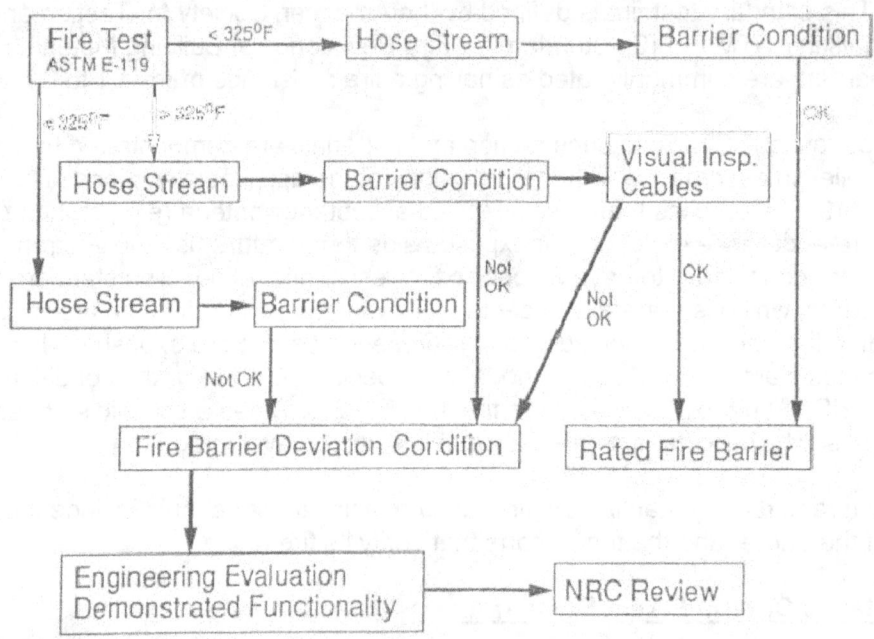

**Figure 4-1. ERFBS Fire Resistance Testing Acceptance Criteria Logic Diagram (GL 86-10 Supplement 1)**

### 4.3.1 American Nuclear Insurers Fire Test Standard

The ANI test standard was enclosed in ANI Information Bulletin #5 (79), dated July 1979. This test standard was to be used by those NPPs insured by ANI to qualify (for insurance purposes only) a Protective Envelope for Redundant Class 1E Cables in NPPs when located in the same fire area. The intent of this qualification standard was to establish the ability of an ERFBS to maintain circuit integrity when exposed to a fire outside of the cabling system, adjacent to the protected cable, or when subjected to the mechanical impact of hose stream or other impact test.

The ANI standard includes a test for exposure fires and subjects the protected cable raceway to an ASTM E-119 standard temperature-time curve. Following the exposure, a hose stream test would be conducted following specific guidelines on line size, pressure, nozzle angle, and flow rate. An energized cable was placed within the ERFBS for monitoring the circuit integrity. The only failure criterion was loss of circuit integrity during the fire exposure or hose stream period. The intent of the test was to identify the onset of fire damage to the cables within the raceway fire barrier test specimen during the fire endurance test period.

Early on during the qualification of ERFBS the NRC accepted the use of the ANI standard from several licensee submittals. However, following review of several test reports, the NRC began to question the applicability of the ANI standard. After completing a more detailed review of the standard and test reports, the NRC determined the ANI monitoring approach to be non-conservative. Specifically, Supplement 1 to GL 86-10 states:

> The use of circuit integrity monitoring during the fire endurance test is not a valid method for demonstrating that protected shutdown circuits are capable of performing their required function during and after the test fire exposure.

### 4.3.2 ASTM E-119 and NFPA 251

GL 86-10 identifies that NRC staff found Chapter 7 of NFPA 251, "Tests of Nonbearing Walls and Partitions" to be an adequate testing acceptance criteria to use for qualifying cable tray fire barrier wraps.

Appendix A to BTP APCSB 9.5-1 Position D.3.(d), states that the design of fire barriers for horizontal and vertical cable trays should, as a minimum, meet the requirements of the ASTM E-119, "Fire Test of Building Construction and Materials," including hose stream test. The technical basis for Section III.M of Appendix R to 10 CFR 50, stipulates that "Fire barriers are 'rated' for fire resistance by being exposed to a 'standard test fire." This standard test fire is defined by the ASTM E-119 test standard. It should also be mentioned that NFPA 251 and ASTM E-119 are nearly identical testing standards.

The following acceptance criteria are found in both the ASTM E-119 and NFPA 251 standards:

- The wall or partition withstood the fire endurance test without the passage of flame or gases hot enough to ignite cotton waste, for a period equal to that for which classification is desired.

- The wall or partition withstood the specified fire and hose stream tests, without the passage of flame, gases hot enough to ignite cotton waste, or the hose stream. The assembly failed the hose stream test if an opening developed that permits the projection of water from the stream beyond the unexposed surface during the hose stream test.

- Transmission of heat through the wall or partition during the fire endurance test did not raise the temperature on the unexposed surfaces more than 139°C (250°F) above their initial temperatures.

These standards specify that the test shall be controlled by the standard temperature-time curve presented in the standards. Table 4.1 and **Error! Reference source not found.** provide reference to the temperature-time values required by this standard. The measurement of these temperatures is the average of no fewer than nine thermocouples symmetrically disposed and distributed near all parts of the sample, at least 6 inches away from the sample.

**Table 4-1. NFPA 251 and ASTM E-119 Temperature Time Curve Values**

| 4.4 Time | 4.5 Temperature °C (°F) |
|---|---|
| 5 minutes | 538 (1000) |
| 10 minutes | 704 (1300) |
| 30 minutes | 843 (1550) |
| 1 hour | 927 (1700) |
| 2 hours | 1010 (1850) |
| 4 hours | 1093 (2000) |
| 8 hours | 1260 (2300) |

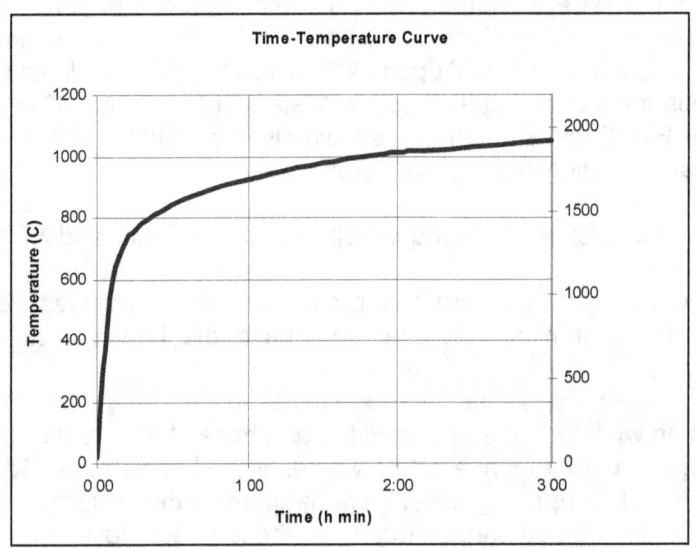

NFPA 251 and ASTM E-119 present acceptance criteria that stipulate transmission of heat through the barrier "shall not be sufficient to raise the temperature on the assembly's unexposed surface more than 140°C (250°F) above the assembly's initial temperature."

Some NRC documentation has referenced a 162°C (325°F) cold side temperature as the acceptance criteria. This 162°C (325°F) criterion is based on the 140°C (250°F) acceptance criteria of NFPA 251, with the assumption that the beginning ambient air temperature is 24°C (75°F) at the start of the fire exposure. The ambient air temperature at the beginning of a fire test is usually between 10°C (50°F) and 32°C (90°F) and is generally recognized that 24°C (75°F) represents an acceptable norm. Therefore, the 140°C (250°F) criterion of NFPA 251, plus the ambient air temperature assumption of 24°C (75°F) is equivalent to the 162°C (325°F) criterion.

Chapter 5 "Conduct of Fire Tests," of NFPA 251 provides information on the qualification time for the fire endurance test along with the Hose Stream Test application. The Hose Stream Tests allows for a duplicate test specimen exposed to half of that indicated as the resistance period immediately after which the specimen shall be subjected to the impact, erosion, and cooling effects of a hose stream.

Additional information on NRC acceptance criteria related to combustibility, ampacity derating, seismic qualification and fire endurance (including test specimen construction, hose stream test, and cable functionality) are located in Appendix H.

### 4.5.1 Purpose and Brief History of the ERFBS Hose Stream Test

Originally, ERFBS were exposed to the same solid-stream hose stream test that applied to primary structural fire walls. As noted above, some standards did allow for a second identical test specimen exposed to only ½ the fire endurance period to be used for hose stream testing, whereas others applied the hose stream to the test article after completion of the full fire exposure duration test.

The solid-stream hose test was derived from very early life safety concerns for firefighting personnel. During certain historical fire incidents (early 1900's), structural walls actually collapsed when a fire hose was played against them putting firefighters at risk. The hose stream test was added to the fire barrier standard (ASTM E119) to reduce the chances of that happening. The standard's specifications were based on a typical solid-stream hand-held fire hose (pressure, line and nozzle characteristics). The ASTM and NFPA standards made no exception to the hose stream test for non-bearing partitions and those protocols were also being applied to ERFBS.

Many of the early ERFBS tests resulted in severe damage to the barrier system itself when the solid-stream hose test was employed. The ERFBS were not, by and large, designed to withstand such impact especially at the end of their fire endurance capacity. Early in industry's Thermo-Lag resolution efforts, industry asserted that the solid-stream hose test was inappropriate to the ERFBS. They argued that the structural integrity issues that drove the standard for a primary fire wall are not of equivalent concern when it comes to ERFBS because structural failure of an ERFBS is unlikely to place firefighters at significant hazard as would collapse of a primary structural element such as a wall. The nuclear industry also argued that industry practice relied on either fog nozzles or adjustable diffusion-type nozzles for interior fire fighting rather than fixed pattern solid-stream type nozzles. The NRC ultimately agreed with this position and endorsed the use of an alternate hose stream test using a diffusion nozzle rather than a solid-stream nozzle. As a result, most of the ERFBS qualification tests performed since the early 1990's have used the diffusion nozzle hose test rather than a solid-stream hose test. The purpose for maintaining the hose stream testing of ERFBS is to evaluate the cooling, impact and erosion aspects of the ERFBS.

# 5. ELECTRICAL RACEWAY FIRE BARRIER SYSTEMS (ERFBSs)

The purpose of an ERFBS is to ensure redundant safe shutdown circuits located in the same fire area are protected and remain operational (i.e., free of fire damage) during a NPP fire. ERFBS can accomplish this in several different methods. The following provides a brief description of each.

- Insulation materials limit the exposure to the heat transfer rate by reducing the conductive heat transfer rate to the protected cable/circuits in accordance with Fourier's Law.

    Insulative ERFBS include:
    - Hemyc
    - Kaowool & FP-60
    - Concrete
    - Mecatiss

- Intumescences materials reduce the heat transfer rate to the protected cable/circuits by chemically absorbing heat energy. This endothermic reaction causes the material to swell, increasing in volume and decreasing in density.

    Intumescent ERFBS include:
    - Pabco
    - 3M Rigid Panel

- Ablation materials reduce the heat transferred to the protected cable/circuits by sublimation.[6] When heated, the ablation material is consumed (sacrificed) through sublimation and mass loss which provides cooling and forms a thermal shield.

    Ablative ERFBS include:
    - Thermo-Lag 330-1

- Hydrate materials contain chemically bound water that is used up during a fire exposure by an endothermic reaction, which maintain temperatures near 100°C (212°F) until the hydrate (water) is converted into steam. Hydrate ERFBS include:

    Hydrate materials include:
    - Darmatt KM-1
    - Promat
    - Pyrocrete

Several barriers use a combination of the heat transfer methods, these ERFBS include:

- Versawrap (hydrate, insulative, and intumescent)
- MT (hydrate and insulative)
- 3M Interam E-50 Series (hydrate and insulative)

This section provides a detailed description of each ERFBS product used in U.S. NPPs in operation at the time this document was written. It provides a description of the barrier, a historical perspective, identification of problems associated with individual barriers, qualification

---

[6] Sublimation is a phase transition from a solid to a gas phase with no intermediate liquid phase

testing, and corrective actions taken to address ERFBS deficiencies. It should be noted that dimensions presented are for the readers understanding of the ERFBS design and doesn't signify an NRC approved design unless specifically stated.

## 5.1 3M Interam™ E-50 Series & Rigid Panel System (Hydrate/Insulative & Intumescent)

Interam E-50 Series is a fire barrier system designed and manufactured by Minnesota Mining and Manufacturing (3M), and supplied by PCI Promatec. 1- and 3-hour rated 3M Interam™ E-54 barrier systems have been designed and tested in accordance with Supplement 1 to GL 86-10.

The 3M Interam™ E-50 Series ERFBS is the most commonly used barrier in operating NPPs. This ERFBS consists of multiple layers of flexible mat that is used to provide 1- and 3-hours of fire resistance rating protection to electrical raceways. Figure 5-1 provides a cut away illustration of a 3M Interam E-54 ERFBS.

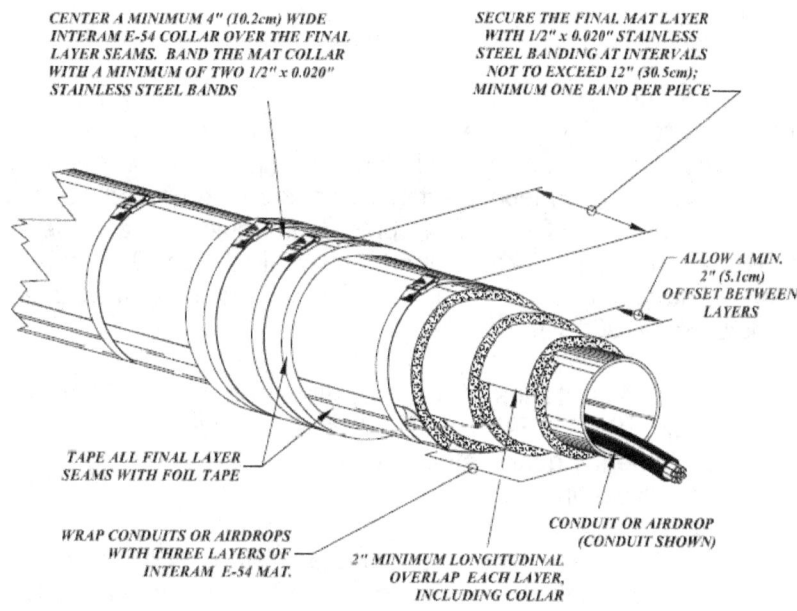

Figure 5-1. 3M Interam E-54 Series ERFBS (PCI Promatec, 2010)

The E-54A ERFBS is manufactured as a mat with a nominal thickness of 10.2 mm (0.4 in). The mat contains aluminosilicate fibers bound in an organic matrix that is sandwiched between a metal foil (aluminum or stainless steel) and a synthetic polymer (nylon) laminated scrim.

The 3M Interam 1- and 3-hour rated ERFBS achieve its fire performance and endurance properties by a combination of chemical and physical properties. The thermal protection is provided by the absorption of heating during an endothermic reaction (from a chemically-bound ingredient that releases chemically bound water), and via the thermal mass (heat sink) of the mat. After the endothermic reaction has gone to completion, remaining ceramic fibers act as a high-temperature insulator. The metallic foil backing is affixed to the outside of the mat to provide a reflective substrate that will reflect radiant energy away from the barrier and reduce the thermal transmission of heat through the barrier.

The manufacture identifies the type of foil backing by the postscript "A" for aluminum backed mats and "C" for stainless steel backed mats. These laminates are 0.076 mm (0.003 in) thick and attached to the base mat by the use of adhesive. Type "C" backing is an annealed Type 304 stainless steel foil and is typically used for inside containment where aluminum is not allowed. UL Test report R10125, 86NK2919 dated May 30, 1986, was conducted to determine any differences in thermal protection among the two barriers. The results indicated that the two barrier are very similar in their thermal conductance, however the stainless steel backed mat did experience a slightly higher internal temperature at the 1-hour time period (approximately 5.6°C (10°F) higher).

WBN Unit 1 use the rigid panel ERFBS (CS-195), while Salem is the only plant that used the FS-195 barrier. Both FS-195 and CS-195 are an intumescent type of material that expands up to eight times its initial volume upon heating. This expansion begins around 250°C (482°F) and exhibits significant expansion in the 350°C (662°F) temperature range. shows a comparison of butt and finger barrier jointing methods.

### 5.1.1 History

When 3M decided to enter into the nuclear fire protection business, they first conducted generic type fire tests at nationally recognized fire testing laboratories. After satisfying results, upgrades and refinements to the installation methods, techniques and materials were obtained by using the 3M internal fire test facility with independent quality assurance procedures and inspections. These tests were conducted prior to GL 86-10 Supplement 1 guidance and the acceptance criteria chosen by 3M was the cable jacket temperature within the ERFBS exceeding 121°C (250°F) above ambient and for structural members a failure criterion of the metal surface temperature reaching 538°C (1000°F).

Originally, 3M manufactured several lines of fire protection products used to protect electrical raceways, including CS-195, FS-195, E-20 and E-50 series materials. However, following successful GL 86-10 Supplement 1 qualification testing of the E-54 series ERFBS, all other series of 3M systems were discontinued.

Peak Seals, Inc. (now PCI Promatec) became the Master Distributer of 3M Interam™ fire wrap system for commercial NPPs on April 24, 1995. Following NRC letter to Peak Seals dated September 5, 1997, Peak Seals agreed to conduct qualification testing of their systems prior to any new installations. Their testing approach was to qualify existing 3M designs to the requirements of Supplement 1 to GL 86-10. When the 3M designs previously qualified to ANI criteria did not meet the more stringent acceptance criteria of GL 86-10 Supplement 1, Peak Seals modified the 3M Interam designs and successfully qualified a range of raceway types, sizes, and configurations for both 1- and 3-hour applications. In their October 3, 1997 letter to NRC, Peak Seals provided a comparison of the barrier design for pre-GL 86-10 Supplement 1 barriers and post GL 86-10 Supplement 1 barriers. This information is reproduced in Table 5-1.

Table 5-1. 3M Interam Design Comparison Old-to-New

| Electric Raceway Configuration | Fire Resistance Rating (hours) | Pre GL 86-10/Supplement 1 Configuration | GL 86-10/Supplement 1 Configuration |
|---|---|---|---|
| Conduit | 1 | 0.9" Thick (3 Layers E-53A) | 1.2" Thick (4 Layers E-54A) |
| Tray | 1 | 0.8" Thick (2 Layers E-54A) | 1.2" Thick (4 Layers E-54A) |
| Junction Box | 1 | 0.9" Thick (3 Layers E-53A) | 1.2" Thick (4 Layers E-54A) |

Table 5-1.  3M Interam Design Comparison Old-to-New (Continued)

| Electric Raceway Configuration | Fire Resistance Rating (hours) | Pre GL 86-10/Supplement 1 Configuration | GL 86-10/Supplement 1 Configuration |
|---|---|---|---|
| Air Drop | 1 | 0.9" Thick (3 Layers E-53A) | 1.2" Thick (4 Layers E-54A) |
| Conduit | 3 | 2.0" Thick (5 Layers E-53A) | 2.8" Thick (5 Layers E-54A with two 0.4" Air Gaps |
| Tray | 3 | 2.0" Thick (5 Layers E-53A) | 2.8" Thick (6 Layers E-54A with one 0.4" Air Gap) |
| Junction Box | 3 | 2.0" Thick (5 Layers E-53A) | 3.2" Thick (6 Layers E-54A 2 Air Gaps) |
| Air Drop | 3 | 2.0" Thick (5 Layers E-53A) | 2.0" Thick (5 Layers E-54A) |

### 5.1.2  Problems

#### 5.1.2.1  Information Notice 93-41

IN 93-41 indentifies an NRC inspection of the testing basis for Salem using a 3M FS-195 fire barrier test report stated,

> "According to the test report, the metal duct temperature on the unexposed side of the fire barrier material exceeded 139°C (250°F) above ambient in about 30 minutes. At 60 minutes the temperature was 326.5°C (620°F). The test specimen was not subject to a hose stream test. The condition of the cables at the end of the test was not reported."

IN 93-41 also identified that a test report issued by Twin City Testing Corporation, dated September 1986, for an Interam™ E-50 Series fire barrier produced by 3M Company didn't adequately document the justification for qualification for this barrier. For this test, circuit integrity acceptance criterion specified by the ANI was used. The temperatures within the fire barrier and the conditions of the cables at the end of the test were not reported. In addition, the fire barrier construction details and methods of fire barrier application for the test specimens were not documented in the test report.

Many of the early test reports did not fully document all of the pertinent information needed by today's guidance to ensure the acceptable qualification of the ERFBS. In addition, Supplement 1 to GL 86-10 did not exist when these early testing was being performed, which resulted in a majority of the testing having not conducted hose stream tests, cable fill or placement of thermocouples, as specified in current NRC guidance documents.

*5.1.2.2 Information Notice 95-52*

IN 95-52, "Fire Endurance Test Results for Electrical Raceway Fire Barrier Systems Constructed from 3M Company Interam Fire Barrier Materials," reported results of full-scale fire endurance tests for ERFBS constructed from 3M Company Interam fire barrier materials. Two 1-hour tests were conducted, one using Interam E-53A mat material that failed the temperature rise criteria towards the end of the fire exposure, while the other test using Interam E-54A mat material resulted in no failures. A 3-hour test was also conducted and resulted in no cable raceway passing the temperature rise acceptance criteria as specified in Supplement 1 to GL 86-10. Section 5.1.3 provides a detailed description of these testing configurations and failures.

*5.1.2.3 Information Notice 95-52, Supplement 1*

IN 95-52, Supplement 1, "Fire Endurance Test Results For Electrical Raceway Fire Barrier Systems Constructed From 3M Company Interim™ Fire Barrier Materials," dated March 17, 1995, documents additional 3-hour fire endurance testing failures identified in test reports forwarded to NRC by the Master Distributer Peak Seals, Inc. In these test reports, only a minimal fraction of the test articles met the temperature rise acceptance criteria of GL 86-10 Supplement 1, with the majority of the articles exceeding the criterion prior to the completion of the tests duration.

*5.1.2.4 Qualifying 3M Interam™ to GL 86-10 Supplement 1*

Prior to GL 86-10 Supplement 1, 3M Interam™ configurations available from the vendor had only been qualified to the ANI/MEARP criteria. To evaluate the 3M barrier performance, Peak Seals (PCI Promatec) performed testing to evaluate their ERFBS design against GL 86-10 Supplement 1 guidance. As documented in IN 95-52 and IN 95-52 Supplement 1, many of the older 3M Interam™ designs did not meet the NRC acceptance criteria. As such the vendor re-engineered the 3M Interam™ ERFBS to achieve a 1- and 3-hour qualification via the GL 86-10 Supplement 1 criteria. In many instances an additional layer of mat was added to the older system designs to achieve the desired fire rating. However, with the additional layer(s), the vendor also had to reevaluate the ampacity derating, and seismic characteristics of the barrier.

## 5.1.3 Testing

Sections 5.1.3.1 and 5.1.3.2, present information on the testing of 3M Interam designs qualified before and after GL 86-10 Supplement 1 guidance, respectively.

*5.1.3.1 3M Interam ERFBS Testing Prior to GL-86-10 Supplement 1 Criteria*

There were numerous 3M Interam ERFBS designs qualified to the ANI/MEARP standard. The following provides an overview of these barrier designs. A typical 3M Interam™ E-50 Series ERFBS of the early vintage consisted of the following;

- 3M Interam™ E-50 series mats (number of layer dependent on rating),
- 3M FireDam 150 Caulk (used as a smoke and flame sealant),
- 3M Interam™ T-49 Aluminum Foil Tape or T-65 Stainless Steel Foil (used as a vapor barrier, radiant heat reflector and installation aid.)
- 3M Scotch® Brand 898 Filament Tape (used as an installation aid)

Older designs may have also used the following two materials finishing the ends of a raceway
- 3M Fire Barrier CS-195 Composite Sheet (used to cover openings and as a collar at the termination of fire protection envelopes)[7], and
- 3M Fire Barrier CP 25N/S Caulk (used as a smoke and flame sealant)[7].

FireDam 150 Caulk was used to seal mat-to-mat assemblies and is a paste version of the E-50 series mat.[8] 3M recommended the use of CP 25N/S Caulk[9] as a sealant whenever the 3M fire barrier terminates at a wall or floor, and caulking is required along the CS-195 product. These two caulks are the only material in the 3M ERFBS which are applied wet.

The 3M Interam™ CS-195 composite sheet is reddish brown in color, 7 mm (0.28-in) thick, with sheet metal on one side and aluminum foil on the other. The sheet metal side faces away from the 3M mat, when installed around a 3M mat. It should be noted that the CS-195 is an intumescent material that will combust if exposed to a heat source that raises its temperature above the materials auto (or pilot with pilot present) ignition temperature. As a result, this material CS-195 should not be used inside containment unprotected.

Although many different methods and configurations existed for installing various 3M components, the manufacturer specified the generic installation requirements shown in Table 5-2 to construct an ERFBS capable of providing the required level of protection.

**Table 5-2. 3M Interam E-50 Series Generic Installation Requirements**

| Electrical Raceway Configuration | Minimum Layers Required for 1-hr Fire-Resistance |
|---|---|
| Cable Trays | |
| < 25% cable fill | 2 layers of E-54A |
| ≥ 25% cable fill | 1 layer of E-54A and 1 layer of E-53A |
| Conduits | |
| Steel | 3 layers of E-53A |
| Aluminum ≥ 5" diameter | 3 layers of E-53A |
| Aluminum < 5" diameter | 1 layer of E-54A and 2 layers of E-53A |
| Air Drops | 3 layers of E-54A |
| Junction Boxes | 3 layers of E-54A |

---

[7] Alternative penetration seal methods have been qualified.
[8] Fire Dam 150 Caulk is no longer manufactured
[9] CP-25 is not used in Supplement 1 to GL 86-10 qualified 3M Interam ERFBS.

Table 5-2. 3M Interam E-50 Series Generic Installation Requirements (Continued)

| Electrical Raceway Configuration | Minimum Layers Required for 1-hr Fire-Resistance |
|---|---|
| Supports and Heat Transfer Items | |
|     Supports underneath cable tray | 2 layers of E-54A |
|     Supports partially protected | 1 layers of E-54A for 12" |
| | 2 layers of E-53A for 9" |
| | 2 layers of E-54A for 6" |
| | 2 layers of E-53A and 1 layer of E-54A for 5" |
| | 3 layers of E-54A for 4" |

| Electrical Raceway Configuration | Minimum Layers Required for 3-hr Fire-Resistance |
|---|---|
| All Raceway Types | 5 layers of E-54A |
| **Other Specifications** | |
|     Weight per unit area: | E-53A = 1.42 psf (6.95kg/sq.m) ±12% |
| | E-54A = 1.81 psf (8.83kg/sq.m) ±12% |
|     Thickness: | E-53A = 0.3 inch (7.6mm) ± 10% |
| | E-54A = 0.4 inch (10.2mm) ± 10% |

3M also recommended a mat seam overlap of 5.1 cm (2.0 in) minimum or a collar with a minimum 10.2 cm (4.0 in) width with butt joint seams caulked with FD-150 caulk, end seams must be coved with T-49 aluminum foil, steel bands spaced no more than 30.5 cm (12.0 in) apart are placed on the surrounding last layer or stainless steel wire mesh to provide structural support during fire conditions.

The selection of the components used in the 3M Interam™ E-50 Series 1-hour and 3-hour fire Protection Systems was varied when the item needing fire protection contained a larger than normal thermal mass, or when an obstruction prevents the use of typical installation techniques. For example, a large bundle of conduits (which represents a thermal mass much greater than a single conduit), can be wrapped inside a single fire protective envelope with one layer less of E-54A in the 3-hour system. In addition whenever the 3M fire protection systems protect an item near a wall or floor, the wall or floor can be used as one or more sides of the fire protective envelope. To accommodate an installation near a wall or floor, half inch by half inch welded wire mesh may be used in place of banding.

Protection of raceway support members is achieved by applying 1-layer of E-54A mat for 1-hour or five layers of E-54A mat for 3-hour rating, for a distance of at least 30.5 cm (12.0 in) from the electrical raceway in both cases.

For cable trays wider than 30.5 cm (12.0 in) the manufacture specified that strapping must be applied around or across the cable tray at a maximum spacing of 30.5 cm (12.0 in) on center and underneath all seams. This strapping is used to minimize sagging of the fire protection mat. Any strapping system with a minimum tensile strength of 227 kg (500 lbs) is satisfactory to the manufacturer. Some options included:

i. Two wraps of three-fourths inch or wider 3M filament Tape #898

ii. Most half inch or wider polyester or nylon strapping
iii. Metal strapping
iv. Metal, plastic, or wood bridging across the top of the cable tray.

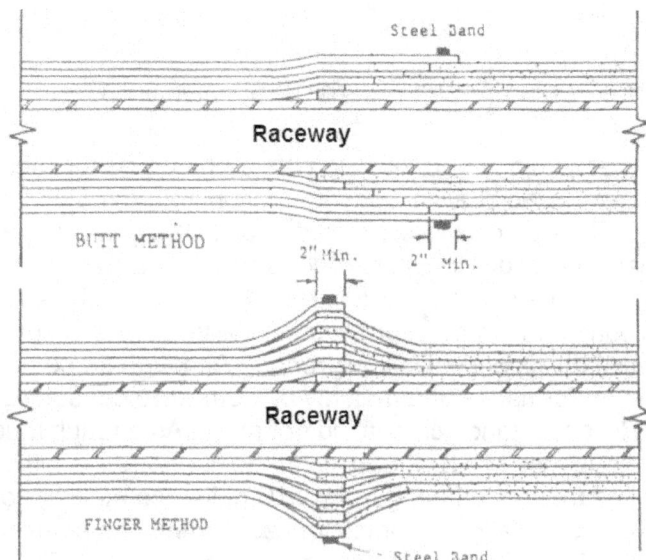

**Figure 5-2. Comparison of Butt and Finger Barrier Jointing Methods
(SwRI Qualification Test – June 1984)**

5.1.3.1.1 NIST Small Scale Testing (Prior to GL 86-10 Supplement 1)

The NIST small scale testing program involved several tests of the 3M Interam E-50 Series barriers. For a description of the NIST test program, please refer to Section 5.2.3.1. The first test (A1-1) consisted of three layers of the Interam E-53A 7.6 mm (0.3 in) thick, green, resilient base material with 0.076 mm (0.003 in) aluminum foil laminated on one face (exposed face) and a reinforcing scrim laminated on the other (unexposed face). This design is considered to be adequate for a 1-hour fire resistance rating. The test assembly was subjected to a 1-hour ASTM E-119 fire exposure. The results indicated that the average unexposed surface temperature rise criterion was met at 1 hour and 12 minutes, while the maximum unexposed single point surface temperature rise was no reached. It was also observed that during removal of the test specimen from the furnace apparatus, a large portion of the exposed surface material fell into the furnace.

The second test (B1-1) consisted of two layers of 10.2 mm (0.4 in) thick blue Interam E-54A, resilient, base material with 0.076 mm (0.003 in) aluminum foil laminated to one face (exposed face) and a reinforcing scrim laminated to the other (unexposed face). This design is representative of a 1-hour rated barrier. The test assembly was subjected to a 1-hour ASTM E-119 fire exposure. The results indicated that this configuration did not provide the 1-hour of protection as it failed to meet the average temperature rise criterion at 48 minutes. The maximum single point temperature rise criterion was not exceeded for the duration of the test.

The third test (B3-1) consisted of five layers of Interam E-54A fire-barrier mat. The test assembly was subjected to the ASTM E-119 standard fire exposure for 3.5 hours. The results indicated that the average and maximum single point temperature rise criterion were reached during the exposure.

As discussed in Section 5.2.3, the small scale testing may not be representative of actual ERFBS fire endurance ratings.

5.1.3.1.2  Fire Endurance Testing (Prior to GL 86-10 Supplement 1)

As part of NRC staffs reverification effort following the problems associated with the Thermo-Lag ERFBS, NRC staff requested qualification information from the 3M to support their review. 3M and Peak Seals (PCI Promatec) willingly provided all of the non-proprietary information to NRC. This included many fire endurance test reports conducted by 3M, UL, and SwRI, along with a hand full of seismic and ampacity derating reports conducted by entities other than 3M. Due to the massive amount of information provided in these reports, a summary of each test report has been presented in Appendix D of this report. Although these test report provide indication of the barriers performance, the majority of them were conducted in the mid- to late-1980's at which time NRC acceptance criteria did not exist. As a result much information required to make a determination of the adequacy of the fire barrier is either not reported or the testing was conducted in a manner differing from NRC guidance making comparison of the results difficult. However, as a matter of completeness, summaries of these test results are provided in Appendix D of this report and the complete reports can be found publically available in NRC NUDOCS system. Another point to make is that a fraction of these tests were completed at 3Ms testing facilities with the observations of staff from nationally recognized testing laboratories (UL, SwRI, etc.). The testing summaries you will find in this section are related to the failed testing reported in the Information Notices discussed above.

IN 95-52, "Fire Endurance Test Results for Electrical Raceway Fire Barrier Systems Constructed from 3M Company Interam Fire Barrier Materials," reported results of full-scale fire endurance tests for ERFBS constructed from 3M Company Interam fire barrier materials. These tests were conducted by OPL and NRC staff members were present to witness the testing. All test assemblies were subjected to an ASTM E-119 exposure for the duration that the barrier was being qualified for, followed by a hose stream test.

The first 1-hour test conducted on April 20, 1995 used three layers of Interam E-53A fire barrier mat materials resulted in failures to meet the temperature rise acceptance criteria for a 61 cm (24.0-in) wide cable tray and three differently sized conduits (7.6 cm, 5.1 cm, and 2.54 cm (3.0 in, 2.0 in, and 1.0 in)), along with the air drop configuration. This assembly passed the hose stream test. The results are shown in Table 5-3.

The second 1-hour test was conducted on May 17, 1995, and used three layers of Interam E-54A mat material to protect a 61-cm (24.0-in-wide) steel cable tray, (2.54-cm and 12.7-cm (1.0-in and 5.0-in, respectively)) diameter steel conduits, and a 5.1-cm (2.0-in) diameter air drop. No cables were contained within any of the test specimens. In all cases, these test specimens met the acceptance criteria specified in Supplement 1 to GL 86-10. The results are shown in Table 5-4.

The last ERFBS qualification test conducted was to qualify a 3-hour Interam barrier system and was conducted on July 7, 1995. This test included a 61.0-cm (24.0-in) wide steel cable tray, 15.2-cm (6.0-in) wide steel cable tray, 2.54-cm (1.0-in) diameter conduit, 3.0-in diameter conduit, 5-in diameter conduit, and 5.1-cm (2.0-in) wide air drop, all arranged in a U-shaped configuration. In addition, a nominal 30.5-cm by 30.5-cm (12.0-in by 12.0-in) by 20.3-cm (8.0-in) steel junction box was included in the test arrangement. The raceways and junction boxes were protected with 5 to 6 layers of 10.0 mm (0.4 in) thick Interam E-54A mat material. All test specimens exceeded the temperature rise acceptance criteria of Supplement 1 to GL 86-10. Table 5-5 summarizes these results.

### Table 5-3. Summary of April 20, 1995 Fire Endurance Test

| Peak Seals – 3M Company 1-hour Interam Fire Barrier<br>Allowable single point unexposed-side temperature criterion = 204°C (399°F)<br>Allowable average unexposed-side temperature criterion = 162°C (324°F)<br>(Shading shows temperatures that exceeded, acceptance criteria of GL 86-10 Supplement 1)<br>Reproduced from IN 95-52 Table 1 ||||| 
|---|---|---|---|---|
| Test Specimen | Thermocouple (TC) Locations | Average TC Temp °C (°F) | Maximum TC TEMP °C (°F) | Remarks |
| 6" Cable Tray | Front side rail | 128 (262) | 170 (338) | Protected with three layers of Interam E-53A.<br>Met acceptance criteria. |
| | Rear side rail | 128 (262) | 169 (337) | |
| | Copper Conductor | 109 (228) | 139 (282) | |
| 24" Cable Tray | Front side rail | 187 (369) | 243 (470) | Protected with four layers of Interam E-53A.<br>Exceeded the maximum single point temperature criterion at 50½ minutes and the average temperature rise criterion at 54½ minutes. |
| | Rear side rail | 197 (387) | 250 (482) | |
| | Copper conductor | 172 (342) | 194 (382) | |
| 5" Conduit | Conduit surface | 136 (277) | 188 (370) | Protected with three layers of Interam E-53A.<br>Met acceptance criteria. |
| | Copper conductor | 103 (217) | 135 (275) | |
| 3" Conduit | Conduit surface | 186 (366) | 206 (402) | Protected with three layers of Interam E-53A.<br>Exceeded the maximum single point temperature criterion at 59½ minutes and the average temperature rise criterion at 53½ minutes. |
| | Copper conductor | 165 (329) | 190 (374) | |
| 2" Conduit | Conduit surface | 181 (357) | 220 (428) | Protected with three layers of Interam E-54A.<br>Exceeded the maximum single point temperature criterion at 55½ minutes and the average temperature rise criterion at 55 minutes. |
| | Copper conductor | 161 (321) | 204 (400) | |
| 1" Conduit | Conduit surface | 183 (361) | 214 (417) | Protected with two layers of Interam E-53A and an outer layer of Interam E-54A.<br>Exceeded maximum single point temperature criterion at 49½ minutes and the average temperature rise criterion at 52 minutes. |
| | Copper conductor | 167 (332) | 203 (397) | |
| 2" Air Drop | Copper conductor | 163 (326) | 201 (393) | Protected with three layers of Interam E-54A.<br>Exceeded average temperature rise criterion at 59 minutes. |
| Junction Box | Metal surface | 125 (257) | 155 (311) | Protected with three layers of Interam E-54A. Met Acceptance. |

### Table 5-4. Summary of May 17, 1995 Fire Endurance Test

Peak Seals – 3M Company 1-hour Interam Fire Barrier
Allowable single point unexposed-side temperature criterion = 207°C (405°F)[1]
Allowable average unexposed-side temperature criterion = 166°C (330°F)
(Shading shows temperatures that exceeded, acceptance criteria of GL 86-10 Supplement 1)
Reproduced from IN 95-52 Table 2

| Test Specimen | Thermocouple (TC)Locations | Average TC TEMP °C (°F) | Maximum TC TEMP °C (°F) | Remarks |
|---|---|---|---|---|
| 24" Cable Tray | Front side rail | 143 (290) | 198 (389) | Protected with three layers of Interam E-54A |
| | Rear side rail | 149 (301) | 179 (354) | |
| | Copper conductor | 108 (226) | 129 (265) | Met acceptance criteria. |
| 5" Conduit | Conduit surface | 107 (224) | 122 (251) | Protected with three layers of E-54A. |
| | Copper conductor | 103 (217) | 118 (244) | Met acceptance criteria. |
| 1" Conduit | Conduit surface | 153 (308) | 190 (374) | Protected with three layers of E-54A |
| | Copper conductor | 141 (286) | 174 (346) | Met acceptance criteria. |
| 2" Air Drop | Copper conductor | 117 (242) | 137 (279) | Protected with three layers of Interam E-54A. Met acceptance criteria. |

### Table 5-5. Summary of July 7, 1995 Fire Endurance Test

Peak Seals – 3M Company 3-hour Interam Fire Barrier
Allowable single point unexposed-side temperature criterion = 208°C (407°F)[1]
Allowable average unexposed-side temperature criterion = 167°C (332°F)
(Shading shows temperatures that exceeded, acceptance criteria of GL 86-10 Supplement 1)
Reproduced from IN 95-52 Table 3

| Test Specimen | Thermocouple (TC)Locations | Average TC TEMP °C (°F) | Maximum TC TEMP °C (°F) | Remarks |
|---|---|---|---|---|
| 6" Cable Tray | Front side rail | 183 (361) | 224 (436) | Protected with four layers of Interam E-53A. Exceeded the maximum single point temperature criterion at 158 minutes and the average temperature rise criterion at 166 minutes. |
| | Rear side rail | 181 (357) | 234 (454) | |
| | Copper conductor | 149 (301) | 173 (343) | |
| 24" Cable Tray | Front side rail | 181 (357) | 214 (417) | Protected with three layers of Interam E-53A. Exceeded the maximum single point temperature criterion at 176 minutes and the average temperature rise criterion at 167 minutes. |
| | Rear side rail | 173 (344) | 208 (406) | |
| | Copper conductor | 117 (243) | 168 (334) | |
| 5" Conduit | Conduit surface | 113 (236) | 233 (451) | Protected with five layers of Interam E-54A. Exceeded the maximum single point temperature criterion at 161 minutes and the average temperature rise criterion at 178 minutes. |
| | Copper conductor | 154 (310) | 211 (411) | |

Table 5-5. Summary of July 7, 1995 Fire Endurance Test (Continued)

| Peak Seals – 3M Company 3-hour Interam Fire Barrier<br>Allowable single point unexposed-side temperature criterion = 208°C (407°F)[1]<br>Allowable average unexposed-side temperature criterion = 167°C (332°F)<br>Reproduced from IN 95-52 Table 3 | | | | |
|---|---|---|---|---|
| Test Specimen | Thermocouple (TC) Locations | Average TC TEMP °C (°F) | Maximum TC TEMP °C (°F) | Remarks |
| 3" Conduit | Conduit surface | 204 (399) | 252 (485) | Protected with five layers of Interam E-54A.<br>Exceeded the maximum single point temperature criterion at 148 minutes and the average temperature rise criterion at 152 minutes. |
| | Copper conductor | 173 (344) | 239 (462) | |
| 1" Conduit | Conduit surface | 185 (365) | 277 (530) | Protected with three layers of Interam E-53A.<br>Exceeded the maximum single point temperature criterion at 126 minutes and the average temperature rise criterion at 167 minutes. |
| | Copper conductor | 174 (345) | 241 (465) | |
| 2" Air drop | Copper conductor | 176 (349) | 219 (426) | Protected with five layers of Interam E-54A.<br>Exceeded the maximum single point temperature criterion and the average temperature rise criterion at 152 minutes. |
| Junction Box | Metal Surface | 188 (370) | 199 (391) | Protected with six layers of Interam E-54A.<br>Exceeded the average temperature rise criterion at 165 minutes. |

\* Shaded cells indicate temperatures that exceeded acceptance criteria of GL 86-10 Supplement 1

In 1995, 3M informed its nuclear costumers that, Peak Seals performed two 1-hour fire tests to NRC Supplement 1 guidance. The following three tables present these results. Table 5-6 identifies testing that used the number of layers and thickness of 3M material that was shown in 3M installation manuals and was conducted on empty (no cable) electrical components, which included cable trays, conduits, air drops, junction boxes and supports.

### Table 5-6. Peak Seals Test Results

| Configuration | Max. Temp. °C (°F) | Avg. Temp. °C (°F) | Rating (min.) |
|---|---|---|---|
| 6" wide Cable Tray | 170 (338) | 128 (262) | 60+ |
| **24" wide Cable Tray** | **250 (482)** | **197 (387)** | **49** |
| 5" diameter Conduit | 188 (370) | 136 (277) | 60+ |
| **3" diameter Conduit** | **206 (402)** | **186 (366)** | **53** |
| 2" diameter Conduit | 220 (428) | 181 (357) | 54 |
| 1" diameter Conduit | 214 (417) | 183 (361) | 54 |

**Bold** text indicates failure to achieve fire resistance rating.

Table 5-7 summarizes the results of the second fire test which used one additional layer of 3M Interam™ mat to each system.

### Table 5-7. Peak Seals Test Results with One Additional Layer

| Configuration | Max. Temp. °C (°F) | Avg. Temp. °C (°F) | Rating (min.) |
|---|---|---|---|
| Cable Tray 24" x 4" | 198 (389) | 149 (301) | 60+ |
| 5" diameter conduit | 122 (251) | 107 (224) | 60+ |
| 1" diameter conduit | 190 (374) | 154 (309) | 60+ |
| Air Drop #8 bare | 137 (279) | 117 (242) | 60+ |

Table 5-8 documents the fire endurance testing results conducted at UL for conduits protected with three layers of Interam™ E-53A and a junction box protected with three layers of 3M Interam™ E-54A ERFBS.

### Table 5-8. Peak Seals Test Results 1-Hour

| Configuration | Barrier Construction | Max. Temp. °C (°F) | Avg. Temp. °C (°F) | Rating (min.) |
|---|---|---|---|---|
| ¾" diameter conduit | 3 layers of E-53A | 167 (333) | 130 (266) | 60+ |
| 2" diameter conduit | 3 layers of E-53A | 157 (314) | 133 (251) | 60+ |
| Junction Box 24" x 24" x 10" | 3 layers of E-54A | 166 (330) | 106 (222) | 60+ |

5.1.3.1.3  Combustibility Testing (Prior to GL 86-10 Supplement 1)

3M documented the combustibility properties of several of its products in its June 9, 1993, response letter to NRC, reproduced in Table 5-9.

Table 5-9.  3M Combustibiltiy Test Results

| Material | Flame Spread | Fuel Contributed | Smoke Development |
|---|---|---|---|
| E-53A & E-54A mat | 0.7 | 0 | 0 |
| FireDam 150 Caulk | 2.2-6.3 | 0 | 0 |
| CS-195 Sheet | 17 | 0 | 0 |
| CP 25, CP 25N/S & CP 25S/L Caulks | 6 | 0 | 0 |

Based on a January 15, 1993, test report on noncombustible Interam™ E-50 Series Mat, the 3M Interam™ products are considered a Category No. 2 noncombustible.

The licensee of Davis Besse Nuclear Power Plant (DBNPP) conducted combustibility testing of E-50 series 3M ERFBS material.  ASTM E-136, "Behavior of Materials in a Vertical Tube Furnace at 750°C (1382°F)," testing requirements for noncombustible materials and ASTM E-84, "Surface Burning Characteristics of Building Materials," were used to test the barrier.  These tests were conducted at OPL and UL.  The 3M Interam E-50 material passed the E-136 test and had a flame spread rating of 0.7 according to the ASTM E-84 standard.  Based on NRC NUREG-0800 Standard Review Plan (SRP) Section 9.5-1 definition of a noncombustible material, the Interam E-50 series ERFBS would be classified as a non-combustible material.

5.1.3.1.4  Ampacity Derating (Prior to GL 86-10 Supplement 1)

The manufacturer reported the following ampacity derating values of the 3M Interam™ E-50 Series 1- and 3-hour Fire Protection Systems.

<u>1-hour System:</u>
Conduit      14-23%
Cable Tray   37-43%

<u>3-hour System:</u>
Conduit      20-30%
Cable Tray   45-52%

Two procedures were used to determine the ampacity derating values:

- "Procedure of the Ampacity Derating of Fire Protected Cables – June 10, 1986," prepared by SwRI.
- UL Subject 1712, "Tests for ampacity of insulated electrical conductors installed in fire protective systems."

In these tests, a baseline of 90°C (194°F) was used and equilibrium current was measured before and after applying the 3M Interam™ E-50 Series 1-hour and 3-hour ERFBS.  The ampacity derating percentages were calculated by dividing the ampere values of the protected system by those of the unprotected system, subtracting from one and then multiplying by 100. This method is presented in further detail in Appendix B to this report. Additional details on numerous 3M ERFBS ampacity testing is found in Appendix D of this report.

#### 5.1.3.1.5 Seismic Analysis

SwRI test reports dated July 1985 and July 1986, document the seismic testing performed on 3M Interam™ E-50 Series Mat ERFBS for generic qualification for use in nuclear and conventional power plants. The actual material tested was the Interam™ E-50A mat in the 1985 test and E-50D 3-hour barrier in the 1986 test. Both seismic testing series were conducted to show that the fire protection system would not break away or act as a missile when subjected to the specified seismic environment. The testing was also performed to demonstrate the effectiveness of the fire protection system would not be impaired as a result of the seismic tests performed.

In 1985, a 1-hour system consisted of five layers of E-50A and the 3-hour system consisted of 10 layers of E-50A mat. The test item was subjected to five Operating Basic Earthquake (OBE) tests and a single Safe Shutdown Earthquake (SSE) test. These were carried out in a five step sequence; ¼ level, ½ level, full level, 1 ½ level and 2x level as specified by the required response spectra (RRS). Test item included a 76.2 cm (30.0-in) wide cable tray, two 5.1 cm (2.0-in) diameter conduits, a Unistrut support, an air drop and 30.5 cm (12.0-in) cube junction box configurations.

The test results indicated that the 1-hour system showed slight physical damage at the highest test levels, confined to the interfaces of the four hangers to cable tray, conduit, and junction box. This damage was limited to tearing of the aluminum foil tape (T-49) at the seams, which is not a critical member of the fire protection system (vapor barrier). The testing of the 3-hour system resulted in physical damage near the joints of the system, where the supports met the raceway. This damage was noted at half level and full level and would required repair to ensure that the ERFBS fire protection characteristics would not be impaired. Based on the acceptance criteria of IEEE 344-1975 "IEEE Recommended Practices for Seismic Qualification of Class 1E Equipment for Nuclear Power Generating Stations" and IEEE 323-1974, "Guide for Qualifying Class 1E Electrical Equipment," these test results conclude that no part of the system broke loose or acted as a missile and therefore the E-50 Series ERFBS seismic performance was acceptable.

The 1986 test was performed on a conduit, air drop and cable tray configuration using a 3-hour E-50D barrier, and conducted in the same manner as the 1985 E-50A seismic tests. The cable tray was tested with a 38.0 Newton-m (28.0 lbs per ft) assembly setup and the conduits and junction box used a 1.84 Newton-m (1.36 lbs per ft) assembly. The seismic qualification testing was performed in successive levels of severity up to twice the SSE level (plus 10-percent margin) as specified by the RRS in the test plan. The test report concluded that no portion of the electrical raceway protection system broke away or acted as missiles for any of the seismic tests performed. Physical damage was observed to the system, mostly confined to the joints of the conduit and junction box (air drop) and where the unistruts and cable tray layers were joined. The damage was first observed following the tests performed at the half and full SSE level tests but was minimal at this point. At successively increasing levels of acceleration, the joints and cracks opened further.

Sargent and Lundy performed a seismic analysis for the 3M Rigid ERFBS, documented in a report dated August 26, 1982 (3M submittal, 1993). This report makes the following conclusion:

"The seismic loads are represented by a system of equivalent static loads corresponding to an acceleration of 5.0 g (0.2 oz) for the systems supported in the middle by the cable tray beams. The systems without a middle support are investigated for an equivalent load of 2.3 g (0.1 oz). The loads are applied simultaneously in three principal directions. The results of this study show that the subject fire barrier system can safely resist an acceleration of 5.0 g (0.2 oz) if it is supported in the middle by cable tray beams and 2.3 g (0.1 oz) without any support."

*5.1.3.2  3M Interam ERFBS Design Following GL 86-10 Supplement 1 Criteria*

The 3M Interam™ E-54A ERFBS design the passed the GL 86-10 Supplement 1 acceptance criteria, based on Peak Seals Procedure Nos. CTP-2003 and CPT-2005, consisted of the materials identified in Table 5-10.

Table 5-10. Components of 3M Interam™ E54A ERFBS

| Component | Application |
| --- | --- |
| Interam™ E54A | Primary Wrap System |
| Dow Corning 732 Adhesive/Sealant | Filling gaps and seams and at barrier terminations |
| FireDam™ FD-150 Caulk | Filling gaps at seams and at supports terminations. |
| T-49 Aluminum Foil Tape | Securing overlap joints, covering exposed mat and caulk at edges and seams. |
| ½" Stainless Banding and Clips | Securement of final mat layer. |

5.1.3.2.1  Fire Endurance Testing (Post GL 86-10 Supplement 1)

The vendor (Peak Seals) performed fire endurance testing in accordance with GL 86-10 Supplemen1 at OPL. Table 5-11 provides a summary of the results and identifies configurations capable of providing the 3-hour fire endurance rating. Table 5-12 provides information on the qualified design developed by Peak Seals. Additional details of these test reports are provided in Appendix D.

Table 5-11.  3M Interam™ E54A Fire Endurance Test Results (OPL Project Nos. 14540-99416 & 14540-99417)

| Thermocouple Location | Fire Endurance Rating (hours) | Max. Single Pt. Temp. (Max Allowable) °C | Max Average Temp. (Avg. Allowable) °C |
| --- | --- | --- | --- |
| 5" Steel Conduit | 3 | 153 (196) | 117 (154) |
| 3" Steel Conduit | 3 | 153 (196) | 153 (154) |
| 1" Steel Conduit | 3 | 192 (196) | 153 (154) |
| Steel Junction Box | 3 | 107 (196) | 101 (154) |
| 6" wide Cable Tray | 3 | 167 (214) | 129 (172) |
| 24" wide Cable Tray | 3 | 163 (214) | 133 (172) |

**Table 5-12. Summary of Peak Seals Upgrade Testing of 3-hr 3M Interam Design to GL 86-10 Supplement 1 Criteria**

| Raceway Configuration | Thermocouple (TC) Location | Average °C (°F) | Maximum °C (°F) | Upgraded Design | Previous Design[1] |
|---|---|---|---|---|---|
| 6" Cable Tray[1] | Front Side Rail | (256) | (319) | 6 layers E54A | 5 layers E54A |
|  | Rear Side Rail | (264) | (333) |  |  |
|  | #8 AWG Conductor | (247) | (303) |  |  |
| 24" Cable Tray[1] | Front Side Rail | (271) | (325) | 6 layers E54A | 5 layers E54A |
|  | Rear Side Rail | (269) | (315) |  |  |
|  | #8 AWG Conductor | (231) | (287) |  |  |
| 5" Conduit[2] | Conduit Surface | (243) | (307) | 5 layers E54A with standoff | 5 layers E54A |
|  | #8 AWG Conductor | (227) | (269) |  |  |
| 3" Conduit[2] | Conduit Surface | (237) | (307) | 5 layers E54A with standoff | 5 layers E54A |
|  | #8 AWG Conductor | (220) | (270) |  |  |
| 1" Conduit[2] | Conduit Surface | (307) | (377) | 5 layers E54A with standoff | 6 layers E54A |
|  | #8 AWG Conductor | (299) | (346) |  |  |
| 2" Airdrop[3] | #8 AWG Conductor | (294) | (338) | 5 layers E54A with standoff | 5 layers E54A |
| Junction Box[2] | Metal Surface | (213) | (224) | 6 layers E54A with standoff | 6 layers E54A |

1 "Previous Design" refers to configurations presented in Table 5-3, Table 5-4, and Table 5-5.
2 OPL Project 14540-99417 (allowed Avg. T=342°F & Max. T=417°F)
3 OPL Project 14540-99416 (allowed Avg. T=310°F & Max. T=385°F)
4 OPL Project 14540-99123 (allowed Avg. T=315°F & Max. T=338°F)

#### 5.1.3.2.2 Combustibility Testing (Post GL 86-10 Supplement 1)

The vendor (Peak Seals) performed ASTM E136, "Standard Method of Test for Behavior of Materials in a vertical tube furnce at 750°C," at OPL on December 14, 1995. The report dated January 15, 1996, only identifies the material tested as "3M E-50 INTERAM™ SERIES MAT," and "Flexible Endothermic Fire Wrap." The report concludes that the specimens met the specific criteria of the standard and are shown in Table 5-13.

**Table 5-13. 3M Interam Combustibility Test Results Conducted at OPL**

| Specimen Number | Initial Wt. (g) | Final Wt. (g) | Wt. Loss (%) | Furnace Temp. at Start of Test (°C) | Max Surface Temp. (°C) | Max. Interior Temp. (°C) |
|---|---|---|---|---|---|---|
| 1 | 60.9 | 49.1 | 19.4 | 750 | 703 | 283 |
| 2 | 59.2 | 47.8 | 19.3 | 750 | 768 | 367 |
| 3 | 63.8 | 51.3 | 19.6 | 750 | 761 | 392 |
| 4 | 64.2 | 52.0 | 19.5 | 750 | 774 | 306 |

5.1.3.2.3  Ampacity Testing (Post GL 86-10 Supplement 1)

The vendor (Peak Seals) performed ampacity tests on the 3M Interam™ E54 ERFBS design qualified to GL 86-10 Supplement 1 criteria at OPL. The results of this testing are shown in Table 5-14. Additional information on this testing is provided in Appendix D to this report.

Table 5-14.  3M Interam™ E54A Ampacity Results

| Test Item | Fire Resistance Rating (hours) | OPL Report No(s). | Ampacity Derating Factor (%) |
|---|---|---|---|
| 1" Conduit | 1 | 14540-99074/75 | 7.86 |
| 4" Conduit | 1 | 14540-99074/75 | 8.96 |
| 24" Wide Cable Tray | 1 | 14540-100770 | 49.88 |
| 1" Conduit | 3 | 8610-102164/65 | 20.29 |
| 4" Conduit | 3 | 8610-102164/65 | 34.92 |
| 24" Wide Cable Tray | 3 | 8610-102166 | 56.62 |

5.1.3.2.4  LOCA Testing (Post GL 86-10 Supplement 1)

In 1998, the vendor (Peak Seals) had loss of coolant accident (LOCA) testing performed on the 3M Interam™ ERFBS, which consisted of radiation exposure, LOCA simulation, and Post-Test Inspection.

The test specimen consisted of two cable trays (24" wide x 4" deep x 24" long) covered with three layers of 3M Interam™ E54C endothermic mat.  One specimen was exposed to gamma radiation from a Cobalt-60 source, receiving a total dose of $2.042 \times 10^8$ rads, while the other was not exposed to any radiation.   Both test specimens were then exposed to elevated temperatures, pressures, and chemical sprays.  The report documents that there was "no disintegration or dislodgement of the mat material or associated components."

### 5.1.4   Resolution and Staff Conclusion

The physical structure and chemical decomposition of 3M Interam™ ERFBS during fire exposures allowed this barrier to perform its specified function.  Fire endurance testing has shown that early designs (pre-1996) of the 3M Interam™ ERFBS were not capable of providing the required 1- or 3-hour fire resistance level of protection required by the regulations; however, continued development and refinement of the ERFBS resulted in the development of an 3M Interam™ E54A ERFBS that has been shown capable of achieving a fire endurance rating as outlined in GL 86-10 Supplement 1 guidance.  Unlike other barrier systems, the manufacture and later distributer performed numerous fire endurance, seismic, LOCA, and ampacity derating testing of which adequately bound the configurations used in NPPs.  As a result of these attributes, the staff has only identified problems with the early 3M Interam™ designs (as identified in IN 95-52 and IN 95-52 Supplement 1).

As a result of this review the staff concluded that Interam 3M Interam™ ERFBS when installed to bound as tested configurations will satisfactory perform its intended design function.

5-20

## 5.2 Thermo-Lag 330-1 (Ablative Material)

Thermo-Lag 330-1 is manufactured by Thermal Science, Incorporated (TSI), of St. Louis Missouri. TSI manufactures a variety of products used in heat transfer applications and the "330-1" class of materials is that typically found in U.S. NPPs. Thermo-Lag can be constructed to provide a fire endurance design rating of 1-hour or 3-hours as required by NRC regulations and guidance.

Thermo-Lag 330-1 is an ablative, water-based material that will volatize at fixed temperatures and change from solid to vapor state. Physically this is achieved by the materials ability to exhibit a volumetric increase through the formation of a multicellular matrix that blocks heat to protect the substrate material to which it is applied. In this process, the ablative agent is consumed through sublimation and mass loss, which when properly designed provides cooling and forms a thermal shield.

During elevated thermal exposures, the Thermo-Lags' highly endothermic reaction takes place above the sublimation layer, in the layer which has been formed by the combined defects of pyrolysis of the binder and other ingredients contained within Thermo-Lag. The Thermo-Lag material composition also includes a specified quantity of glass fiber that strengthens the virgin material and also enhances the physical retention properties of the char layer when formed. The formed char layer has a high emissivity that makes its surface an effective retardant to heat. The char layer further serves as an effective mass and heat exchanger as well as a transport medium for the volatiles leaving the subliming surface.

The original use of Thermo-Lag was in spray-on applications to provide fire protection of structural steel members. When the need to protect electrical cables presented itself, the TSI proposed using the spray-on application of Thermo-Lag as a rated ERFBS. During the initial installation of Thermo-Lag on NPPs electrical raceway, the installers found that an excessive amount of off spray was being applied to adjacent raceways and other equipment that did not require protection. This problem caused TSI to develop a new method of using Thermo-Lag to protect the electrical raceways. What they developed was a prefabricated panel and half-round conduit system that used the base Thermo-Lag material and some additional structural members. This new product is what has come to be known and recognized in the nuclear industry as Thermo-Lag 330-1.

The Thermo-Lag 330-1 material is manufactured in nominal 1.588±0.318 cm (0.625±0.125 in) flat panels (for use on cable trays and junction boxes) and in half round prefabricated sections (sized for use on conduits). Most utilities use Thermo-Lag 330-1 ERFBS to satisfy NRC's fire protection requirements for safe shutdown capability and in some cases licensees use Thermo-Lag to achieve physical independence of electrical systems per Regulatory Guide (RG) 1.75. Thermo-Lag 330-1 ERFBS is used in NPPs to protect cable trays, conduit, air drops (cable in free space), junction boxes and structural supports and hangers. Thermo-Lag 330-1 material has also been used as components of penetration seals and fire barrier walls. In addition, Thermo-Lag ERFBSs were used by some plants to construct a Radiant Energy Shield (RES) for cables located in containment; however, when it was identified that Thermo-Lag is a combustible material, the utilization of Thermo-Lag as a RES was typically eliminated or modified to eliminate combustible materials within containment.

The Thermo-Lag 330-1 ERFBS exists in several basic designs for use in NPPs, including:

- Pre-fabricated Panel Design
- Pre-shaped Conduit Section Design
- Direct Spray Over Stress Skin Design
- Direct Spray-on Design.

The first three consist of the same material components—a Thermo-Lag Stress Skin and a Thermo-Lag 330-1 ablative material—the only difference being a prefabricated product versus a spray-over application.

The stress skin is a steel mesh[10] used in conjunction with the Thermo-Lag 330-1 ERFBS. It is composed of an open-weave, self-stiffened steel mesh and is used to provide an enclosure and mechanical base for the Thermo-Lag 330-1 ablative material. The stress skin was originally designed to be placed over cable trays, conduits, and other items (i.e., beneath the ERFBS), but some licensees have also applied the stress skin around the exterior of a Thermo-Lag ERFBS with a top coat of trowel grade material to help reinforce and upgrade the barrier system.

The trowel grade Thermo-Lag 330-1 material is the same material used to fabricate the prefabricated panels and preshaped or preformed conduit section. It can also be supplied by the vendor in a sprayable form. The trowel grade material is typically applied to seal the joints between adjacent Thermo-Lag panels but, as discussed later, was also used to reinforce and upgrade the Thermo-Lag ERFBS. Common terms used in the trade were "pre-butter" or "post-butter" thermo-lag assemblies, meaning the trowel grade Thermo-Lag 330-1 was applied prior to assembly (i.e., pre-butter) or applied after assembly to fill joints (i.e., post-butter). The trowel grade Thermo-Lag 330-1 requires a minimum of 72 hours to cure or a moisture content of less than 100 when using a meter with a scale of 0-100.

The direct spray method installations are limited to Susquehanna Steam Electric Station Unit 1 and 2, and limited applications at Washington Nuclear Project, Unit 2 (now Columbia Generating Station). Most Thermo-Lag fire barriers installed in the field are constructed of prefabricated Thermo-Lag 330-1 panels and preshaped conduit sections that have been cut to size and shape and fastened together with either stainless steel wires or bands. When securing the half round preshaped conduit pieces to conduit, the manufacture recommends as a minimum, an 18 gauge standard stainless steel wire and/or a 0.05 cm (0.02 in) thick by 1.27 cm (0.50 in) wide standard stainless steel banding be used.

In addition to protecting the raceway with the Thermo-Lag system, the vendor also recommends that all penetrations into the ERFBS should be fire protected for a distance of at least 45 cm (18 in) measured from the outer surface of the fire barriers (to prevent thermal shorts). That is, any raceways support members of adjoining raceways also need to be protected by the ERFBS for a particular distance. Thermal short members, such as support members, may be covered (or protected) for various distances depending on the qualified fire tested configurations.

Although design and construction of Thermo-Lag 330-1 ERFBS vary by plant, Figure 5-1 and Figure 5-2 show a breakdown of the components used to construct a Thermo-Lag conduit and cable tray ERFBS, respectively.

---

[10] Stress Skin physical parameters 0.043 mcm (0.017 inch] minimum diam; 56 holes/sq. in. minimum; 1.75 lbs/sq yd min)

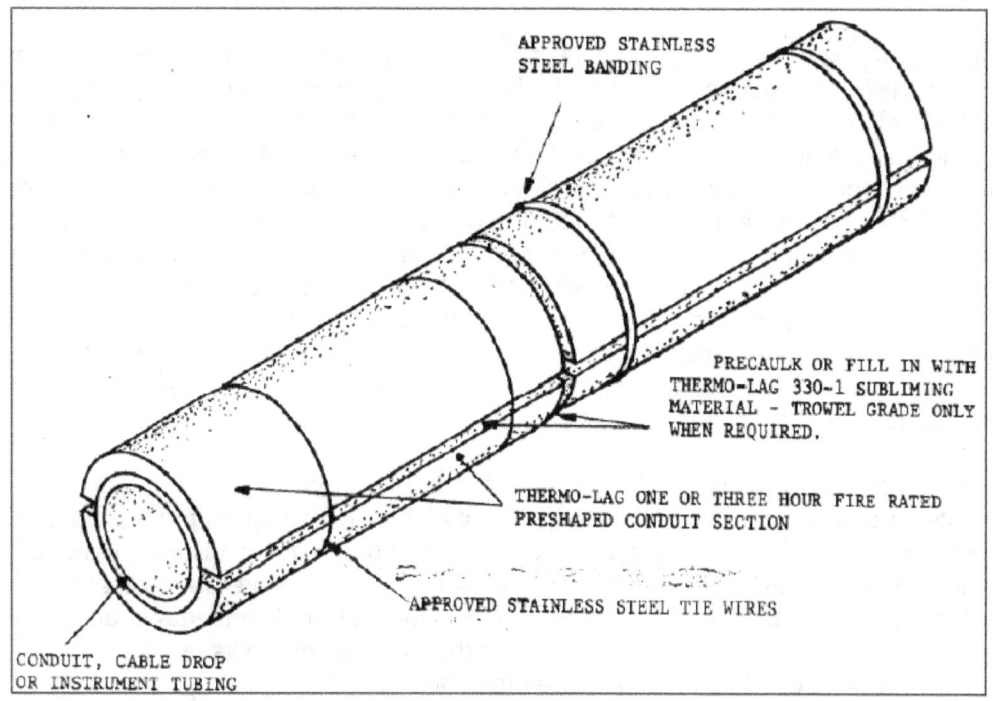

**Figure 5-3.** Thermo-Lag 330-1 ERFBS Conduit Application 1- or 3-hour (TSI Technical Note)

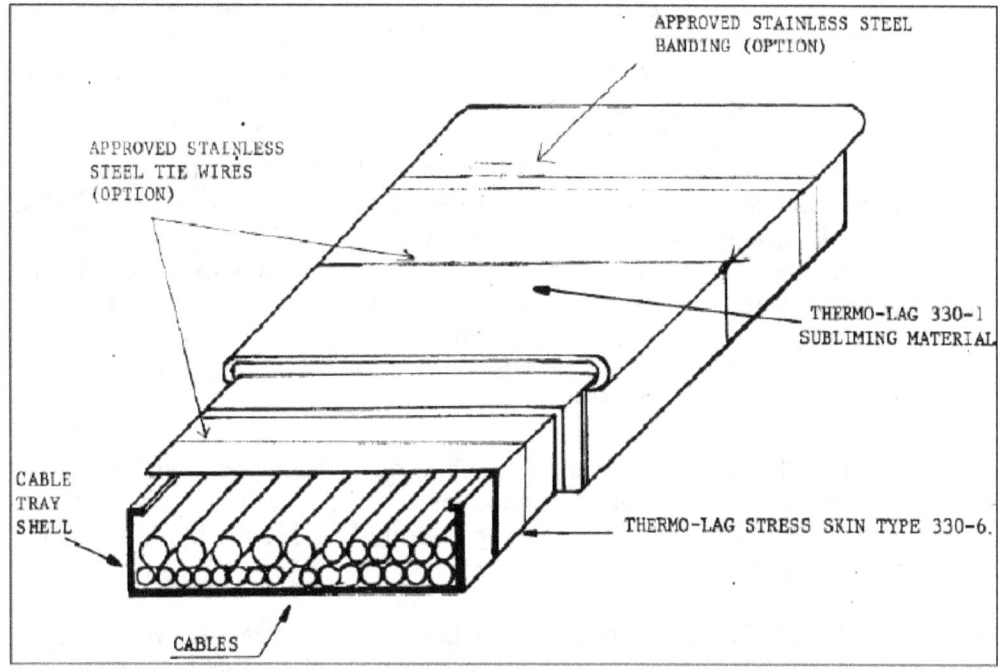

**Figure 5-4.** Thermo-Lag 330-1 ERFBS Cable Tray Application 1- or 3-hour (TSI Technical Note)

## 5.2.1 History

Thermo-Lag 330-1 has had a long and contentious history in the NPP industry. Licensees who made the first NPP installations of this material did not fully recognize the physical properties and limitations of the material nor did they understand its potential benefits. When Appendix R was published as a regulation, nuclear utilities unable to meet the requirement for 6.1 m (20 ft) of separation between redundant equipment needed to quickly correct their problem, and TSI's Thermo-Lag 330-1 ERFBS appeared to meet their need. As a result of Thermo-Lag's being readily available and having test reports documenting its performance (which would later be questioned), Thermo-Lag became the predominant ERFBS used in the industry for compliance with Appendix R. However, as discussed below, when the true nature and performance of this material became more widely understood, the licensees had to expend considerable resources (i.e., a multimillion dollar fire testing project was conducted) to bring their plants into compliance with Appendix R's requirements.

Because Thermo-Lag 330-1 had no history of use in NPPs to protect safe shutdown cable/circuits, prior to 1980 utilities proposing to install this fire barrier material sought NRC staff acceptance. Along with their proposals to use Thermo-Lag 330-1, the utilities submitted test reports and other documentation to qualify Thermo-Lag 330-1 as a fire barrier that met NRC's fire protection requirements. NRC began receiving requests from licensees for acceptance of Thermo-Lag 330-1 in 1981, but it wasn't until after they first accepted its use that numerous additional proposals to use this material were submitted to NRC. Within a few short years over three-fourths of the nation's commercial NPPs had Thermo-Lag installed for Appendix R compliance.

NRC's concerns regarding Thermo-Lag 330-1 ERFBS began after receiving several licensee event reports (LERs) from Gulf States Utilities (GSU) between 1987 and 1990, citing failed qualification fire tests and installation problems. The LERs stated that the ASTM E-119 fire endurance testing GSU had performed at Southwest Research Institute (SwRI) showed the 3-hour Thermo-Lag ERFBS installed on wide aluminum cable trays resulted in a complete failure within about 60 minutes (i.e., one-third of the 3-hour requirement). GSU conducted this confirmatory testing after identifying that the fire barriers had not been installed at its River Bend Station in accordance with the manufacturer's specifications. NRC issued IN 91-47, "Failure of Thermo-Lag Fire Barrier Material to Pass Fire Endurance Test," dated August 6, 1991, to inform NPP licensees of this issue. At the time of issuance, NRC knew of at least 40 plants that had used Thermo-Lag to construct fire barrier assemblies with 1-hour and 3-hour ratings to enclose electrical raceways and other safe shutdown equipment. The amount of Thermo-Lag used at each plant varied from only two conduits at Monticello to over 1858 $m^2$ (20,000 $ft^2$) at Comanche Peak Steam Electric Station, Unit 1 (CPSES).

On December 6, 1991, NRC issued IN 91-79, "Deficiencies in the Procedures for Installing Thermo-Lag Fire Barrier Materials," which provided information on deficiencies in procedures that the manufacturer (TSI) provided for installation Thermo-Lag 330 fire barrier material, along with details of the TU Electric test failures.

In response to GSUs operating experience, NRC established a special review team in June 1991 to review the safety significance and generic applicability of the technical issues regarding the use of Thermo-Lag. As part of the teams' effort, about 40 fire endurance test reports and nine ampacity derating test reports were reviewed. Based on this review, the team determined that the fire endurance rating of the Thermo-Lag 330-1 system to be indeterminate and the ampacity derating tests indicated conflicting results. In addition, the team found that some licensees did not adequately review and evaluate the test results, did not adequately review their configurations, were not bounded by tested configurations, and used inadequate or incomplete installation procedures. Based on these findings, the review team issued IN 91-47, "Failure of Thermo-Lag Fire Barrier Material to Pass Fire Endurance Test," August 5, 1991, and IN 91-79, "Deficiencies in the Procedures for Installing Thermo-Lag Fire Barrier Materials," dated December 6, 1991. Following the completion of this effort, NRC issued IN 92-46, "Thermo-Lag Fire Barrier Material Special Review Team Final Report Findings, Current Fire Endurance Testing and Ampacity Calculations Errors," on June 23, 1992. This report informed the nuclear power utilities of the recent raceway barrier fire endurance testing failures completed by TU Electric and identified an ampacity calculation error in a test report published by Industrial Testing Laboratories (ITL) Incorporated Test Report ITL-82-5-355C.

Attachment 1 to IN 92-46 contained the Final Report of an NRC Special Review Teams finding on the Review of Thermo-Lag Fire Barrier Performance. The final report concluded that the fire resistance ratings and ampacity derating factors for Thermo-Lag 330-1 ERFBS were indeterminate[11] and that some licensees had not adequately reviewed and evaluated the fire endurance and ampacity test results for applicability to the Thermo-Lag ERFBS installed in their facilities. In addition, the special review team found that some licensees had used inadequate installations procedures to construct their Thermo-Lag fire barriers.

Following issuance of IN 91-47 and IN 91-79 and due to its wide use, TU Electric conducted a series of full-scale fire endurance tests to qualify the Thermo-Lag 330-1 electrical raceway fire barrier configurations it had installed at its Comanche Peak Steam Electric Station. These tests produced additional fire endurance failure results on wide cable trays and small conduits.

On the basis of these findings, NRC issued Bulletin 92-01, "Failure of Thermo-Lag 330 Fire Barrier Systems to Maintain Cabling in Wide Cable Trays and Small Conduits Free From Fire Damage," and Supplement 1, "Failure of Thermo-Lag 330 Fire Barrier Systems to Perform its Specified Fire Endurance Function," in June and August 1992, respectively. The bulletin identified that NRC had made the determination that the 1- and 3-hour Thermo-Lag 330-1 preformed assemblies installed on conduits smaller than 10 cm (4 in) diameter and cable trays wider than 36 cm (14 in) did not provide the level of safety needed to meet NRC requirements. Bulletin 92-01 required licensees to identify areas that contained such constructions, implement the appropriate compensatory measures, and provide NRC with written notification of its use of Thermo-Lag 330-1 fire barrier systems. NRC required the licensees compensatory measures to remain in place until the licensee could declare the fire barriers operable on the basis of applicable tests that demonstrate successful 1- or 3-hour barrier performance.

Following receipt of all licensee responses related to NRC Bulletin 92-01, NRC staff determined that 83 operating plants had Thermo-Lag installed and 28 operating plants did not (based on September 21, 1992, data). The staff also determined that most of the licensees with Thermo-Lag 330-1 installed did take the proper corrective actions, (i.e., they had declared the barriers inoperable and implemented compensatory measures consistent with their Technical Specifications or licensing conditions for an inoperable barrier). However, some licensees that had declared their barriers inoperable also provided arguments to support a determination of operability. In most cases, the arguments were that the low fire loading in the area, control of transient combustibles, or other administrative controls provided adequate assurance that the

---

[11] Indeterminate test results meant that the details of the testing were not sufficient for the staff to conclude that those tests served as an acceptable regulatory basis for Appendix R compliance.

Thermo-Lag would remain operable for the limited amount of time needed to perform its functions.[12]

The staffs' positions with regard to Thermo-Lag installations within the scope of Bulletin 92-01 were;

1. The staff considers Thermo-Lag barriers inoperable unless the licensee has specific test data that would demonstrate otherwise. Operability determinations made on the basis of functionality of the protected system are not acceptable. To be operable, the barrier must be capable of performing its specified function for 1- or 3-hours as required.

2. GL 86-10 interpretations regarding fire area boundaries and deviations from tested configurations are not applicable since the supporting engineering analysis assumes that a qualified tested configuration that has successfully passed the test acceptance criteria is being used as a basis for the analysis.

The bulletin was followed by GL 92-08, "Thermo-Lag 330-1 Fire Barriers," in December 1992, requesting information from licensees on their use of Thermo-Lag 330-1 to verify compliance with NRC requirements. GL 92-08 addressed three areas of NRC concern: (1) fire endurance capability of Thermo-Lag 330-1 barriers, (2) ampacity derating of cables enclosed in Thermo-Lag 330-1 barriers, and (3) evaluation and application of the results of the endurance and ampacity tests.

GL 92-08 provided for the mechanism for NRC to evaluate the specific details of each licensee's use of Thermo-Lag ERFBS and to prompt the nuclear industry to resolve issues related to deficient barriers. Section 6 provides a brief description on each plants resolution to problems associated with various barriers used at that plant. NRC closeout of GL 92-08 was a significant effort, involving hundreds of requests of additional information (RAI) and several site verification visits. GL 92-08 requested that the licensee provide the following information:

- Chemical composition.
- Material thickness.
- Material weight and density.
- Presence of voids, cracks, and delimitations.
- Fire endurance capabilities.
- Combustibility.
- Flame spread rating.
- Ampacity derating.
- Mechanical properties such as tensile strength, compressive strength, shear strength, and flexural strength.

Following the numerous response to GL 92-08 and subsequent RAIs, NRC staff met with licensees to discuss their plans and schedules for implementing GL 92-08. NRC staff became concerned with the licensees completing their commitments when a number of licensees reported that they had already passed their completion dates without complete resolution. Some licensees informed NRC that their completion dates had slipped by as much as three

---

[12] Plants not initially declaring barrier inoperable following Bulletin 92-01 included: Oyster Creek; Three Mile Island1; Beaver Valley 2; Vermont Yankee; St Lucie 1 and 2; Browns Ferry 1, 2 and 3; Sequoyah 1 and 2; Davis Besse; Zion 1 and 2.

years. In 1998, NRC issued Confirmatory Orders to each plant with outstanding Thermo-Lag resolution, modifying their license. These Orders required the plants to complete their Thermo-Lag modification by the dates previously committed to NRC. Table 5-15 provides a list of those plants issued Confirmatory Orders, along with the NRC Agencywide Document Access and Management System (ADAMS) Accession Numbers for those documents.

Table 5-15. Thermo-Lag 330-1 Confirmatory Order Documentation

| Plant Name | Docket No. | Date (ADAMS Accession No.) | |
|---|---|---|---|
| | | Confirmation Order | Order Completion |
| St. Lucie Plant, Unit 1 | 50-335 | 07/13/98 (ML013580124) | 04/07/00 (ML003703549) |
| Three Mile Island Nuclear Station, Unit 1 | 50-289 | 05/22/98 (ML003765653) 08/11/99 (ML003766024) | 12/30/99 (ML003676460) 03/12/00 (ML003693928) |
| Columbia Generating Station | 50-397 | 03/25/98 (ML022130143) | 01/19/00 (ML003678400) |
| Peach Bottom Atomic Power Station, Units, 2 and 3 | 50-277,278 | 05/19/98 (ML040990313) | 10/12/99 (ML040990314) |
| Limerick Generating Station, Units, 1 and 2 | 50-352,353 | 05/19/98 (ML011560778) | 09/17/99 (ML040990326) |
| Crystal River Unit 3 Nuclear Generating Plant | 50-302 | 05/21/98 (ML020670496) | 05/25/00 (ML003722384) |
| Sequoyah Nuclear Plant, Units, 1 and 2 | 50-387,388 | 07/02/98 (ML010160064) | 04/28/00 (ML003711917) |
| North Anna Power Station, Units, 1 and 2 | 50-338 | 06/15/98 (ML013530026) | 02/01/99 (ML040990189) |
| Sequoyah Nuclear Plant, Units, 1 and 2 | 50-327,328 | 06/18/98 (ML013320074) | 06/30/99 (ML040990478) |
| Davis Besse Nuclear Power Station, Unit 1 | 50-346 | 06/22/98 (ML021210216) | 01/25/99 (ML040990274) |
| Clinton Power Station, Unit 1 | 50-461 | 06/26/98 (ML020990547) | 04/27/99 (ML040990340) |
| Comanche Peak Steam Electric Station, Units, 1 and 2 | 50-445,446 | 07/28/98 (ML021820291) | 12/22/98 (ML040990491) |
| Turkey Point Nuclear Generating, Units, 3 and 4 | 50-250,251 | 07/09/99 (ML013390600) | 06/18/01 (ML011770240) |
| Oyster Creek Nuclear Generating Station | 50-219 | 05/22/98 (ML040990167) | 01/30/01 (ML010370267) |
| Edwin I. Hatch Nuclear Plant | 50-321,366 | 06/24/98 (ML013030297) | 10/16/98 (ML040990196) |
| Surry Power Station, Units, 1 and 2 | 50-280,281 | 07/09/98 (ML012700090) | 02/01/99 (ML040990189) |
| South Texas Project, Units, 1 and 2 | 50-498,499 | 10/02/98 (ML040990301) | 02/08/99 (ML040990180) |

On May 20, 1994, NRC staff briefed the Commission on the status of Thermo-Lag issues. As a result of this meeting the staff was directed to provide details on which plants had achieved compliance with Appendix R, how much Thermo-Lag material was previously used in these plants, and the corrective actions performed. Section 6 of this report, provides plant-specific information related to resolution of Thermo-Lag ERFBS issues.

In addition to providing NRC Information Notices on numerous deficiencies with Thermo-Lag ERFBS, NRC special technical review team, the U.S. Attorney's Office, NRC Office of Inspector General (OIG), and NRC Office of investigations conducted an investigation as a result of numerous anomalies with the reviewed test report. On March 30, 1994, the testing laboratory that certified the original Thermo-Lag fire tests, ITL of St. Louis, Missouri, and Alan M. Siegel, the president of the company, pleaded guilty to five counts of making and aiding and abetting the making of false statements within the jurisdiction of NRC, in violation of Title 18, US Code, Section 1001 and 1002. More than 30 false reports transmitted from TSI to NRC and other entities. ITL, Inc. was fined $150,000 and agreed to fully cooperate in the criminal investigation and prosecution of organizations and individuals associated with the Thermo-Lag fire barrier material. On September 29, 1994, TSI was charged by a federal grand jury in Maryland with seven counts of wrongdoing, including conspiracy and fraud. On August 1, 1995, a Federal jury found TSI and its president not guilty of making false statements about the role of ITL in the qualification testing of Thermo-Lag ERFBS.

### 5.2.2 Problems

#### NRC Staff Findings

Following issuance of IN 91-47 NRC staff visited several sites to inspect the as installed Thermo-Lag ERFBS and associated documentation. During those site visits, NRC staff found a number of field installations that were not constructed in accordance with the vendor recommended installation procedures. The staff also found that the vendor had revised its recommended installation procedures without notifying the licensees, and that the vendors' installation procedures were not complete. These two issues were a major cause of Thermo-Lag ERFBS variations among plants because the installers would construct the barriers following either the old procedures or their own judgment when the procedures didn't provide specific instruction regarding a particular aspect of the installation. As a result, the qualification of all barriers so constructed was brought into question.

Upon further review, the NRC staff identified some configurations that did not appear to be qualified by fire endurance testing, and installations that deviated from the tested configurations without adequate engineering justification. From these findings, it was clear to the staff that further regulatory oversight was needed to ensure that issues identified in the field were brought to resolution and all licensees who used ERFBS had qualified and properly installed barriers for the configurations in their plants.

#### Acceptable Test Report Become Unacceptable

Beginning in 1981, NRC had received numerous reports documenting fire tests of Thermo-Lag 330-1 that were conducted by TSI and witnessed and documented by ITL. In the early 1990s, an NRC review of a number of these reports disclosed that the TSI tests had not been performed in accordance with the applicable standards. For example, the test furnace and temperature measuring devices used by TSI during the tests did not meet the ASTM E-119 standard. Also, NRC requirements state that a fire endurance test on barrier materials must be conducted by a nationally recognized fire testing laboratory. Although it was later learned that neither ITL nor TSI had acceptable fire testing experience, NRC staff (erroneously) accepted the

ITL test reports in the 1980s for the TSI tests[13], and those reports were subsequently used throughout the industry to qualify Thermo-Lag 330-1 for use in NPPs.

A later OIG inspection found that although the ITL test reports state the fire tests were supervised and controlled entirely by ITL, the ITL representative was present only as a witness to verify that a test was conducted. The test reports were actually written by TSI and then signed by the President of ITL with no substantive verification that the data in the reports reflected the actual tests. In some instances, the ITL President merely signed test report cover sheets without seeing the test report. OIG identified about 25 tests of Thermo-Lag 330-1 that were conducted by TSI with ITL "acting as a witness." Since neither TSI nor ITL were qualified per NRC requirements to conduct the tests, further discussion of who ran and witnessed the tests is important only for legal or administrative issues.

## Installation Errors & Procedure Issues

The most prominent problem involving Thermo-Lag 330-1 ERFBS was the differing and changing installation requirements. NRC staff found that although the Thermo-Lag 330-1 materials performed adequately in laboratory test furnaces, field installations introduced uncertainties due to variations in the training and abilities of installation personnel. In several instances, NRC staff found that the protection provided did not qualify as a 1- or 3-hour fire barrier because the licensee applied the material improperly and in untested configurations. When these configurations were tested, results showed that the 1-hour barriers would actually only provide a nominal 32-minute to 50-minute fire rating while the 3-hour application might provide a 150-minute to 160-minute fire rating.

While conducting site visits after issuing IN 91-47, NRC staff observed that the vendor had revised its recommended installation procedures without notifying the licensees, that the vendor installation procedures were incomplete, that a number of field installations were not constructed in accordance with the vendor recommended installation procedures, that some installations did not appear to be qualified by fire endurance testing, and that some installations deviated from the tested configurations without justification. All of these issues resulted in wide variation in the barriers' performance among the plants.

Simple material parameters, such as, inadequate Thermo-Lag thickness also resulted in fire barrier degradation. One of the larger problems associated with installation of the Thermo-Lag 330-1 fire barrier assemblies resulted from the product's not coming from the vendor as a complete assembly (such as a fire door assembly). Instead, assemblies were often "custom built" to meet variations in the actual in-plant installations as compared to the tested configurations, these variations commonly resulted in plant-to-plant dissimilarities in the barriers' performance.

---

[13] NRC staff review of the test reports consisted of an audit of the paperwork submitted by the utilities. NRC staff considered it to be the responsibility of the utilities to provide accurate information concerning the conduct of the qualification tests. The licensees' submittals were under oath and affirmation per 10 CFR 50.9, "Completeness and Accuracy of Information."

## Ampacity Derating

When current flows in a conductor, heat is produced because every conductor offers some resistance to the flow of current. The current-carrying capacity of a particular conductor is dictated by its "ampacity" (that is, how many amperes of current it can handle based on its design). ERFBS inhibit electrical cable conductors from dissipating resistive heat energy to the environment and thus cables protected by ERFBS must be derated to ensure sufficient capacity and capability to perform their intended safety functions.

The special review team reviewed nine ampacity derating test reports and found conflicting test results. For example, the vendor has reported derating factors for cable trays that range from 7 to 28 percent for 1-hour fire barriers and from 16 to 31 percent for 3-hour barriers. In addition, ampacity derating tests of Thermo-Lag materials conducted for 3M found the ampacity derating to be 37 percent for a 1-hour barrier, 9 percent higher than what had been previously reported by the vendor. There are similar inconsistencies for conduit barriers. In addition, Sandia National Laboratories (SNL) conducted Ampacity testing of a Thermo-Lag 330-1 "U"-shaped configuration and found Ampacity Derating factors to be even higher than that specified by the previous testing. (See Section 5.2.3.1 below for more information on the SNL ampacity testing.)

The results of an OIG inspection identified the root cause of the inconsistencies, excerpt follows:

> Originally, TSI reported to CPSES that Thermo-Lag 330-1 would require a 10-percent ampacity derating. In 1982, TSI conducted an ampacity derating test with ITL as the witness and produced a derating factor of about 17 percent. During this same time period, manufactures of other fire barrier materials conducted ampacity derating tests and reported ampacity derating figures far higher than those reported by TSI, some as high as 40 percent.
>
> In 1986, an ampacity derating test on Thermo-Lag 330-1 was conducted at a nationally recognized laboratory—Underwriters Laboratories (UL). However, TSI refused to follow the UL ampacity derating testing procedure and these non-standard tests resulted in ampacity derating figures of about 31-percent for the 3-hour Thermo-Lag 330-1 and about 28-percent for the 1-hour Thermo-Lag 330-1 ERFBS. These figures were significantly larger than those previously reported by TSI. Following TSI representative leaving the UL testing facility, UL performed an additional ampacity test on Thermo-Lag 330-1 following UL procedures, resulting in ampacity derating factors of nearly 40-percent for the 3-hour barrier and 36-percent for the 1-hour.

Unfortunately, these results were not reported to NRC at the time they were discovered and were only identified during an OIG inspection six years later, in 1992.

## Licensee Review Evaluation

When licensee performed independent testing to verify an installed barriers capability, they typically found two physical deficiencies (1) for conduits less then 10.2 cm (4 in) the temperatures recorded during testing exceeded the maximum allowable limits, and (2) joints on the barriers where two sections of material butted would open during the fire test.

Initial confirmatory and plant-specific testing raised numerous questions associated with the capability of Thermo-Lag 330-1 ERFBS to perform its design function. For example, during tests did the Thermo-Lag 330-1 structurally remain intact for the fire exposure? Is there a sufficient quantity of Thermo-Lag 330-1 material (i.e., thickness) to protect electrical raceways of differing mass and materials?

## Bounding Plant Installations

A few instances were identified where facilities had installed fire barriers without a basis for their fire rating such as an UL Listing or testing conducted by a nationally recognized testing laboratory for the configurations installed in the plant. Some licensees did not adequately reviewed and evaluated the fire endurance test results and ampacity derating test results used as the licensing basis for their Thermo-Lag 330-1 barriers to determine the validity of the tests and the applicability of the test results to their plant designs. Some licensees did not adequately review installed fire barrier configurations to ensure that they either replicate the tested configuration or provide an equivalent level of protection.

## Combustibility

The National Institute of Standards and Technology (NIST) testing (Section 5.2.3.1) provided results that allowed NRC to conclude that the Thermo-Lag 330-1 ERFBS is a combustible material. NRC fire protection requirements (Section III.G, Appendix R to 10 CFR 50) preclude the use of combustible material to (1) enclose other combustibles, such as cables, between redundant safe shutdown trains to eliminate the combustibles as a fire hazard or (2) provide radiant energy heat shield protection form shutdown components inside containments.

## OIG Inspection Report

In August 1992, an OIG investigation determined that NRC staff had accepted manufacturer fire qualification test results for Thermo-Lag that were reported to have met required standards but were later found to have been falsified.

The NRC OIG, in its Inspection Report entitled, "Adequacy of NRC Staff's Acceptance and Review of Thermo-Lag 330-1 Fire Barrier Material," of August 12, 1992, found that NRC staff did not conduct an adequate review of fire endurance and ampacity derating information concerning the ability of Thermo-Lag fire barrier material. The findings suggest that had NRC staff conducted a thorough review, they would have found that the TSI test furnace was not adequate along with the inadequacy of the TSI quality assurance procedures. Moreover the report shows that had a vendor inspection been conducted, NRC would have determined that the tests were not conducted, as required by a nationally recognized testing laboratory and that the vendor had falsified the test reports. However, because these review and inspections were not conducted, it was not until 1992 when the staff determined that the performance of Thermo-Lag 330-1 with respect to fire resistance ratings and ampacity derating was indeterminate. The

OIG report concludes that in seven instances between 1982 and 1991, NRC did not pursue reports of problems with Thermo-Lag 330-1.

Former NRC Chairman Ivan Selin responded[14] to the OIG report by directing the staff to address the following three matters:

(1) the reasons the initial review process did not identify the problems with Thermo-Lag 330-1 and the causes of deficiencies in NRC's response to later indications of problems that were brought to the agency's attention;

(2) whether the problems identified with respect to the initial review and the lack of follow-up to latter indications of problems represented a systematic weakness with our review and response programs; and

(3) what corrective actions are necessary to rectify the deficiencies identified with respect to the review and response processes.

### 5.2.3 Testing

Attachment 2 to IN 92-46, "The Final Report of the Special Review Team for the Review of Thermo-Lag Fire Barrier Performance," documented the teams finding on 34 of the available 49 Thermo-Lag fire test reports. This effort by NRC raised several concerns regarding compliance with NRC requirements and guidance, compliance with ASTM E119, and adherence to good engineering practices. The team's specific concerns involved test procedures, test facilities, test equipment and personnel, methods of assembly, quality assurance, and acceptance criteria. The team also found that the configurations of the test specimens for many of the previously performed tests are atypical of the field installations observed during the special review teams site visit to the plant. The Final Report concluded that many of the tests did not meet NRC requirements and guidance and, therefore, may not provide adequate technical bases for establishing fire resistance ratings of Thermo-Lag fire barriers.

GL 92-08, "Thermo-Lag 330-1 Fire Barriers," required all licensees to individually confirm that Thermo-Lag systems have been qualified by representative fire endurance tests, ampacity derating values have been derived from valid tests, and barriers have been installed with appropriate procedures and quality controls to ensure that they comply with NRC's requirements. The following discusses the various testing completed by NRC, Nuclear Energy Institute (NEI), and licensees.

*5.2.3.1 NRC Fire Endurance Testing*

NRC conducted two testing programs at separate national laboratories to independently evaluate the performance of the Thermo-Lag 330-1 ERFBS material. NIST performed the initial small-scale testing which resulted in the need for full scale testing, subsequently conducted by SNL. The following provides a brief description of these tests and the results.

---

[14] Memorandum dated August 17, 1992

**NIST Small-Scale Testing**

NIST conducted pilot-scale fire-endurance testing on 1- and 3-hour Thermo-Lag 330-1 panels. This type of testing is limited to provide insights on materials performance such as determining the thermal-transmission characteristics of fire barriers, often under non-conservative edge-loss conditions. The report cautioned the use of the results to assess the potential fire performance of full-scale cable-tray fire barriers. In full-scale testing the fire barrier enclosure is exposed to elevated temperatures on all sides, which is typically more severe than the pilot scale testing that only exposes one planar surface to the ASTM E-119 thermal exposure.

The results for the small-scale 1-hour and 3-hour fire endurance testing of Thermo-Lag 330-1 materials, conducted at NIST, indicated that the 1-hour test exceeded the temperature rise criteria within 22 minutes, while the 3-hour barrier failed at 2 hours 20 minutes. In addition, one thermocouple on the unexposed side recorded a temperature of 935°C (1716°F), exceeding the corresponding furnace temperature of 923°C (1694°F). These data indicates that Thermo-Lag may be supplying energy to the fire, (i.e., combustible). The small-scale testing at NIST also consisted of combustibility testing following ASTM E-136 and ASTM E-1354 test standards. The NIST tests revealed that Thermo-Lag 330-1 fire barrier material is combustible. NRC viewed the results of the NIST tests as indication of an inability of the Thermo-Lag material itself to provide protection according to its specified fire resistive rating. The tests conducted at NIST were not considered definitive in that the tests were not full scale and only panels were tested. However, the information gained from the NIST tests provided enough evidence to NRC to confirm doubts raised during the TU Electric tests, such as the bare stress skin observed following the TU 76 cm (30 in) wide cable tray test on August 21, 1992, leading to a conclusion that Thermo-Lag 330-1 fire barriers should be treated as inoperable in the absence of successful, applicable plant-specific tests.

**SNL Testing**

Following the small-scale testing conducted at NIST, NRC contracted with SNL to conduct three full-scale 3-hour fire endurance tests and one full-scale ampacity derating test of the Thermo-Lag 330-1 ERFBS. Of the three fire endurance tests conducted, two used the procedure-based TSI installation instructions (as presented in TSI Technical Note 20684, Rev. V) and a third test was a full reproduction of one of the original manufacturer's fire endurance qualification tests articles (a configuration typically not found at NPPs). These tests were conducted to evaluate the performance of the barrier against the results of tests previously reported by TSI (the vendor). The program evaluated performance using both the ANI standard and the ASTM E-119 temperature rise limits. All tests consisted of a "U" shaped cable tray raceway protected with two layers of half inch thick Thermo-Lag 330-1 preformed panels designed to achieve a 3-hour fire endurance rating. Stainless steel 18 gauge wire ties were used to secure the panels to the test article, with trowel grade Thermo-Lag 330-1 applied to any gaps and to pre-butter all material joints. The fire endurance test assemblies were exposed to the standard time-temperature as described in ASTM E-119. As discussed below, all three fire endurance tests resulted in prematurely failing both acceptance criteria methods. IN 94-22 documents and inform the utilities of these results. In addition, the ampacity derating test results indicated larger ampacity derating factors (ADFs) than those specified by the vendor.

Figure 5-5 shows the test assembly used in the SNL tests. The top middle picture presents the base test assembly structure, loaded with a single layer of cables. Bottom left photo shows a completely protected testing assembly ready to be tested. The bottom right photo shows the test assembly immediately after being pulled from the test furnace (note that the Thermo-Lag 330-1 material is combustible and missing in some areas).

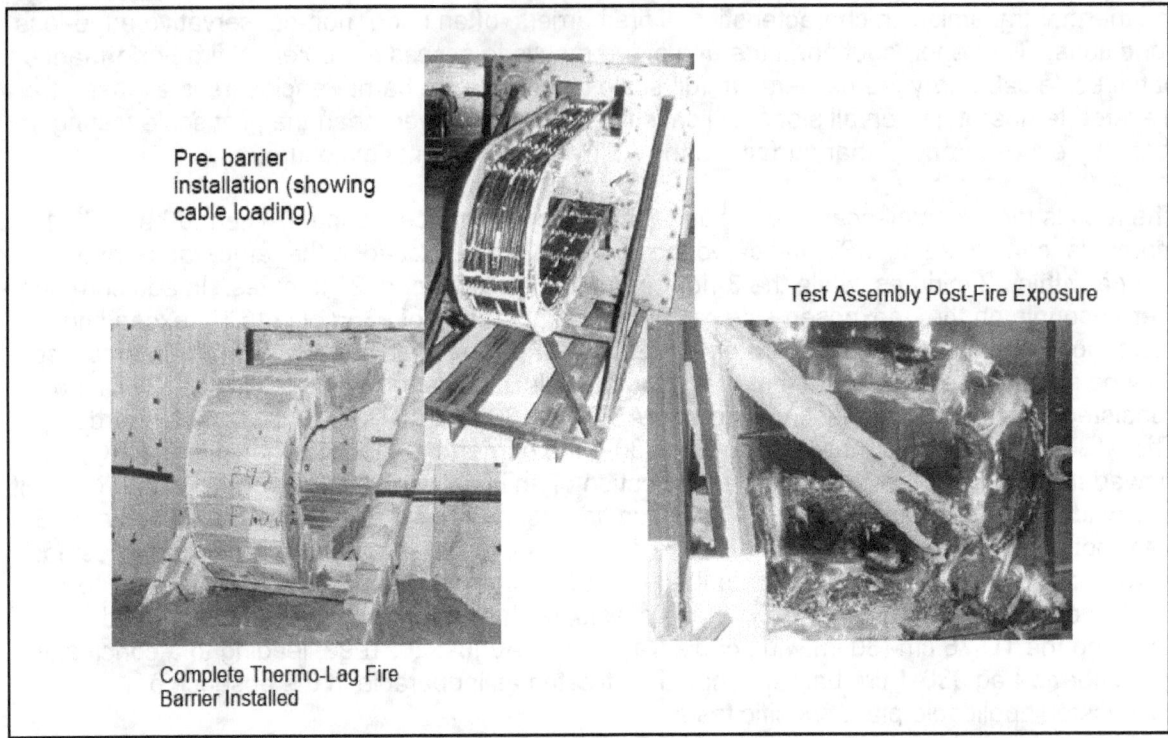

Figure 5-5.   SNL Full-Scale Thermo-Lag Test Article Shown in Various States (SAND94-0146)

All three fire endurance test assemblies failure prematurely and the failure criteria was exceeded for both circuit integrity and temperature rise acceptance criteria. The procedure-based installation articles failed electrically 1:15:39 and 0:58:32 into the fire test, while the upgraded reproduction assembly failed at 1:58:52. All were well short of the 3-hour design rating. The assemblies exceeded the minimum temperature rise (typically single point first) at 65, 55, and 110 minutes, respectively, again well short of the 180 minutes design rating.

Two failure modes were noted for the two procedure-based assemblies; namely, very early in the test (several minutes) the seams were observed to open in regions between where the tie wires were used, and after heating, the tie wires themselves were noted to stretch, allowing the protective panels to sag, eventually the tie wires failed allowing the panels to fall away. During testing of the reproduction assembly, the failure modes were quite different. Here the seams remained closed; however the panel began to sag and eventually the stress skin tore resulting in the panel material falling away to allow direct heat exposure to the protected raceway.

The last test assembly was used for an ampacity derating test. Both protected and unprotected ampacity testing was performed on the "U"-shaped test assembly and the testing was conducted at UL using the "High Ambient" environmental test chamber. The final ampacity derating factors indicated by the SNL test set were 46.4, 36.6, and 35.3 percent for the 8 American Wire Gauge (AWG), 4 AWG and 2/0 cables respectively. These factors were slightly higher than those determined by TSI at 38.1, 32.4, and 35.6 percent, respectively.

### 5.2.3.2 NRC Combustibility Testing

As part of the small-scale testing program of Thermo-Lag 330-1 ERFBS, the NRC staff contracted NIST to perform combustibility tests using two industry standards: (1) ASTM E-136, "Standard Test Method for Behavior of Material in a Vertical Tube Furnace at 750°C (1382°F)," and (2) ASTM E-1354, "Standard Test Method for Heat and Visible Smoke Release Rates for Materials and Products using an Oxygen Consumption Calorimeter."

Both nominal 1.27 cm (0.50 in) and 2.54 cm (1 in) thick Thermo-Lag 330-1 boards were tested with the half-inch thick board having a stainless steel wire mesh on only one side and the 1-inch board containing the stress skin on both sides.

The results of the ASTM E-136 testing indicated that the Thermo-Lag 330-1 material is combustible. Under this testing standard the material is considered to be "combustible" if three out of four samples tested exceed the following criteria:

- The recorded temperature of the specimen's surface and interior thermocouples, during the test rise 30°C (54°F) above the initial furnace temperature.

- There is flaming from the specimen after the first 30 seconds of irradiance.

- The weight loss of the specimen, due to combustion, during the testing exceeds 50 percent.

Of the four specimens tested, all experienced weight loss greater than 50 percent and flaming continued in excess of 30 seconds.

ASTM E-136 test standard assesses the material combustibility as either pass or fail (noncombustible or combustible). To access the level of combustibility of the Thermo-Lag 330-1 material (i.e., to provide a quantitative measurement scale), the material was additionally tested in accordance with ASTM E-1354. ASTM E-1354 is a heat release rate (HRR) test, which provides time-resolved information on the combustion of a specimen. Eight ASTM E-1354 tests were conducted on 100 cm x 100 cm (39 in x 39 in) specimens irradiated at 50kW/m$^2$ for four tests and 75kW/m$^2$ for the remaining four. Table 5-16 shows the results of these tests. Comparing the total heat released from Thermo-Lag 330-1 to the heat released from other building products, Thermo-Lag contributes more heat than Gypsum Board (12 mm (0.47 in) thick / unpainted – 4.2MJ/m$^2$ Total heat release) and is about equivalent to fire retardant plywood (12mm thick – 35.8 MJ/m$^2$ total heat release). Therefore, from a combustibility stand point, protecting cable raceways with one layer of Thermo-Lag material would be equivalent to enclosing that same raceway with one layer of fire retardant plywood. NRC staff concluded that Thermo-Lag 330-1 fire barrier material has combustible characteristics similar to those of other NPP combustible materials, such as fire-retardant plywood and cable jackets.

Table 5-16. NIST Results of ASTM E-1354 Thermo-Lag 330-1 Testing

| Test No. | Test condition | Irradiance kW/m$^2$ | Peak HRR KW/m$^2$ | Total heat in 600s MJ/m$^2$ | Total heat in 900s MJ/m$^2$ |
|---|---|---|---|---|---|
| 5489 | Grid used; wire mesh up | 50 | 74 | | 34.1 |
| 5490 | Grid used; wire mesh up | 50 | 83 | | 44.1 |
| 5491 | Grid used; wire mesh down | 50 | 74 | | 28.2 |
| 5492 | Grid used; wire mesh down | 50 | 76 | | 25.9 |
| 5466 | Grid not used; wire mesh up | 75 | 120 | 28.5 | |
| 5486 | Grid used; wire mesh up | 75 | 107 | 46.9 | |
| 5487 | Grid used; wire mesh down | 75 | 110 | 38.9 | |
| 5467 | Grid used; wire mesh down | 75 | 100 | 35.3 | |

*5.2.3.3 Vendor Testing (TSI)*

UL report dated June 16, 1981 evaluated the surface burning characteristics of Thermo-Lag 330-1. The testing was conducted in accordance with UL 723, "Surface Burning Characteristics of Building Materials." The UL 723 standard provides a method to classify flame spread by determining the area under the flame spread distance versus time curve, ignoring any flame front regression, and using one of the calculation methods as described below:

1. If the total area ($A_T$) is less than or equal to 97.5 min-ft (m-minute x 3.3) the flame spread classification (FSC) shall be 0.515 times the total area (FSC = 0.515 x $A_T$).

2. If the total area (AT) is greater than 97.5 min-ft (m-minute x 3.3) the flame spread classification is to be 4900 divided by 195 minus the total area ($A_T$) (FSC=4900/(195-$A_T$).

These UL tests determined that the FSC for thermo-lag 330-1 to be 2.9 – 3.1. This standard provides for the determination of the fuel contribution which was determined to be zero for Thermo-Lag 330-1. For the smoke development classification, a photoelectric circuit output operating across the furnace flue pike was used. A curve is then developed by plotting values of light absorption (decrease in cell output) against time. The classification of the smoke development is derived by expressing the net area under the curve for the tested material as a percentage of the net area under the curve for untreated red oak. The Thermo-Lag 330-1 tests indicated a 20.3 and 12.9 smoke development classification for the two tests conducted.

It should be noted that the test report is unclear as to how the Thermo-Lag 330-1 material was tested and doesn't appear to have tested actual pre-fabricated board but rather an asbestos cement board coated in 6.35 mm (0.25 in) thick trowel grade Thermo-Lag 330-1 material.

*5.2.3.4 NUMARC Fire Endurance Testing*

An industry Thermo-Lag fire endurance test program was subsequently established by the Nuclear Management and Resources Council (NUMARC), which later became NEI. The purpose of the nuclear industry program was to assess material performance and to provide a basis for evaluation of installed Thermo-Lag fire barriers. Specifically, the industry program, assessed current industry configuration through the use of survey data, conducted a number of fire tests to document performance of various baselines and upgraded Thermo-Lag fire barrier assemblies, and developed a guideline to assist utilities in evaluating their Thermo-Lag fire barrier configurations for compliance with respect to the guidance provided by the series of NRC Bulletins and GLs.

NUMARC testing program aimed to test bounding Thermo-Lag installations related to those utilities that supported the testing program. Licensees who had not addressed the issues raised in GL 92-08 and didn't plan to do so on an individual basis, planned to use the results of the NUMARC testing to determine any plant specific configuration modifications. A majority of the licensees using Thermo-Lag choose to wait for industry testing results prior to taking any actions to permanently resolve the issues. This is one of the causes for the delay in closing out GL 92-08 and ultimately resulted in NRC issuing more RAIs to confirm the licensees' closure plans.

In December 1993, NRC staff became concerned that the scope of the NUMARC testing would not be sufficient to resolve all Thermo-Lag barrier issues identified in the GL 92-08.
The principal concern of the staff was that the limited number of tests would not yield enough data for extrapolation of the large number of specific configurations needing evaluation.
In response to NRC concerns, NEI developed an Application Guide to assist utilities evaluation of their existing Thermo-Lag installations. The NEI Application Guide was used to evaluate Thermo-Lag enclosures. The application guide provided a methodology for evaluating equivalency between tested and installed Thermo-Lag configurations and is consistent with the process previously established by GL 86-10 Supplement 1. By letter dated October 16, 1995, NRC staff notified NEI of its review of this method as unacceptable as replacement to the guidance provided in GL 86-10 Supplement 1, and that NRC would not generically endorse the use of this guide. Instead, NRC staff would review the application of the guide by individual licensees on a plant-specific basis.

The testing was conducted in three phases. The first phase occurred between September and October 1993 and tested articles using a minimum material thickness and other construction attributes to provide a conservative baseline representation of the Thermo-Lag barrier.
The second phase of testing involved constructing upgraded Thermo-Lag 330-1 ERFBS to evaluate the effectiveness of various upgrading techniques to meet NRC requirements. The final phase assessed the performance of baseline and upgraded Thermo-Lag ERFBS on cable and raceway "box" design configurations (e.g., junction boxes, cable tray boxes, etc.) using both 1-hour and 3-hour prefabricated panels. The following presents an overview of the results with a summary of all the NEI testing provided in Appendix D of this report.

Of the phase 1 tests, four configurations passed all NRC acceptance criteria, these included;

- 1-hour conduits and junction box.
- 3-hour straight steel cable tray.
- 3-hour junction box.
- 3-hour small conduit.

Configurations not meeting all aspects of NRC criteria included:

- 3-hour cable tray with "T" section, 91.4 cm (36.0-in) wide cable tray,
- 3-hour medium and large conduits,
- 3-hour air drop, and
- a 1-hour 91.4 cm (36.0-in) wide cable tray.

Temperature rise exceeded the criteria with in 1 hour and 13 minutes of 3-hour test completion. Although the temperature limit was exceeded, post-test evaluation of the cables determined that no cable damage was visible. These results demonstrate the differences in testing standards acceptance criteria, with NRC acceptance criteria being more conservative than that of the ANI or UL standard.

Phase 2 of the NUMARC testing emphasized the need to test upgraded barriers versus the baseline configurations which showed poor resistance to thermal heat up within the barrier. Appendix D also provides a summary of these test results.

### 5.2.3.5 TVA and TU

In 1981, TU Electric conducted a full-scale fire endurance qualification test of the TSI Thermo-Lag 330-1 at SwRI. The purpose of the testing was to obtain a 1-hour fire rating for Thermo-Lag 330-1 in accordance with ANI Bulletin No. 5. Based on the ANI criteria all test specimens passed and the NRC accepted the qualification in a letter dated December 1, 1981.

Following issues surrounding the validity of the ANI standard as an appropriate method to qualify ERFBS, TU Electric proposed an alternate, site-specific fire test methodology and acceptance criteria that did not rely on the ANI approach. In October 1992, NRC concurred with the alternative site-specific fire test methods that TU Electric would utilize for conduct of future fire tests.

TU Electric performed Thermo-Lag testing in 1992 and 1993 to qualify ERFBS configurations for CPSES Unit 2 which was undergoing the plant licensing process at that time. Testing performed to qualify Unit 1 ERFBS configurations was performed later in 1993 and in 1998. The results of the TU Electric test program were consistent with the TVA results (discussed below), in that small diameter conduits required an additional layer of material, and joints on cable tray and other "box" design enclosures required external stress skin reinforcement.

In the mid-1990s TVA undertook extensive testing program to design, engineer, test, and qualify a series of Thermo-Lag 330-1 ERFBS for use in its NPPs. The enclosed DVD contains a video presentation summarizing the types of testing and process TVA used to qualify the use of Thermo-Lag 330-1 ERFBS for use in its plants, specifically the near term operating license of Watts Bar Nuclear Plant (WBN). TVA presented this video summary of its Thermo-Lag test program at a public licensee meeting for WBN Unit 1. Thermo-Lag 1-hour and 3-hour designs were qualified by TVA testing. The results of the testing confirmed that conduits 10.2 cm (4 in) and larger could be successfully protected with a single layer of material. However, smaller conduits required an additional layer of Thermo-Lag to pass the ASTM E-119 test. During their testing, TVA also found that a simple modification of adding an external layer of stainless steel mesh and trowel grade Thermo-Lag 330-1 material could alleviate the joint failure problem. The external layer controlled and directed the Thermo-Lag 330-1 subliming reaction to that of a more effective ablative shielding process.

Because of the amount of Thermo-Lag 330-1 materials used at CPSES, its licensee also performed an extensive testing and qualification program to ensure any Thermo-Lag 330-1 barrier used would provide the acceptable endurance rating. The NRC staff witnessed CPSES testing conducted at Omega Point Laboratories (OPL) in San Antonio, Texas.

TVA and TU entered into agreement to share information containing Thermo-Lag ERFBS. TVA in an effort to license Watts Bar Unit 1, has researched, developed, and rewritten the installation procedures, re-engineered the design, and performed the fire resistance qualifying tests for Thermo-Lag 330-1, 1-hour ERFBS. Thermo-Lag has undergone extensive testing by both TU and TVA. These tests were developed consistent with the guidance contained in the applicable codes, standards and regulatory guidance.

*5.2.3.6 Chemical Testing*

In a letter of November 7, 1992, TSI informed NRC that pre-shaped Thermo-Lag conduit sections received by TU for CPSES, Unit 2 showed signs of delaminating and voids. The NRC staff concern was that the use of such materials could affect the results of TU fire tests and the performance the Thermo-Lag barriers installed in CPSES 2.

GL 92-08 requested that licensees take actions to fully address the technical issues, independent of information and data supplied by TSI, before the staff makes any determination regarding whether the use of Thermo-Lag fire barriers complies with NRC regulations.

To support industries assessment of the Thermo-Lag materials, NEI (NUMARC) initiated a Thermo-Lag fire barrier material chemical testing program. NUCON International, Inc. performed pyrolosis gas chromatography testing consistent with ASTM D3452 on samples collected from 25 plants (169 samples total). Based on the test results of the chemical testing program sponsored by NEI, the results showed that all samples were consistent with one another in terms of organic and inorganic chemical composition.

NEI transmitted the results of this Thermo-Lag chemical testing program to NRC. The overall NEI test program (including 169 utility provided samples) demonstrated that the composition of Thermo-Lag fire barrier materials has remained consistent throughout the production dates of 1984 – 1995.

*5.2.3.7 Industry Combustibility Evaluation Method*

On October 12, 1993, NEI (NUMARC) submitted its "Thermo-Lag 330-1 Combustibility Evaluation Methodology Plant Screening Guide". In that report, NEI stated that Thermo-Lag 330-1 may not necessarily be considered a combustible material from a generic standpoint and recommended a performance-based approach, using fire modeling techniques, to evaluate the combustibility hazards presented by Thermo-Lag 330-1 installations.

The NEI guide provided a screening method that determines those areas where the potential ignition and subsequent flame propagation of Thermo-Lag is a concern. This method provides guidance on determining where Thermo-Lag is located in the plant, calculating the combustible loading attributed to the Thermo-Lag, and determining if this combustible loading will impact the fire hazards analysis (FHA)/fire safe-shutdown (FSSD) analysis of the plant. Any areas not screened out would be analyzed using fire modeling methods developed with the NEI guide. This fire modeling analysis would presumably result in additional screening of fire areas.

Following an NRC staff review, based on existing NRC requirements and guidance, the staff determined that the NEI method was unacceptable to justify the use of Thermo-Lag materials, or other materials such as fire retardant plywood or cable jackets, as noncombustible where noncombustible materials are specified by NRC fire protection requirements. NRC concluded that the NEI method does not provide a level of fire safety equivalent to that specified by existing NRC fire protection regulations and guidelines. As an alternative to the NEI guidance, NRC staff recommended that licensees re-evaluate their use of Thermo-Lag as radiant energy heat shields inside the containment or as an enclosure to create a 6.1 m (20 foot) combustible-free zone between redundant trains and seek other solutions. Possible solutions include:

- Reanalyze post-fire safe shutdown circuits inside containment and their separation to determine if the Thermo-Lag RES is needed.
- Replace Thermo-Lag barriers installed inside the containment with noncombustible barrier materials.
- Replace Thermo-Lag barriers used to create combustible-free zones with noncombustible barrier materials.
- Reroute cables or relocated other protected components.
- Request plant-specific exemptions where technically justified.

*5.2.3.8 Seismic Testing*

TVA performed shake-table testing of some typical cable tray and conduit configurations to address the seismic adequacy concern related to the Thermo-Lag material at WNB Unit 1. This testing showed that Thermo-Lag ERFBSs will not impact the functionality of cables, cable trays, and other components during and following a seismic event, when designed to TVA standards. The testing included single and ganged cable trays and conduits along with air drop configurations. The DVD included with this report provides a video recording of the seismic testing conducted on Thermo-Lag ERFBS. Although TVA made the determination that the Thermo-Lag 330-1 ERFBS would not impact functionality, the results did indicated significantly lower mechanical properties values than those used by the TSI consultant to evaluate seismic adequacy of the Thermo-Lag configurations.

Following the notification of differences between the vendors analysis and the licensees, NRC sent follow-up letter to GL 92-08 (pursuant to 10 CFR 50.54(f)) to the licensees and construction permit holders that had used Thermo-Lag 330-1 ERFBS in their plants. The letter included a request for additional information and some background information. In particular, item 1(9) requested information related to the mechanical properties of the Thermo-Lag material. In response, a number of licensees stated that they relied on the vendors analysis performed by the TSI consultant.

As a result of their review, NRC issued Information Notice 95-49 that informed the addresses of two specific NRC concerns related to the possibility of varying physical composition of the Thermo-Lag barrier in use across industry and the actual weights of Thermo-Lag use in plants. IN 95-49 was followed up with a supplement in 1997 which documented NRC's evaluation of the Thermo-Lag 330-1 seismic properties. The results of the testing indicated significantly lower mechanical properties than those used by the vendor to demonstrate the seismic adequacy of Thermo-Lag 330-1 panels.

In addition to the differences associated with the vendor reported material parameters and industry and NRC testing results, a review of the as received weights of the Thermo-Lag 330-1 panels, prefabricated conduit sections, and 330-660 flexi-blanket fire barriers indicated that there could be a variation of as much as 45 percent in the unit weights of the fire barriers when calculated as a percentage of the weight associated with the thickest panel (maximum weight). The variation is primarily related to the variations in the thickness tolerances. For example, the thickness of a 1-hour-rated panel could vary between 1.27 cm (0.50 in) and 1.91 cm (0.75 in), thus indicating a variation of 33 percent. The remaining 12 percent variation could result from the density variation of the material. The weight of one layer of stress skin, staples, steel bands, and trowel-grade material (applied during the installation) can increase the average weight of a panel by about 10 percent. Depending on the method used by the licensees to incorporate the weight of the Thermo-Lag ERFBS in seismic analysis of the raceways and their supports and anchorages, the effects of the variations could be non-conservative when the maximum unit weight of the fire barrier and its accessories (wire mesh, staples, bands, etc.) is higher then nominal values considered in determining the loads on the raceways and their supports and anchorages.

The seismic adequacy of various configurations of Thermo-Lag panels attached to the raceways has been determined by static analyses, subjected to simultaneous horizontal and vertical accelerations of up to 7.5 g (0.3 oz) and 5.0 g (0.2 oz), respectively. The TSI consultant performed their analysis using the mechanical properties (i.e., tensile strength, shear strength, and corresponding moduli) at various temperatures specified by TSI. Based on this analysis the TSI consultant concluded that the panels and conduit wraps were seismically adequate. After review this analysis, NRC staff determined that properly installed Thermo-Lag panels and conduit wraps would not undergo appreciable damage during the postulated seismic events at the nuclear reactor sites.

*5.2.3.9 Ampacity Derating Test Results*

Table 5-17 lists results from ampacity derating test conducted by TSI and UL. Appendix B provides a description of ampacity derating and testing methods.

**Table 5-17. Ampacity Derating Test Results - Thermo-Lag 330-1**

| Test Report | Description | TSI Derating Value | UL Derating Value |
|---|---|---|---|
| ITL No. 82-355-F-1 | 1-hour Cable Tray Test | 12.5% | |
| ITL No. 84-3-275-A | 3-hour Cable Tray Test | 20.55% | |
| Technical Note No. 111781 | 1-hour Conduit Test | 7.2% | |
| ITL No. 84-10-5 | 3-hour Conduit Test | 9.72% | |
| UL 86NK23826, File R6802 | 1-hour Cable Tray Test | | 28.0% |
| UL 86NK23826, File R6802 | 3-hour Cable tray Test | | 31.2% |
| UL 86NK23826, File R6802 | 3-hour Conduit Test | | 9.4% |

## 5.2.4 Resolution and Staff Conclusion

As a result of GL 92-08, licensee from 71 operating reactor units indicated that actions necessary to restore the operability of these barriers would be based on the results of the industry testing program being coordinated by the NUMARC. Based on meetings between NRC and NUMARC discussing the scope of the testing, the limited success of Phase 1 testing and a description of Phase 2 testing, NRC determined that the scope of this testing program would not address all issues associated with Thermo-Lag and informed those licensee affected. As a result, NRC based its acceptance on plant specific resolution paths, including those using the NUMARC results on plant-specific bases.

Section 6, "Plant Specific Usage and Resolution of ERFBS Issues," discusses the specific resolution each NPP took. In general plants undertook several tasks in the process to resolving compliancy issues.

These activities included re-analyzing the plants FSSD analysis to take credit for other methods of plant post-fire safe shutdown paths that wouldn't rely on redundant trains located within the same fire area. This method would typically eliminate a fraction of instances where the Thermo-Lag ERFBS was previously used. Those locations that still required protection would be resolved by either re-routing the cables through another fire area not containing the other train, or the Thermo-Lag ERFBS would be replaced by a qualified barrier or upgraded by applying more Thermo-Lag material per qualified configurations or by applying a different ERFBS atop of the existing Thermo-Lag ERFBS.

In addition, where barriers were found to no longer be needed, some licensees conducted destructive examinations of those barriers to determine the exact installation methods used when the barriers were initially installed. The information collected from this initiative was then compared to documented installation procedures to provide the licenses with a level of

confidence that the remaining barriers were constructed to a specific level of performance and future work would provide a barrier what can endure a fire for the required duration.
If deviations from installation manuals were identified during the destructive examinations, then the licensee would evaluate the effects of these deficiencies and determine what corrective actions would be required for the remaining barriers relied on for protection. These destructive examinations reviewed the following information.

- Material type
- V-rib orientation
- Stress skin location and use
- Joint gap
- Fastener spacing
- Joint reinforcement mechanisms
- Box and conduit interface

- Material thickness
- V-rib flattening
- Joint type
- Fastener size and material
- Fastener distance from joints
- Structural support and intervening steel protection

Methods to upgrade Thermo-Lag 330-1 ERFBS to meet 1-hour fire resistance rating varied somewhat based upon the specific testing; however the following example provides a typical upgrade that was found acceptable.

> An additional layer of Thermo-Lag 330-1 1.27cm (½ in) preformed conduit sections were pre-buttered and encompassed the baseline 1-hour barrier. Tie wires spaced 30.5 cm (12 in) apart were used to secure additional layer and all joints were offset and covered with a skim coat of trowel grade material. In addition, a layer of stress skin was applied to cover the joints of the barrier and a final skim coat of trowel grade material was applied atop of all stress skin and fasteners. (TVA 1995)

Although a number of different methods of upgrading the Thermo-Lag 330-1 ERFBS were qualified and used, the example mentioned above could be considered a typical upgrade. The specific geometry of the configuration also plays a roll as to the performance of the material and how it needs to be constructed to perform the intended design function. Both TU and TVA conducted extensive test programs to develop and qualify their unique upgrades.

From its review of documentation and applications of Thermo-Lag 330-1 as an ERFBS, the staff concludes Thermo-Lag 330-1 can be used as an acceptable ERFBS, provided that the proper testing and engineering assessments of their plant specific applications for the barrier to perform its intended design function.

## 5.3 DARMATT KM-1 (Hydrate) ERFBS

Darmatt KM-1 is a fire barrier system designed and manufactured by Darchem Engineering Ltd., and supplied by Transco Products, Inc. 1- and 3-hour rated Darmatt KM-1 barrier systems have been designed and tested in accordance with Supplement 1 to GL 86-10. Darmatt KM-1 consists of a semi rigid endothermic reactive insulating board, expanding paper gaskets, a silicon rubber cloth, and conduit mix. Under fire conditions, the Darmatt panels undergo multiple endothermic reactions. At the same time a refractory chain interspersed with pockets of carbon dioxide ($CO_2$) are produced. These processes reduce the thermal conductivity of the material and absorb the heat (by the endothermic process) transmitted into the barrier.

The Darmatt KM-1 boards are nominally 16mm (5/8-in) thick with a surface density of 13.0 kilograms per $m^2$ (3.1 lbs per $ft^2$) and a thermal conductivity of 0.113 Watts per m Kelvin (0.783 Btu inches per hour *$ft^2$ * °F) (at 68.9°C (156°F)). The Darmatt KM-1 layers are manufactured from a mix of commercially available raw material and cut into pieces or panels as needed for installation. Preformed half-round sections are also available for use in protecting conduits. The Darmatt KM-1 barrier consists of a multi-layer system that is placed around cable tray. The Darmatt KM-1 panels are secured to the raceway by the use of J-hooks attached to the Darmatt KM-1 insulating board. These J-hooks are typically spaced 15.2 cm (6.0-in) apart near the edge of a panel and lacing wire (18 gauge stainless steel wire) is used to secure panels together. Individual panels butt up to the preceding piece and joints are offset.

Expanding paper gaskets (nominally 3.2 mm (0.1-in) thick) are installed at panel joints. During fire conditions the paper gasket expands to fill any remaining joint gaps that are formed during installation. Expanding paper gaskets are used along the joints between adjacent panels and between the panels and the concrete. For irregular gaps between panels, a conduit mix, known as KM-1 Thermal Filler, is used to fill those gaps. The conduit mix has the same density, composition, and reaction under fire conditions as the Darmatt KM-1 panels. The exterior most panels are wrapped in a wire mesh reinforced silicon rubber cloth (also known as inconnel reinforced silicone fabric) to increase the resistance to abrasion of the system during normal conditions and to maintain barrier structural integrity during and after the fire exposure.
The silicon rubber cloth has no fire resistive properties and will burn off leaving behind the inconnel wire mesh. Figure 5-6 shows a sample of the Darmatt KM-1 ERFBS.

For a 1-hour rated barrier, typically two layers of Darmatt KM-1 panels are placed around cable trays and a single layer two-piece (half-round) sections are secured around conduits. 3-hour Darmatt KM-1 cable tray barriers consist of four layers with the exterior layer pre-wrapped with the silicon rubber coated glass cloth with a double layer of half-round sections for the 3-hour conduit protection. Air drops are either two-piece pre-molded conduit type sections, or four piece cable tray type panels which may have integral steel angle frame sections. Table 5-18 provides the nominal weight and thickness for the various KM-1 applications.

Table 5-18. Darmatt KM-1 Specifications

| System Type | Weight kg/m² (lbs/ft²) | Thickness cm (in) |
|---|---|---|
| Cable Tray | | |
| 1-hr replacement | 26.02 (5.33) | 3.20 (1.26) |
| 1-hr upgrade | 6.83 (1.40) | 0.79 (0.31) |
| 3-hr replacement | 55.17 (11.3) | 6.71 (2.64) |
| 3-hr upgrade | 13.67 (2.80) | 1.57 (0.62) |
| Conduit – Replacement only | | |
| 1-hr ¾-inch dia. | 21.97 (4.5) | 3.20 (1.26) |
| 1-hr 2 inch dia. | 34.17 (7.0) | 3.20 (1.26) |
| 1-hr 4 inch dia. | 53.70 (11.0) | 3.20 (1.26) |
| 1-hr 6 inch dia. | 73.23 (15.0) | 3.20 (1.26) |
| 3-hr ¾ inch dia. | 86.41 (17.7) | 6.40 (2.52) |
| 3-hr 2 inch dia. | 113.75 (23.3) | 6.40 (2.52) |
| 3-hr 4 inch dia. | 110.82 (22.7) | 6.40 (2.52) |
| 3-hr 6 inch dia. | 203.09 (41.6) | 6.40 (2.52) |

Figure 5-6. Photo of Two Layer Darmatt KM-1 System (Author, 2009)

Unlike other ERFBS where removal of the barrier requires destruction of that material, the Darmatt barrier was designed for future removal by the use of J-hooks which allows for a disassembly method that will not damage the barrier. This feature would be useful for evaluation of as installed barrier construction.

### 5.3.1 History

Darmatt KM-1 material came into use in the U.S. operating commercial NPP applications after deficiencies documented in GL 92-08 involving Thermo-Lag 330-1 were identified. Many licensees used Darmatt KM-1 ERFBS as a replacement or upgrade to their existing Thermo-Lag 330-1 installations.

### 5.3.2 Problems

No generic problems with the current use of Darmatt KM-1 ERFBS in the U.S. nuclear fleet have come to the attention of NRC. Test reports and test observations conducted by NRC staff have documented that Darmatt KM-1 is capable of achieving the fire endurance temperature rise acceptance criteria and hose steam tests as specified by NRC in Supplement 1 to GL 86-10. Seismic and ampacity derating reviews by NRC staff have also shown that the licensees who use this material are adequately accounting for these ERFBS engineering design aspects.

### 5.3.3 Testing

In its brochure[15], the vendor states that,

> "The system has successfully passed 1 and 3 hour fire tests in a wide range of boundary and site specific configurations. Boundary conditions include zero percent cable fill as well as free fall (air drop) single and grouped cables. Fire tests are fully compliant with NRC Generic Letter 86-10, Supplement 1, and include configurations for both upgrades of existing systems as well as new applications."

The vendor has tested the Darmatt KM-1 Barrier to the following testing standards at independent laboratories with acceptable performance results. They include,

- ampacity derating per IEEE P848 Draft 16,
- ageing per ASTM E1027, c
- combustibility per ASTM E136,
- corrodibility per US Reg. 1.36,
- surface spread of flames per ASTM E84 and
- UV resistance testing.

Testing Laboratories include, Faverdale Technology Centre (NAMAS accredited), in-house UKAS approved testing facility, and independent laboratories in the USA, such as OPL and Wyle Laboratories.

In addition to the vendor testing, the following licensees also evaluated performance of Darmatt KM-1 ERFBS.

#### 5.3.3.1 Lasalle County Station 1-hour Fire Resistance Testing

A 1-hour fire endurance test, including hose stream test following ASTM E-119 and GL 86-10 Supplement 1 acceptance criteria was carried out on the Darmatt KM-1 ERFBS for Lasalle County Station (LCS) Units 1 and 2. The test furnace was 3.7 m (12.0 ft) long by 2.1 m (7.0 ft) wide and 2.1 m (7.0 ft) high, fired by 8 gas burners and controlled by a total of 13 thermocouples. The test protected various electrical raceway assemblies including a 76.2 cm by 10.1 cm (30.0 in by 4.0 in) galvanized steel cable tray, four small 1.9 cm (0.75 in) conduits,

---

[15] http://www.esterline.com/Portals/8/Darchem/PDF/DTP_Brochure.pdf

an air drop and a 30.5 cm by 30.5 cm by 7.6 cm (12.0 in by 12.0 in by 3.0 in) galvanized steel junction box all protected with two layers of Darmatt KM-1 material representing a composite of LCS Unit 1 and 2 conditions. The raceways only contained a single bare copper #8AWG conductor instrumented with thermocouples every 6 inches. The raceways were monitored for temperature by placement of thermocouples as specified in GL 86-10 Supplement 1.
In addition, the effects of thermal shorts on conduits were evaluated by attaching 1.9 cm (0.75 in) long copper pieces to the ends of the 1.9 cm (0.75 in) conduit.

The tests were conducted on June 16, 1994 at Faverdale Technology Centre in Darlington (England – UK). Following the ASTM E-119 exposure and hose stream test, post hose stream test visual inspections found no instances of barrier failure. Data provided in the report indicated that the highest average temperature rise ($\Delta T$) reached in any raceway was 65°C (149°F) and a maximum single point temperature rise ($\Delta T$) of 84°C (183°F). By safety evaluation report (SER) dated November 20, 1995, the NRC notified the licensee that the Darmatt KM-1 fire endurance test was conducted in accordance with the methodology and acceptance criteria specified in GL 86-10 Supplement 1. NRC staff also acknowledged that the 1-hour fire-rated Darmatt KM-1 fire barriers installed at LaSalle were bounded by the test data. However, this SE didn't address seismic or ampacity derating acceptance related to Darmatt KM-1, but indicated that a follow-up action would address these matters generically.

*5.3.3.2 Carolina Power and Light (CPL) Company*

On December 20, 1995, an NRC staff member witnessed a 3-hour fire endurance test of Darmatt KM-1 performed at Faverdale Technology Center, Darlington, England, by CPL and IES Utilities, Inc. (IES). The Darmatt KM-1 is used by CPL and IES to replace certain Thermo-Lag barriers at Brunswick Steam Electric Station and Duane Arnold Energy Center. The test plan, installation procedure, and quality control were provided by Transco Products, Inc., Chicago, the sole sub vendor of Darmatt KM-1 in the United States.

The 3-hour test consisted of a large junction box 73.7 cm by 83.8 cm by 135.9 cm (29.0 in by 33.0 in by 53.5 in), a 61.0 cm (24-in) wide steel ladder back cable tray with air drop, four rigid steel conduits (two 1.9 cm (0.75 in), one 10.1 cm (4.0 in), and one 12.7 cm (5.0 in) diameter) and two flexible steel conduits (1.9 cm (0.75 in) and 10.1 cm (4.0 in) diameter). The test assembly was subjected to the ASTM E-119 standard fire endurance test for three hours. The maximum average temperature for any raceway was 142°C (287°F) (on the 1.9 cm (0.75 in) conduit) while the maximum single point temperature was also observed on the same conduit as being 184°C (363°F). The ambient conditions at the start of the test were 20°C (68°F), resulting in acceptance criteria of 159°C (318°F) and 201°C (393°F), respectively. The test specimens all met, with margin, the thermal and hose stream acceptance criteria specified in GL 86-10, Supplement 1.

*5.3.3.3 Prairie Island Nuclear Generating Plant, Units 1 and 2*

Fire endurance testing of Darmatt KM-1 was performed for Commonwealth Edison by Faverdale Technology Centre on March 29, 1994, in accordance with NRC GL 86-10 Supplement 1. The testing involved a 91.4 cm (36.0-in) and 15.2 cm (6.0-in) wide cable trays, along with a 1.9 cm (0.75-in) conduit protected with a 1-hour rated KM-1 Darmatt barrier. The raceways were empty except for the instrumented bare copper conductor. Table 5-19 provides the results this testing. A hose stream test was conducted separate from the fire endurance test. The hose stream testing followed the guidance from NFPA 251, thus using an assembly identical to the fire endurance assembly subjected to one-half of the fire endurance duration and then a hose stream was applied for 5 minutes.

Table 5-19. KM-1 1-hr Fire Endurance Results (FTCR/94/0060)

| Raceway | Failure Time (minutes) | Failure Criteria |
|---|---|---|
| 36 x 6 inch cable tray | 79 | Avg. Temp. Rise |
| 12 x 3.5 inch cable tray | 81 | Avg. Temp. Rise |
| ¾ inch RSC | 70 | Single Point Temp. Rise |

*5.3.3.4 Callaway Plant Ampacity Derating*

Union Electric (the licensee of Callaway Plant) submitted a Darchem Engineering Limited. Ampacity derating report conducted at the Faverdale Technology Centre. The testing was in accordance with IEEE P848 and carried out between August 29 and September 12, 1996. Test Report S-1064-00011-00 evaluated 2.54 cm and 10.1 cm (1.0-in and 4.0-in) diameter conduits each insulated with one layer of pre-formed half-round conduit sections (i.e., 1-hour barrier). The results are presented in Table 5-20.

Table 5-20. Ampacity Results Faverdale (Test Report FTCR/96/0077)

| Test | Average Room Temperature °C (°F) | Conductor Temperature °C (°F) | Normalized Current (Amps) | Ampacity Derating Factor % |
|---|---|---|---|---|
| 1" diameter conduit baseline | 39.96 (103.9) | 89.8 (193.6) | 38.88 | - - - |
| 1" diameter conduit insulation | 39.82 (103.7) | 90.26 (194.5) | 35.56 | 8.54 |
| 4" diameter conduit baseline | 39.86 (103.7) | 89.99 (193.9) | 18.03 | - - - |
| 4" diameter conduit insulation | 40.31 (104.6) | 89.83 (193.7) | 15.84 | 12.15 |

Test Report S-1064-00012-00 documents the ampacity derating factors for 2.54 cm (1.0 in) and 10.1 cm (4.0 in) diameter rigid steel conduits encapsulated with two layers of KM-1 Darmatt material to provide a 3-hour barrier. The tests were conducted at Faverdale Technology Centre on September 3, 9, 11, and 20, 1996. The results are presented in Table 5-21.

**Table 5-21. One & four-inch RSC Ampacity Results 3-hour KM-1 (Test Report FTCR/96/0099)**

| Test | Average Room Temperature °C (°F) | Conductor Temperature °C (°F) | Normalized Current (amps) | Ampacity Derating Factor % |
|---|---|---|---|---|
| 1" diameter conduit baseline | 39.96 (103.9) | 89.80 (193.6) | 38.88 | --- |
| 1" diameter conduit insulation | 39.91 (103.8) | 89.74 (193.5) | 32.00 | 17.7 |
| 4" diameter conduit baseline | 39.863 (103.8) | 89.99 (193.9) | 18.03 | --- |
| 4" diameter conduit insulation | 40.11 (104.2) | 90.03 (194.1) | 14.46 | 19.8 |

Test Report S-1064-00014-00 provided ampacity derating factors for 600 mm by 101mm by 3650 mm (24-in by 4-in by 144-in) long cable tray protected with four layers of KM-1 Darmatt 3-hour replacement material. The testing was performed on August 4 and December 6, 1996 at Faverdale Technology Centre. The results are presented in Table 5-22.

**Table 5-22. 600mm Cable Tray Ampacity Results (Test Report FTCR/96/0108)**

| Test | Average Room Temperature °C (°F) | Conductor Temperature °C (°F) | Normalized Current (amps) | Ampacity Derating Factor % |
|---|---|---|---|---|
| 600mm tray baseline | 39.58 (103.2) | 90.45 (194.8) | 13.63 | --- |
| 600mm tray insulation | 40.40 (104.7) | 90.39 (194.7) | 6.74 | 50.55 |

### 5.3.4 Resolution and Staff Conclusion

No generic issues or deficiencies have been identified and associated with this ERFBS; there has been no need for generic resolution related to Darmatt KM-1 ERFBS. Darmatt KM-1 has been used as replacement materials to other deficient ERFBS.

As a result of its late entry into NPP applications and testing in accordance with GL 86-10 Supplement 1, the staff concluded that Darmatt KM-1 when installed to bound as tested configurations will satisfactory perform its intended design function.

## 5.4 Hemyc and MT (Insulative and Insulative/Hydrate ERFBS)

The Hemyc and MT ERFBS were initially products fabricated by B&B Insulation, Inc. an affiliated company of INSULCO Inc., but subsequently manufactured and typically installed by PCI Promatec, Inc. This transition happened sometime in the early 1980's. These ERFBS have been installed at NPPs to protect circuits in accordance with regulatory requirements and plant-specific commitments. Hemyc ERFBS is utilized in 1-hour fire barrier applications and as radiant energy shields, while MT can be used as a 1-hour or 3-hour barrier depending on how it is constructed. Both ERFBS are basically an assembly of common industrial materials. The enclosed DVD contains video footage of NRC testing of Hemyc and MT ERFBS.

Hemyc is a simple thermal insulator consisting of ceramic blankets constructed of 38.0 mm (1.5 in) or 50.0 mm (2.0 in) thick, 128.0 kg per m$^3$ (8.0 lbs per ft$^3$) or (6.0 lbs per ft$^3$) ceramic blanket manufactured by any of the following:

- Kaowool® blanket manufactured by Thermal Ceramics (formally Babcock and Wilcox),
- Cerablanket® blanket manufactured by Thermal Ceramics (formally JM Manville), or
- Durablanket manufactured by Carborundum Fiberfrax.

The ceramic fiber blanket is covered with a Siltemp®[16] mesh fabric to produce what is called the Hemyc mat. The primary purpose of the cover materials is to protect the ceramic fiber core from physical damage. The materials are sewn together with "Astroquartz" thread (high temperature thread ~3300°C (5972°F)). The fireproof mats are pre-manufactured to fit the specific cable tray or conduit where it is to be installed.

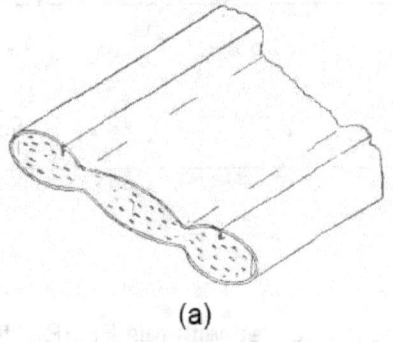
(a)

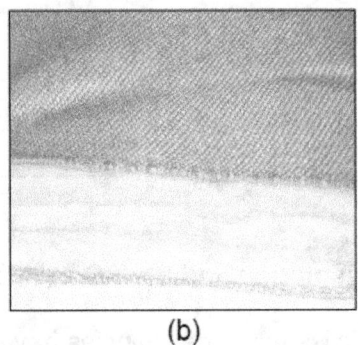
(b)

**Figure 5-7.** (a) Sketch and (b) Photo of Hemyc Mat (Installation Procedure 1985; Author, 2009)

Hemyc can be applied directly onto a raceway using 50.0 mm (2.0-in) thick wraps (i.e., direct attachment) or an air gap construction can be used (i.e., air gap attachment). The Hemyc air gap attachment consisted of two basic components; a light weight metal framework attached to the raceway and the Hemyc mat that surrounds the framework. The metal framework supports the Hemyc mat and provides the required off-set from the raceway to allow a dead air space. For air gap attachment, a 38.0-mm (1.5 in) thick Hemyc mat is used.

---

[16] The Promatec vendor manual references either Siltemp®, Refrasil®, or Alpha 600 as equivalent materials for the outer fabric mesh covering on all surfaces exposed to the fire.

Originally, the attachment of the Hemyc blanket is accomplished using threaded studs connected to the metal framework for cable trays or with finger straps (as shown in
Figure 5-8) for conduit. The blanket is impaled onto the threaded studs or strap fingers using nuts and washers, or clips. Experience with this method of assembly showed signs of rips occurring at the point of impalement and latter installations used stainless steel banding wrapped and compressed around the circumference of the barrier.

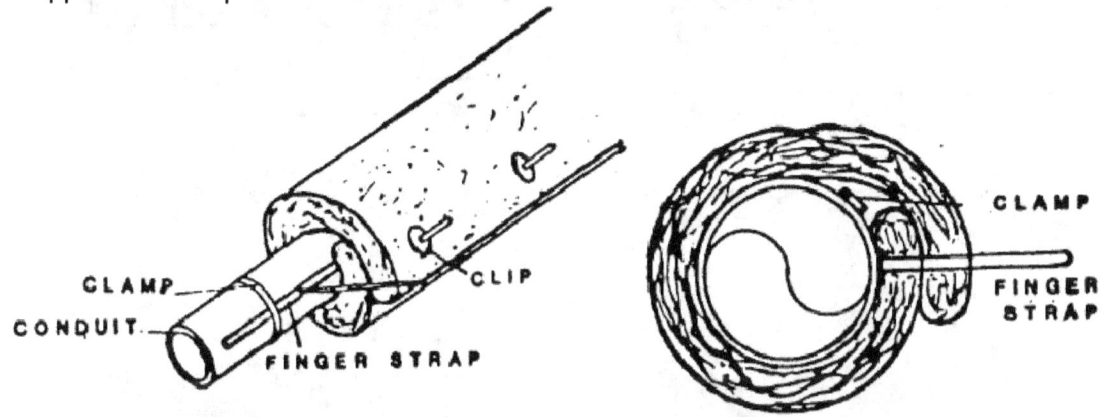

Figure 5-8.   Hemyc 1-hour ERFBS Conduit Construction (Installation Procedure 1985)

Figure 5-8 shows the construction of a typical Hemyc ERFBS used on conduit. However, instead of the finger straps shown, licensee tended to use stainless steel banding to secure the Hemyc wrap to the raceways. This method of attachment is shown in Figure 5-9.

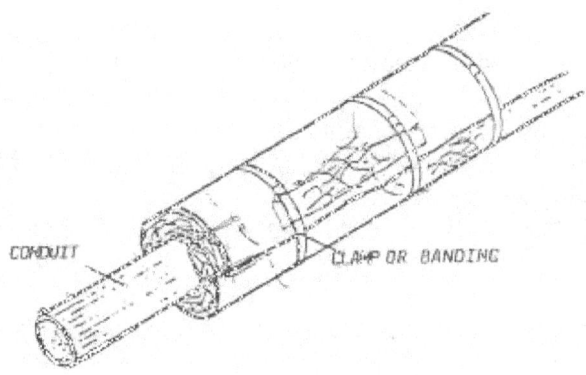

Figure 5-9.   Hemyc 1-hour ERFBS Banding(Installation Procedure 1985)

Hemyc mats were manufactured in sections which needed to be joined together to maintain the overall barrier integrity. Two methods predominantly used in the industry were the overlap and collar type joints. The two predominant joint techniques are shown in Figure 5-12.

In the overlap joint, an end of an installed Hemyc mat section is overlapped by a minimum 5.1 cm (2-in) overlap of the next section of Hemyc to be installed. Typically, at the overlap joint, several steel bands are used to secure the joint. In collar type joints, two individual Hemyc mats are butt jointed against each other and then a minimum 15.2 cm (6 in) collar made out of Hemyc mat is secured around the butt joint seam. As will be discussed later, the joints of the Hemyc

ERFBS tend to be the weakest link in maintaining the barriers integrity during elevated fire conditions.

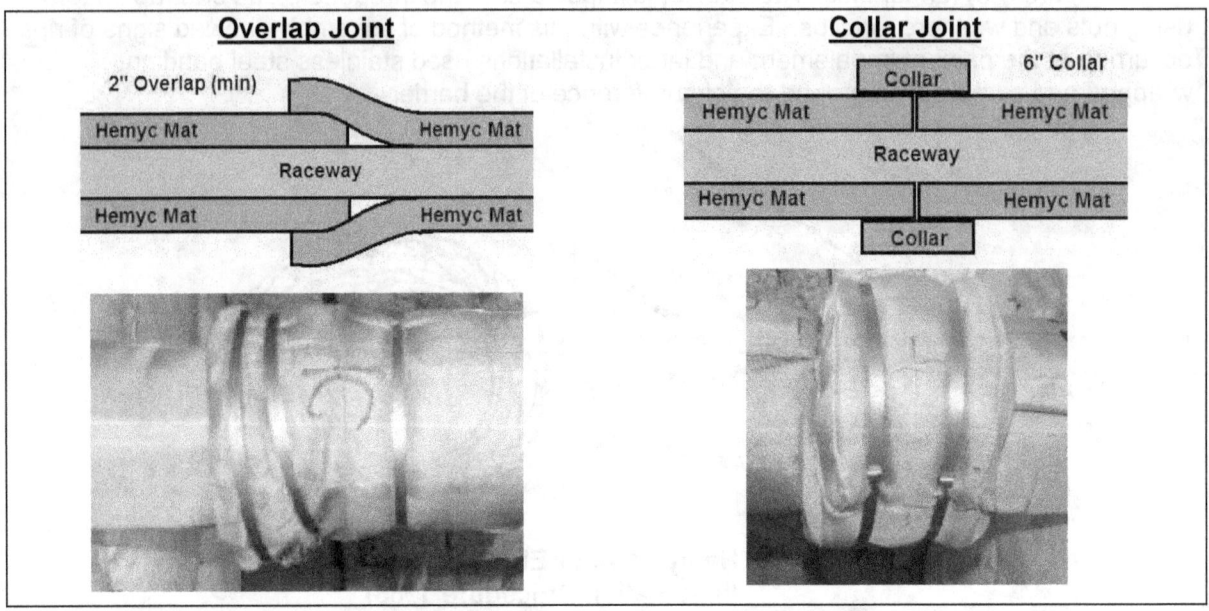

Figure 5-10. Sectional View and Photos of Hemyc Joint Techniques (Author, 2009; NRC Hemyc Testing, 2005)

The MT ERFBS is an upgraded version of the Hemyc ERFBS and is installed as a 3-hour fire barrier. MT consists of four layers, with one of the layers being a heat activated compound:

Closest to Raceway
    Layer 1: 2.54 cm (1.0 in) of Kaowool® ceramic fiber blanket wrapped in a fiberglass fabric
    Layer 2: 2 mm (0.08 in) sheet of stainless steel (moisture barrier)
    Layer 3: hydrate packet made by stitching together packets of aluminum trihydrate in a fiberglass-coated fabric
    Layer 4: 3.8 cm (1.5 in) Kaowool® blanket wrapped in a fiberglass fabric
Farthest from Raceway

Exceptions include air drops which consist of a 7.6 cm (3.0 in) thick blanket of Kaowool® as the inner layer and structural supports which do not have the hydrating packet layer or the stainless steel sheet (Layers 2 and 3). Some licensees may use MT as a 1-hour ERFBS.

Installation of the MT barriers included the use of lacing hooks, lacing washers and tie wire to securely hold the barrier together.

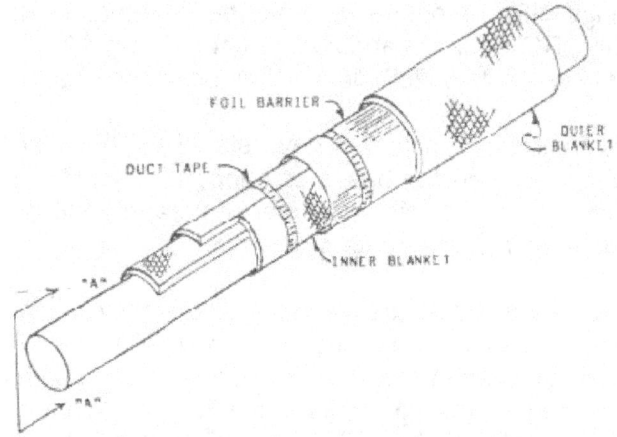

Figure 5-11. MT ERFBS Construction (Installation Procedure 1985)

### 5.4.1 History

Hemyc ERFBS was tested and qualified in the early 1980's using the protocols outlined by ANI/MAERP Bulletin No. 5. Supplement 1 to GL 86-10 did not exist at this time, but NRC staff in several instances approved the use of Hemyc as a qualified ERFBS based on these ANI test results. These acceptances documented in the plants licensing basis provides the applicability of Hemyc as a 1-hour fire barrier for specific applications in the plants.

Following the issuance and licensee responses to GL 92-08, NRC staff believed that some technical concerns identified in the GL remained unanswered. However, the staff believed that since Hemyc was a 1-hour rated ERFBS only used in application where protection is provided by the levels of defense-in-depth, the indeterminate fire resistance rating of Hemyc was not seen to be an area of significant risk. At the time 1993 through 1998 the staff was consumed with addressing the performance deficiencies of Thermo-Lag 330-1, Kaowool, and FP-60 ERFBS. In 1994, NRC conducted a series of small-scale, two dimensional fire tests on fire barrier materials. The tests were performed as scoping tests to evaluate the generic fire-endurance characteristics of available materials. Barriers tested, included 3M Interam E-53A, E-54A, FP-60, Promat-H, Thermo-Lag, and Hemyc. Unfortunately these tests couldn't be used to qualify the barriers but provided qualitative insights on the performance of each barrier.

In the late 1990's, NRC inspection staff raised concerns with the adequacy of Hemyc ERFBS. At the turn of the century, NRC inspections issued several plant-specific findings related to the performance of the Hemyc and MT fire barriers. The inspections revealed that the licensee's fire testing and acceptance criteria used to determine the fire resistance of Hemyc required further NRC review to determine their acceptability. This review was documented in TIA 99-028 which concluded that Hemyc was not qualified for use as a fire barrier in NPPs.

As a result of these findings, NRC recommended that the industry conduct testing to qualify the barrier in accordance with NRC guidance and that NEI coordinate this effort. In NEI's review of the matter and as stated in its April 25, 2001 letter, their positions was,

> "Licensees using the Hemyc materials have stated to NEI their belief that they are complying with their licensing bases as supported by prior NRC acceptance of the test protocol and use of these fire wrap applications, and that their licensing bases demonstrate adequate protection."

As a result of industry reluctance to address the safety and technical issues related to the acceptability of Hemyc to provide the 1-hour of protection, in 2001 the NRR within NRC planned to conduct a series of full scale confirmatory fire endurance tests on both Hemyc and MT ERFBS. NRR later transferred the testing program to RES. NRC contracted SNL to conduct two full scale Hemyc and one full scale MT ERFBS test.

Throughout the planning phases of NRC Hemyc/MT testing program, NEI provided comments and information to NRC to improve and bound the actual plant configurations. In addition, NRC held a public meeting on October 31, 2002 with several licensees and NEI to discuss the proposed testing plan and acquire additional feedback from the licensees on the proposed testing to assure that the testing contained a representative sample of configurations found in plants. It should be noted that NRC tests were not conducted to qualify the Hemyc or MT ERFBS for use in any NPP, but were undertaken to provide confirmatory evidence of the barriers ability or lack thereof to provide the required protection.

NEI provided formal comment on the testing plan by letter dated December 6, 2002, commenting on the licensing basis, program plan additions, including bounding conditions for non-tested configurations and a mathematical calculation, information and representation during the construction of the barrier materials, along with suggested guidance on the reporting and interpretation of results. A detailed discussion of NRC testing program is provided in Section 5.4.3.

The results of NRC Hemyc ERFBS testing were communicated via IN 2005-07, "Results of Hemyc Electrical Raceway Fire Barrier System Full Scale Fire Testing," dated April 1, 2005. The Information Notice indicated that the Hemyc ERFBS didn't perform for the 1-hour period as designed because shrinkage of the external cloth covering and thermal shorts. The IN concluded that, "… the Hemyc ERFBS does not provide the level of protection expected for a rated 1-hour fire barrier."

The results of NRC MT ERFBS testing were made publically available on May 23, 2005. The test report documented that no raceways protected with a 3-hour MT ERFBS meet the acceptance criteria to be rated a 3-hour barrier. More information on these results is provided in Section 5.4.3.3 NRC Testing below.

As a follow-up to IN 2005-07, NRC issued GL 2006-03, "Potentially Nonconforming Hemyc and MT Fire Barrier Configurations," dated April 10, 2006, requesting the addresses to determine whether or not Hemyc or MT fire barrier material is installed and relied upon for separation and/or safe shutdown purposes to satisfy applicable regulatory requirements. GL 2006-03 also requested a description of the controls used to ensure other fire barrier types are capable of providing the necessary level of protection. The responses were compiled into a list of ERFBS used for each operating NPP as presented in Memorandum to Alex Klein dated December 21, 2007 and reproduced in Appendix F.

The GL 06-03 guidance further stated, "If licensees identify nonconforming conditions, they have several options. (1) replace existing ERFBS with a qualified one, (2) upgrade existing barrier to fire rated one, (3) reroute cables or instrumentation lines through another fire area, or (4) voluntarily transition to the risk-informed approach to fire protection (NFPA 805)." Although other solutions may be acceptable, these are the most popular methods that NRC believes the industry would take to resolve the issues identified by the confirmatory testing results.

### 5.4.2 Problems

As noted above, NRC concern with Hemyc ERFBS was brought to light after several inspection findings were identified in the 1999 and 2000 time frame.

During a fire protection inspection at Shearon Harris NPP, NRC inspectors identified that the licensee's fire testing and acceptance criteria used to determine the fire resistive performance of the Hemyc/MT ERFBS installed to separate safe shutdown functions within the same fire area required further NRC review to determine its acceptability. Shearon Harris Nuclear Plant (HNP) based their acceptance criteria for the use of Hemyc/MT ERFBS on that reflected by the ANI Information Bulletin 5(79) standard, which NRC considers a non-conservative qualification approach in that cable damage can occur without indication of excessive temperature on the cables. The inspectors identified that the licensee was unable to provide the inspectors with engineering evaluation documentation which demonstrated that the shutdown capability is protected. Additionally, the inspectors were unable to confirm that the licensee had established an acceptable design basis for the Hemyc/MT ERFBS used to separate safe shutdown functions within the same fire area. The inspectors concluded that the actual fire resistive performance of the Hemyc/MT ERFBS installed to separate safe shutdown functions within the same fire area was indeterminate. This issue was documented as an unresolved item (URI) 50-400/99-13-03 in NRC Inspection Report No. 50-400/99-13, dated February 3, 2000.

An inspection at Waterford Steam Electric Station, Unit 3, identified that equipment required for safe shutdown of the plant following a fire were not separated by 1-hour fire barriers. Specifically, several cables from the redundant Train A/B of the chilled water system had either missing or damaged 1-hour fire wrap. This Green safety significance non-cited violation was documented in NRC Inspection Report No. 50-328/00-07, dated November 29, 2000.

A triennial fire protection inspection at McGuire Nuclear Station, Units 1 and 2 resulted in the inspection team finding that the licensee was unable to provide documentation to demonstrate that an adequate design basis had been established for the Hemyc ERFBS in use to protect the Train B service water control cable, which was located within the same fire area as Train A. This finding was identified as an unresolved issue in the inspection report dated December 15, 2000.

These three inspection findings revealed that the licensee's fire testing and acceptance criteria used to determine the fire resistive performance of the Hemyc fire barrier systems installed to separate safe shutdown functions within the same fire area required further NRC review to determine their acceptability. NRC was concerned that the Hemyc ERFBS may not fulfill the requirement of a 1-hour rated fire barrier as required by Appendix R to 10 CFR Part 50 in all applications.

On November 23, 1999, NRC Region II offices requested assistance from NRR in Task Interface Agreement (TIA) 99-028. A TIA is a process used to address questions or concerns raised within NRC regarding nuclear reactor safety and the related regulatory and oversight programs.

NRR provided its response to TIA 99-028 on August 1, 2000. In its response, the NRR staff evaluated the three fire endurance tests used by HNP to qualify the Hemyc and MT barriers used at the plant.

<u>1-hour Hemyc Barrier Final Report CTP 1026</u>
The staff determined that information documented in Final Report CTP 1026 is insufficient to qualify the Hemyc fire barrier as a 1-hour-rated ERFBS. Its determination was based on the staffs concern with the size of the furnace, its accuracy, and the type, location and number of thermocouples used both in the furnace and on the testing assembly to provide this data, along with the fact that the assemblies were not bounding (due to heavy and unrealistic cable loading).

<u>3-hour MT Barrier Report No. 1100A</u>
NRC staff determined that, although the acceptance criteria used (ANI continuity criteria and thermocouples attached to cables) deviated from the acceptable method identified by NRC guidance, the test may be used to qualify cable configurations protected with the "MT" ERFBS, provided that they met the conditions identified in the TIA response.

<u>3-hour MT Barrier Report No. 1071</u>
The staff determined that information documented in Final Report CTP 1071 is insufficient to qualify the "MT" fire barrier as a 3-hour-rated conduit fire barrier system. While the thermocouple may appear to meet qualifying temperatures, there is a concern as to the type, location, and number of these thermocouples used on the test assembly to provide the data. In addition, these results were not bounding in the fact that the 10.1 cm (4.0 in) conduits were heavily loaded in some cases with a non-realistic arrangement of cable.

TIA 99-028 concluded that CTP 1026 results were "inconclusive to qualify" Hemyc as a 1-hour rated fire barrier; CTP 1071 results were "inconclusive to qualify" MT/Hemyc as a 3-hour rated fire barrier; and CTP 1100A results could be used to qualify MT/Hemyc as a 3-hour rated fire barrier only if the specific configuration of MT/Hemyc installed in NPPs met the criteria (i.e., cable tray sizes and cable masses) in CTP 1100A. The TIA response also identified the series of NRC small-scale Hemyc tests conducted at NIST, which resulted in failure to meet the cold-side temperature rise criteria within 25 minutes.

Subsequent to the August 2000 TIA determination that Hemyc was not qualified for use as a fire barrier in NPPs, NRC staff requested that licensees address the Hemyc concerns as a voluntary initiative and that NEI assist with the coordination of this initiative. Industry later decided not to pursue voluntary testing to resolve Hemyc issues. NEI stated to NRC that there was insufficient evidence to indicate a safety concern that would warrant an industry initiative and in the opinion of NEI it believed that the nuclear industry was in compliance with NRC requirements based on the fact that NRC had previously accepted the original Hemyc manufacturer fire qualification tests. Therefore, NRC through its Office of Research preformed confirmatory testing to identify potential safety problems with the Hemyc and MT ERFBS. As discussed below, these NRC tests resulted in failure of Hemyc and MT to meet their respective fire endurance rating.

NRC informed licensees that, notwithstanding that the Hemyc material may be part of a plant's licensing basis, the test results that were cited do not fully address the contemporary technical concerns regarding the adequacy of this material in satisfying the intent of Commission's regulations. Specifically, technical concerns which were identified in GL 92-08 remain unresolved. These issues involve testing adequacy, for example, minimum and maximum fill were not performed for all configurations, cable damage occurred in some configurations, and energized cables were not included in all tests. Application and bounding questions concerned the staff, such as, which sizes and configurations bound what other sizes and configurations.

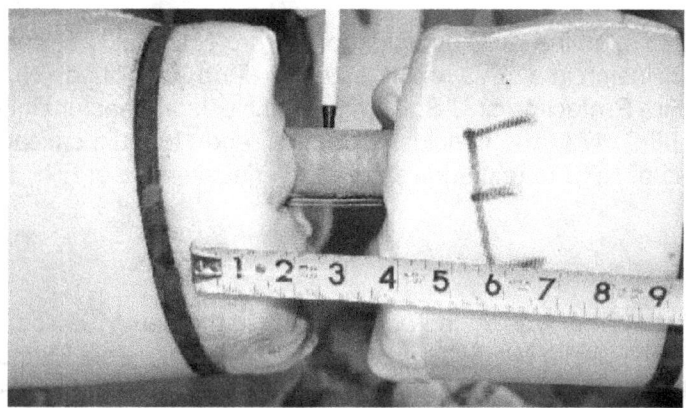

Figure 5-12. Post-test Photo of Hemyc ERFBS Showing Shrinkage at Junction (NRC Hemyc testing, 2005)

*5.4.2.1 NRC Office of Inspector General Special Inquiry*

The Office of Inspector General (OIG), NRC, initiated a special inquiry in response to concerns pertaining to Hemyc fire barriers. This inquiry intended to evaluate the failure of Hemyc during NRC 2005 tests and whether NRC staff acted appropriately to address the problem.

In the OIG report, "NRC's Oversight of Hemyc Fire Barriers," dated January 22, 2008, the following findings were identified:

- NRC did not communicate the results of the failed 1994 NIST testing of Hemyc to the licensees, nor did NRC conduct any follow-up to the NIST small-scale testing.
- Via its response to TIA 99-028, NRC determined that qualification tests used previously to supporting licensing of Waterford were subsequently determined inadequate to qualify; however, NRC did not require licensees to take corrective action.
- NRC confirmatory testing of Hemyc and MT resulted in failures of the barrier. NRC informed the industry of the results through an information notice, but required no follow-up (take action or written response) by the industry.
- GL 2006-03 required licensees to resolve Hemyc / MT issues by December 1, 2007, however NRC did not schedule or budget for any inspections to review licensees' resolution of the Hemyc fire barrier issues.
- In 1993 the former NRC Chairman provided testimony to the House of Representatives Subcommittee on Oversight and Investigations. This testimony included a commitment to conduct assessments of all fire barriers used to protect

electrical cables in NPPs to identify what improvements were needed to have these fire barriers meet NRC requirements.

Letter dated July 15, 2008 from NRC Chairman Dale E. Klein, addressed the finding of the OIG report. Chairman Klein agreed with most of the OIG findings and provided clarification to the findings related to budgeting of inspections and purpose of the issued information notice and generic letter.

*5.4.2.2 U.S. Government Accountability Office Report to Congressional Requesters*

The GAO was asked to examine NRC oversight of fire protection at U.S. commercial NPPs and documented their conclusion in a GAO report issued in June 2008 titled, "Nuclear Safety – NRC's Oversight of Fire Protection at U.S. Commercial Nuclear Reactor Units Could Be Strengthened, GAO-08-0747." The conclusions of this report found it critical, in the opinion of the GAO, for the need of NRC to test and resolve the effectiveness of ERFBS at nuclear units.

### 5.4.3 Testing

The Hemyc fire barrier system was tested and qualified in the early 1980's using the protocol outlined by ANI/MAERP bulletin No. 5, "Standard Fire Endurance Test Method to Qualify a Protective Envelope for Class 1E Electrical Circuits." NRC performed follow-on confirmatory testing of Hemyc ERFBS in the spring of 2005. This full-scale testing was performed on typical NPP conduit, cable tray, junction box and support configurations protected with the Hemyc ERFBS. During this testing NRC discovered a previously unidentified failure mode; the outer layer of high temperature cloth covering experienced thermal shrinkage resulting in the opening of joints in the ERFBS and exposing the protected raceway. Subsequent testing was pursued by an informal Hemyc users group, consisting of several utilities, with results consistent with those documented in NRC testing. The following provides a detailed review of the testing of Hemyc and MT ERFBS

*5.4.3.1 Vendor Testing*

Final Report CTP 1026, "HEMYC Cable Wrap System – One Hour Test," dated June 1, 1982, records the results of a 1-hour Hemyc tested conducted by Central Nuclear de Asco in Tarragona, Spain. The purpose of the test was to qualify the barrier for acceptance by ANI and as such, the ANI test standard was used, including the ASTM E-119 standard time/temperature curve. Three individual tests were conducted, involving a variation of 30.5 cm (12 in) wide cable trays, 10.2 cm (4 in) diameter conduits, and air drop configurations. Test acceptance criteria included maintaining circuit integrity and having no pass through during the hose stream test. Cable tray fill varied from 100 percent visual fill to a single layer, while all conduits were tested with 100% visual fill and the air drops consisted of 10 cables bundled together. Water hose test consisted of 6.4 cm (2 ½ in) hose provided with 2.86 cm (1 1/8 in) nozzle, from a distance of 6.1m (20.0 ft) at 206842 Pascal (30 lbs per in$^2$) pressure for 2 minutes and 30 seconds, while maintaining power to monitored cables. The results of the Test 1 indicated no cable damage was observed but following the hose stream test, a portion of material was damaged in some of the areas of direct impact, having lost 10 to 15 percent of the fiber material. Test 2 also indicated that portions of the barrier were slightly torn by the hose stream and again no cable damage was observed during post test examinations. Test 3 showed that some of the cable located in the cable tray lost continuity and insulation, portions for the barrier mesh fabric were torn and some permanent deformation of the barrier covers was observed. Although these three tests passed the ANI criteria, they do not pass NRCs GL 86-10 Supplement 1 criteria. In response to TIA 99-028, NRC staff concluded that information documented in Final Report CTP 1026 is insufficient to qualify the Hemyc fire barrier as a 1-hour-rated ERFBS.

Final Report CTP 1071, "Three Hour fire Qualification Test of PROMATEC 'MT' Barrier Wrap System Electrical Conduit Circuits," dated January 6, 1986 documents a qualification test conducted by SwRI in San Antonio, Texas on July 30, 1985 (SwRI Project No. 01-8305-049). The test assembly consisted of several 10.1 cm (4.0 in) diameter conduits, a 25.4 cm by 25.4 cm by 61.0 cm (10.0 in by 10.0 in by 24.0 in) junction box, 10.1 cm (4.0 in) diameter pull boxes, and a 10.1 cm (4.0-in) diameter conduit tee. Cable loading ranged from 40 percent actual (100 percent visual) to a single layer of polyethylene (PE) insulated, polyvinyl chloride (PVC) jacketed cables in 1/3 power, 1/3 control, and 1/3 instrumentation mix. All raceways were protected by a three layered system consisting of (1) an inner blanket assembly (7.6 cm (3-in) thickness of alumina silica blanket enveloped with fiberglass cloth, (2) stainless steel foil moisture barrier, and (3) a multi-layered outer blanket assembly consisting of a 3.8 cm (1.5-in) alumina silica blanket and a fiberglass assembly containing a powdered ingredient enveloped with a fire resistant outer fabric. The MT ERFBS was installed by the vendor in accordance with procedure number CTP-1071. Test acceptance criteria were based on the ANI/MAERP test standard and based on these criteria the MT barriers all passed, with the exception of one conduit, which failed at 177 minutes into the 180 minute test. NRC staff reviewed this test report for its response to TIA 99-28 and determined that, "Final Report CTP 1071 is insufficient to qualify the "MT" fire barrier as a 3-hour-rated conduit fire barrier system."

Report CTP 1077, "HEMYC Cable Wrap System – One Hour," dated October 29, 1984, was performed by SwRI in San Antonio, Texas. The testing was for engineering purposes only and not intended to qualify the barrier. The test placed 100 percent visually filled 7.6 cm (3.0-in) conduits straight through a furnace controlled to the standard ASTM E-119 time/temperature curve. At the center of the conduit the Hemyc ERFBS was butt jointed with a collar surrounding the joint. The results indicated that the barrier failed at the joint, as the thermocouple nearest the joint location exceeded the 163°C (325°F) temperature rise approximately 55 minutes into the test.

Final Report CTP 1100A, "Three Hour Fire Qualification Test of PROMATAC 'MT' Barrier Wrap System Electrical Cable Tray Circuits," dated June 4, 1986, documents a qualification test conducted by SwRI on February 19, 1986 (SwRI Project No. 01-8821-016). The test assembly consisted of four raceways, two 45.7 cm (18.0 inch) wide and two 61.0 cm (24.0 in) wide by 10.1 cm (4.0 in) high cable trays. Each raceway configuration consisted of one-half the length ladder back type construction and the remaining half solid back construction. Single layer cable tray fill and 100 percent visual (50 percent actual) tray fill were used in this testing. The MT ERFBS was constructed with the same three layer system used in CTP 1071. The test assembly was exposed to the standard ASTM E-119 standard time temperature curve for three hour duration. Based on the acceptance criteria of ANI/MAERP, no test assembly exceeded the temperature rise or lost circuit integrity during or after the test. However, the test report does state,

> "Post-test examination immediately after the hose stream tests showed that some of the PROMATEC, Incorporated, protective envelope was dislodged... but none of the seals were penetrated by the hose stream."

NRC staff reviewed this test report for its review of TIA 99-028 and determined that, although the acceptance criteria used (ANI continuity criteria and thermocouples attached to cables) deviated from the acceptable method identified by NRC guidance, the test may be used to qualify cable configurations protected with the "MT" ERFBS, provided that they met the conditions identified in the TIA response.

### 5.4.3.1.1 Vendor Ampacity Derating Tests

For the materials and configurations tested, the vendor supplied the following ampacity derating values:

| | |
|---|---|
| MT 3-hour Cable Tray | 73.57% |
| MT 3-hour Conduit | 42.08% |
| Hemyc 1-hour Cable Tray | 54.06% |
| Hemyc 1-hour Conduit | 39.58% |

These values were derived from conducting tests in accordance with the original IEEE 848 draft standard.

### 5.4.3.1.2 Vendor Surface Burning Characteristics

The vendor performed testing for flame spread, fuel contribution and smoke development for both one and three hour systems. Testing was conducted in accordance with ASTM E-84. The values were the same for both barriers and were reported as follows:

| | |
|---|---|
| Flame Spread Index | 5 |
| Fuel Contribution | 0 |
| Smoke Developed | 0 |

### 5.4.3.2 NIST Testing

In 1994, NRC conducted a series of small-scale, two dimensional fire tests on numerous fire barrier materials used in NPPs at the time. The testing used a small scale furnace with one side of the sample exposed to the furnace environment; the other side was exposed to the open laboratory environment (unexposed side of the barrier). Five thermocouples were placed on the unexposed side of the barrier, covered with 7.6 cm (3.0 in) square insulation material and the exposed side was subjected to the ASTM E119 standard time/temperature curve. The tests were performed as scoping tests to evaluate the generic fire-endurance characteristics of available materials. Materials tested included, Hemyc 1-hour and MT 3-hour, among others.

The 1-hour Hemyc test results indicated that the average temperature rise criterion was exceeded at 23.2 minutes into the test, while the maximum temperature rise criterion was exceeded at 24.8 minutes. NIST determined that the maximum uncertainty for this test was +30/-24 seconds (+0.5/-0.4 minutes).

The MT 3-hour test assembly consisted of multiple layers: a fire-blanket, a sheet of stainless steel foil, a layer of encapsulated hydrated powder material, a fire-barrier blanket encapsulated within a glass cloth (e.g., Siltemp). The assembly was subjected to 3.5 hours of the ASTM E-119 standard time/temperature exposure. At no point during the test did the unexposed surface temperature exceed NRC 325°C (250°F) criteria. At the end of the 3.5 hour exposure the peak temperature recorded on the unexposed side was 77°C (171°F).

Although the NIST tests provided insights into the performance of several different barrier systems, the NIST report emphasized that this type of testing is limited to assessing the thermal-transmission characteristics of fire-barrier materials, often under non-conservative edge-loss conditions. As such, NRC believed that the NIST tests were not sufficient to make a final determination regarding the capability of fire barrier. Although the Hemyc tests indicated a less robust material among other 1-hour barriers, NRC staff didn't use these results to determine whether additional testing or review was necessary. NRC also didn't communicate the results of the failures identified in the NIST testing to the industry through its typical means (e.g., Information Notices, Generic Letter), as these small-scale tests were not qualification tests and industries use of such information was indeterminate.

*5.4.3.3 NRC Testing*

As a result of Industries reluctance to undertake the responsibility of performing testing of Hemyc and MT, NRC performed three ASTM E-119 furnace tests on a number of cable raceway types protected by the Hemyc ERFBS (with and without air gaps) and MT ERFBS at the OPL in San Antonio, Texas. The Hemyc and MT ERFBS were manufactured and installed by qualified Promatec employees to the manufactures vendor manual and procedures. A bare No. 8 stranded copper conductor, instrumented with thermocouples every 15.2 cm (6.0 in) along its length, was routed through each of the conduit and cable tray test specimens. To expand on the testing methodology and understand how various aspects of raceway configurations affect the heat transfer characteristics of this particular barrier, NRC testing included both empty and fully loaded conduits and it also tested supports independently. The Hemyc ERFBS tests were performed for a period of 60-mintutes each and 180-minutes for the MT testing, followed by a hose stream test and post-test visual inspection of the ERFBS. In other words, this testing was performed in accordance with Supplement 1 to GL 86-10 guidance. An average temperature rise of ≤121°C (250°F), maximum single point temperature rise of ≤162°C (325°F) and hose stream testing were the acceptable criteria for qualification. Ampacity derating and seismic position retention testing was beyond the scope of this testing program. Table 5-23 provides a summary of the raceway configurations tested.

Table 5-23. Hemyc and MT Test Matrix (NRC)

| Raceway Type | Hemyc (1-hour, Direct Attachment) Test #1 | Hemyc (1-hour, Framed For Air Gap And Direct Attachment) Test #2 | M.T. (3-hour, Direct Attachment) Test #3 |
|---|---|---|---|
| 27-mm (1 in) Conduit[17] | X | | X |
| 63-mm (2.5 in) Conduit[9] | X | | X |
| 103-mm (4 in) Conduit[9] | X | | X |
| 305-mm (12 in) Tray | | X | |
| 914-mm (12 in) Tray | | X | |
| Junction Box | X | X | X |
| Cable Drop | | X | |
| Unistrut Support | X | | X |
| Tube Steel Support | X | | X |

During the development of the test plan, the NRC interfaced with industry to better understand what was installed in NPPs, however, no test articles were constructed to conform to a specific site installation. Although NRC Hemyc and MT tests were not intended to address all issues with the limited number of tests, NEI did provide, in its letter dated December 28, 2001, a list of typical installation practices used at commercial NPPs. This letter provided NRC with information to help develop the test plan that was representative of configurations found in NPPs. However, it should be emphasized that NRC test program was to evaluate the conformance of the Hemyc and MT barriers to perform their intended function and not to qualify any particular barrier configuration.

Test 1 consisted of 2.5, 6.4, and 10.1 cm (1.0, 2.5 and 4.0 in) conduits empty and with significant cable fill, junction box, and structural steel supports, with all Hemyc material directly attached to the raceways (no air gap). Test 2 consisted of 30.5 and 91.4 cm (12.0 and 36.0 in) cable trays and cable airdrop configurations with direct attachment and 5.1 cm (2.0 in) air gap Hemyc attached over special frames, and a junction box with direct attachment. Test 3 was identical to Test 1 with the use of MT ERFBS instead of Hemyc. The DVD enclosed, contains video footage of the NRC Hemyc and MT testing.

**Hemyc Tests**

The Hemyc mats were constructed of 5.1 cm (2.0 in) Kaowool insulation inside an outer covering of Refrasil® high temperature fabric. The mats are custom sized for the particular application and machine stitched at the factory. Where the 5.1 cm (2.0 in) air gap configuration was used, 3.8 cm (1.5 in) Kaowool mat was used instead of 5.1 cm (2.0 in) material. Refrasil® was used for the outer covering of the Hemyc ERFBS mats (the vendor manual referenced Siltemp, Refrasil or Alpha 600 as equivalent materials for the outer covering of the Hemyc

---

[17] Conduit test specimens were tested under both "empty" and "loaded-with-cable" conditions.

EFRBS mats). At the time of NRC testing Siltemp was not available for purchase and the only remaining quantities of this material were new-old-stock remaining in some licensees warehouses.

After construction of the Hemyc ERFBS, the test assembly was lowered into the test furnace and exposed to the standard fire endurance test for 1-hour as specified in ASTM E-119. The results of the testing indicated gross Hemyc ERFBS shrinkage and opening of joints which resulted in none of the protected raceways passing the 1-hour test. Thermal shorting of the raceway support members was also identified as a root cause of the barrier failure. During the testing, the Refrasil mesh consistently experience a phenomenon of thermal shrinkage and change of color from tan to white. This shrinkage led to the mats contracting and opening gaps in the ERFBS. The temperature rise acceptance criteria was exceed in all raceways between 15 and 57 minutes, with the average failure time of approximately 30 minutes.

NRC testing demonstrated that when the Hemyc ERFBS is constructed per vendor procedures and evaluated against NRC acceptance criteria, it is unable to meet the required fire endurance rating of 1-hour. Table 5-24 provides a list of the Hemyc configurations tested and the final fire endurance rating.

Table 5-24. Summary of NRC 1-hour Hemyc ERFBS Tests

| Raceway ID | Raceway | Time to $\Delta T_{avg} \geq 250°F$ (min) | Time to $\Delta T_{ind} \geq 325°F$ (min) | Max. Temp Bare #8 @1h (°C) | Burn-Through/ Structural Failure Yes/No | Pass Hose Stream Yes/No | Final Fire Endurance (min) |
|---|---|---|---|---|---|---|---|
| 1E | 1" Conduit (Empty) | 46 | 42 | 545 | Yes | Yes | 42 |
| 1F | 1" Conduit (1.02 lb./lin.ft. Cable Fill) | 44 | 34 | 636 | Yes | Yes | 34 |
| 1C | 2 ½" Conduit (Empty) | 48 | 41 | 376 | Yes | Yes | 41 |
| 1D | 2 ½" Conduit (5.85 lb./lin.ft. Cable Fill) | 51 | 38 | 230 | Yes | Yes | 38 |
| 1A | 4" Conduit (Empty) | 49 | 33 | 463 | Yes | Yes | 33 |
| 1B | 4" Conduit (14.84 lb./lin.ft. Cable Fill) | 57 | 43 | 93 | Yes | Yes | 43 |
| 1I | Junction Box 18" x 24" x 8" (Empty) | 17 | 15 | N/A | Yes | Yes | 15 |
| 1G | Unistrut | N/A | 22 – 32 | N/A | N/A | Yes | 22 – 32 |
| 1H | 2" Tube Steel Support | N/A | 13 – 25 | N/A | N/A | Yes | 13 – 25 |
| 2A | 12" Cable Tray (Empty, Direct Attach) | 27 | 18 | 682 | Yes | Yes | 18 |

Table 5-24. Summary of NRC 1-hour Hemyc ERFBS Tests (Continued)

| Raceway ID | Raceway | Time to $\Delta T_{avg} \geq 250°F$ (min) | Time to $\Delta T_{ind} \geq 325°F$ (min) | Max. Temp Bare #8 @1h (°C) | Burn-Through/ Structural Failure Yes/No | Pass Hose Stream Yes/No | Final Fire Endurance (min) |
|---|---|---|---|---|---|---|---|
| 2B | 12" Cable Tray (Empty, 2" air gap) | 33 | 35 | 539 | Yes | Yes | 35 |
| 2C | 36" Cable Tray (Empty, Direct Attach) | 34 | 33 | 721 | Yes | Yes | 33 |
| 2D | 36" Cable Tray (Empty, 2" air gap) | 28 | 31 | 603 | Yes | Yes | 31 |
| 2E | Air Drop (Direct Attach) | 35 | 32 | 933 | Yes | Yes | 32 |
| 2F | Air Drop (2" air gap) | 32 | 28 | 766 | Yes | Yes | 28 |
| 2G | Junction Box 18" x 24" x 8" (Direct Attach, with Bands) | 31 | 28 | N/A | Yes | Yes | 31 |

Failures are usually manifested by the opening of a gap in the outer covering material at its weakest point. The weakest point is most often at a seam between two pieces of the material, or at a fastening where the material is connected to the underlying electric raceway. If there are no seams, or the existing seams are exceptionally strong and connections to the raceway are made so forces generated by the thermal shrinkage are distributed over a large area, the outer covering material itself will rip. Shrinkage also causes extreme compression of the Kaowool insulation material under the outer covering, which decreased the ERFBS heat transfer resistance sufficient to exceed acceptance criteria.

NRC testing also examined the four most common methods of joining the Hemyc material into a complete ERFBS, namely stitched joints, minimum 15.25 cm (6 in) collars over a joint, minimum 5.1 cm (2.0 in) overlapping of the mats, and through bolts/fender washers for cable trays and junction boxes using the 5.08 cm (2 in) air gap space frames. The shrinkage led to failure of each of the joint systems. At the time, NRC was uncertain if this shrinkage effect was solely a result of using the Refrasil or if the other Hemyc outer coverings experienced the same shrinkage phenomena.

As a result of the shrinkage experienced during NRC testing of Hemyc, which used Refrasil® as the fabric mesh covering, NRC contracted with SNL who performed testing on both Siltemp® and Refrasil® to determine any differences between the two materials thermal properties. The testing was conducted by SNL on March 24, 2005. The Refrasil sample was taken from the actual bolt on material used to construct insulating pads for the SNL/NRC fire endurance testing, while the Siltemp® was provided as new-old-stock from licensees own on hand stock. Siltemp® is no longer manufactured and can no longer be purchased on open market.

The two materials were placed on a thin insulating board, and placed inside a cylindrical radiant heating chamber. The radiant chamber shroud temperature began at 300°C (572°F) and was increased in 50°C (122°F) increments to a maximum temperature of 800°C (1472°F). Total test duration was 90 minutes.

The test results indicated that the two materials behaved in a virtually identical manner with shrinkage on the order of 5% and the physical aspects of the two materials are similar enough to be considered essentially the same. The first visible signs of shrinkage were noted at a shroud temperature of 450°C (842°F), where the shrinkage was visually estimated at 0.16 cm to 0.32 cm (1/16 in to 1/8 in) total (or about 2%). At 600°C (1112°F) the total shrinkage looked to be roughly 3% and at 800°C (1472°F) the material turned stark white in appearance. Post test measurements revealed that both materials experienced a total shrinkage of about 5% with uniform shrinkage in both directions. These results indicate that no substantial differences in either the timing or extent of the material shrinkage behavior between Refrasil and Siltemp should be expected. It should be noted that, Siltemp and Refrasil were both available from the manufacture in standard (as tested) and pre-shrunk versions. The typical installation in the nuclear industry is with the standard (not pre-shrunk) version of the outer covering. Pre-shrunk can be identified by its stark white appearance, while the standard (non pre-shrunk) version is tan in color.

## MT TESTS

As stated above, Test #3 was identical to Test #1, with the exception of a 3-hour MT ERFBS being used instead of the 1-hour Hemyc wrap. Qualified Promatec workers installed the MT ERFBS in accordance with the vendor installation manual. Figure 5-13 provides pictures and annotation of the four layers used to construct the MT barrier.

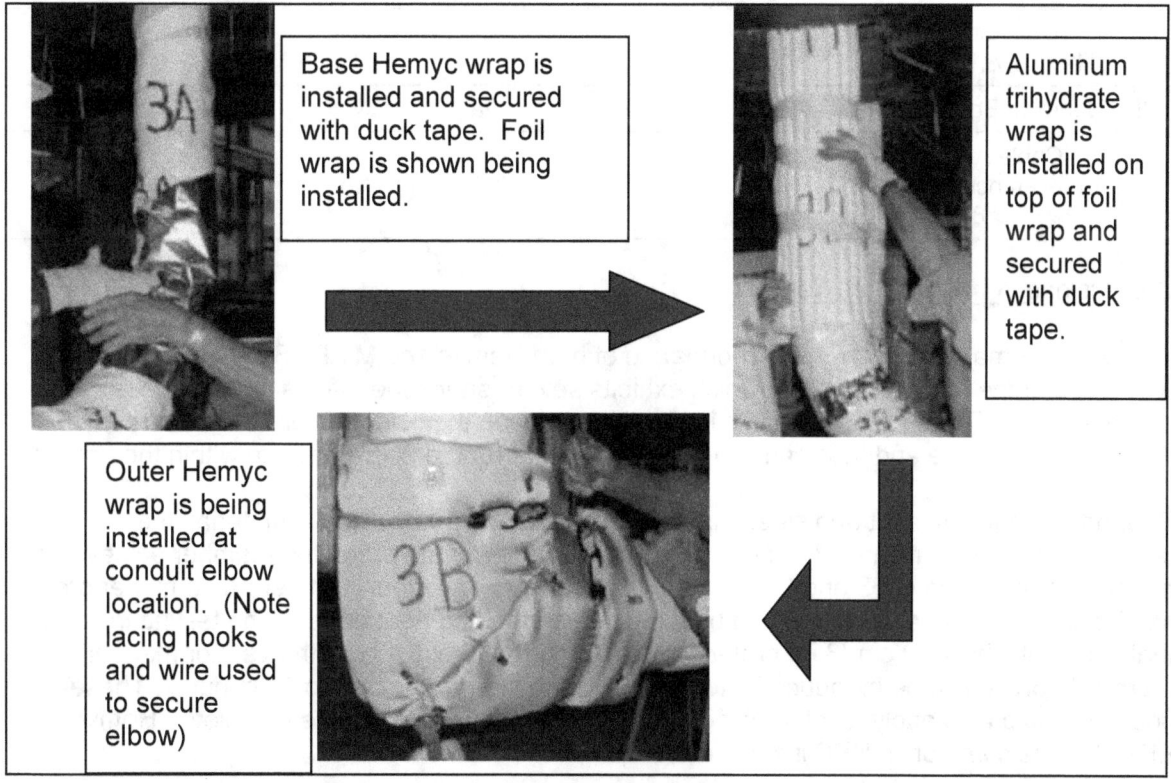

**Figure 5-13. MT Installations Process for NRC Testing (NRC MT testing, 2005)**

After the raceway barrier had been installed on all raceways, the entire test assembly was hoisted into the test furnace and subjected to the standard time/temperature curve exposure as specified in ASTM E-119. Thermocouples were located on the exterior of the raceways with 15.2 cm (6.0 in) spacing and on a bare # 8 AWG conductor located within the raceway. Acceptance criteria were based on GL 86-10 Supplement 1 guidance. Table 5-25, provides the results of the 3-hour test for each raceway. For the 3M tests, all raceways systems failed to meet the 3-hour fire endurance period.

### Table 5-25. Summary of NRC 3-hour MT ERFBS Tests

| Raceway ID | Raceway | Time to $\Delta T_{avg} \geq 250°F$ (min) | Time to $\Delta T_{ind} \geq 325°F$ (min) | Max. Temp Bare #8 @3h (°C) | Turn-Through/ Structural Failure Yes/No | Pass Hose Stream Yes/No | Final Fire Endurance (min) |
|---|---|---|---|---|---|---|---|
| 3A | 4" Conduit (empty) | 121 | 110 | 516 | No | Yes | 110 |
| 3B | 4" Conduit (loaded: fill=30%, 14.48 lb/ft) | 143 | 113 | 190 | No | Yes | 113 |
| 3C | 2 ½" Conduit (Empty) | 119 | 103 | 604 | No | Yes | 103 |
| 3D | 2 ½" Conduit (loaded: fill=29.8%, 5.85 lb/ft) | 126 | 112 | 303 | No | Yes | 112 |
| 3E | 1" Conduit (empty) | 98 | 87 | 712 | No | Yes | 87 |
| 3F | 1" Conduit (loaded: fill=29.7%, 1.02 lb/ft) | 108 | 96 | 584 | No | Yes | 96 |
| 3I | Junction Box (empty) | 122 | 134 | n/a | No | Yes | 122 |
| 3J | Cable Air Drop (seven pcs of bare #8) | 169 | 159 | 319 | No | Yes | 159 |

<u>Conclusions from NRC/SNL Testing</u>

Shrinkage: A major defect in the performance of both Hemyc and MT ERFBS is the physical properties of the exterior covering which exhibits severe shrinkage effects when exposed to a thermal insult. This shrinkage results in junctions to open exposing the protected raceway to direct heat exposure and thus causing unacceptable abrupt temperature rise within the barrier.

Supports: Where Unistrut and steel supports were protected with Hemyc material, the temperature at a location 45.7 cm (18.0 in) from the edge of the protected member achieved a fire endurance rating of 58 and 56 minutes respectively. The steel member had a higher cross-sectional area of metal and as would be expected conducted heat better. The testing indicated that, with only the 7.62 cm (3 in) protection on supports as required by the vendor manual, thermal shorts could be introduced into the ERFBS in the range of 13 to 32 minutes. These findings would also apply to intervening metallic items that penetrate the completed Hemyc ERFBS as is common in NPP installations.

Raceway Loading: Although both loaded and empty raceways were tested, it is believed that the barriers failure by shrinkage altered the effects associated with cable fill and therefore effects of cable fill could not be obtained from the MT tests.

*5.4.3.4 Industry Testing*

Most utilities using Hemyc and MT relied on qualification testing conducted and supplied by the vendor (Promatec). As a result of the failures identified through NRC testing, an informal Hemyc users group conducted testing of the Hemyc ERFBS on August 23, 2005, in configuration similar to Test 1 conducted by NRC. The objective of the industry testing was to identify performance differences between new-old-stock Siltemp fabric mesh and the Refrasil fabric used in NRC testing. The industry test program also used more robust barrier construction design, some of which were considered to be more representative of what is used in NPPs. The Siltemp® used for the testing was taken from two different licensees stock and used for comparison purposes to NRC test results. The industry results are proprietary to the licensees involved with the testing and cannot be discussed in this report. However, what can be said is that with the design changes and use of Siltemp fabric, the Hemyc barrier failure times were consistent with what was observed during NRC testing, as were the modes of failure.

## 5.4.4 Resolution and Staff Conclusion

The specific resolution each NPP took is discussed in detail in Section 6.

Out of the 17 units that use Hemyc, 12 have informed NRC of their intent to transition to the risk-informed method per 10 CFR 50.48(c). A discussion of NFPA 805 is provided above in Section 3, "ERFBS Regulations."

NRC staff has confirmed through inspection that licensees not transitioning to NFPA 805, where their fire protection program includes Hemyc and MT fire barrier materials, have resolved issues with the Hemyc and MT materials by completing plant modifications and/or requesting and receiving NRC staff approval for changes to their licensing bases.

Based on its review of Hemyc, the staff believes that the licensees are taking the adequate actions to bring their plants into compliance. The NRC reactor oversight process allows for periodic review of licensee fire protection programs, which can identify future deficiencies. The NRC staff review of each NFPA 805 application will ensure that the performance-based methods using Hemyc as part of it fire protection plan will adequately protect plant safe shutdown capability in the event of a fire.

## 5.5 Mecatiss (Insulative ERFBS)

Mecatiss ERFBS is manufactured by Mecatiss of Morestal, France. According to the Mecatiss website (www.mecatiss.com), Mecatiss specializes in passive fire barrier systems, watertight, airtight, and biological protections, but can also provide private laboratory and testing facilities.

The Mecatiss ERFBS used in US NPPs consists of several layers, including a silicon fabric, a mineral wool insulation, a silicon based mastic, and an adhesive. The silicon fabric, called Silco cloth is a nominal 0.05 cm (0.02 in) thick woven glass silicon fabric. This material is claimed to be gas and water tight at normal pressures and chemically inert. It is applied around the cable raceway and again around the exterior of the completed barrier and held together by the use of an adhesive identified as Silicone Glue Mastic Type 75A. The 75A adhesive is cold application silicon-based mastic used in thin layers for filling, coating, insulating, bonding, and joining work. It is used to seal the SILCO fabric and bounds Silco to itself, concrete, metal, etc. The MPF-A and MPF-B refractory mineral wool insulation provided that actual thermal insulation of the system. The type and number of layers used depend on the design of the ERFBS, but all mineral wool mats are held together with Mecatiss refractory glue Type F-active adhesive. This adhesive is an air-hardening adhesive component and exhibits adhesive characteristics up to 1302ºC (2375°F). Figure 5-14 shows a conduit test assembly protected with Mecatiss prior to testing.

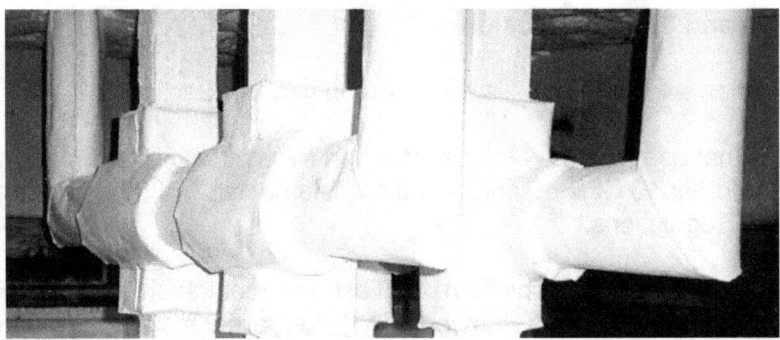

**Figure 5-14. Photo of Mecatiss ERFBS (NRC Photograph)**

### 5.5.1 History

Mecatiss found its application as an ERFBS upgrade and standalone product approximately 14 years following the issuance of the fire regulation (Appendix R). When it was introduced in the late 1990's the ERFBS testing criteria specified in NRC guidance documents had been used and understood for several years. As a result, the Mecatiss ERFBS was designed and tested to pass the GL 86-10 Supplement 1 testing criteria. In addition, licensee who elected to use Mecatiss learned from previous failure to install a barrier system per the tested configurations and were more vigilant to install Mecatiss ERFBS in plant applications that were bounded by tested configurations.

## 5.5.2 Problems

As of the date this report is being written, there have been no generic problems or issues identified related to the plants use of Mecatiss as an ERFBS. As with all ERFBS, NRC inspection staff inspects numerous ERFBS systems during plants triennial fire protection inspections and during routine resident inspector inspections.

## 5.5.3 Testing

Florida Power Corporation (FPC), the licensee of Crystal River 3, notified NRC during a February 28, 1995 meeting of its intent to use Mecatiss as an upgrade and replacement ERFBS for resolution of the extensive amount of Thermo-Lag used in the plant. At that time, this licensee noted 2,345m (7,700 linear feet) of Thermo-Lag protecting conduits and cable trays and another 4,732m (15,526 feet) of Thermo-Lag protecting raceway supports, were used at CR-3. FPC also informed NRC of the recently completed fire barrier testing conducted in France and their planned follow-up testing of the replacement barriers at a US testing laboratory. These two testing programs are described below. The French test was conducted at the Mecatiss vendors' site as a viability test, while the UL test provided the qualification of the barrier.

### 5.5.3.1 Mecatiss over Thermo-Lag Testing (Morestel, France)

Test report file No. NE0016 documents 1- and 3-hour fire endurance and hose stream tests conducted on a combination of Thermo-Lag and Mecatiss barriers, tested in Morestel, France. The raceway test assemblies were identical and included, eight "U" shaped aluminum conduits consisting of a single group of six conduits (two 1.9 cm (0.75 in) and four 2.54 cm (1.0 in) aluminum conduits all oriented side-by-side) and two singular conduits (1.9 cm (0.75 in) diameter). The six conduit group was enclosed in a single ERFBS, while the singular conduits were protected by individual ERFBS. Figure 5-15, shows the configurations of the test assemblies. Note that the individually enclosed conduits had condulets on one bend while the group of six used radial bends exclusively. All conduits were supported by Unistrut and Unistrut pipe hanger hardware. The two test articles (1- and 3-hour) were designed and constructed to represent CR-3 existing thermo-lag barrier system.

Temperature measurements were taken using Type K, No. 24 AWG thermocouples installed on surface of fire conduits at FPC, and on a single bare #8 AWG stranded copper conductor, all spaced approximately 15.2 cm (6.0 in) on center.

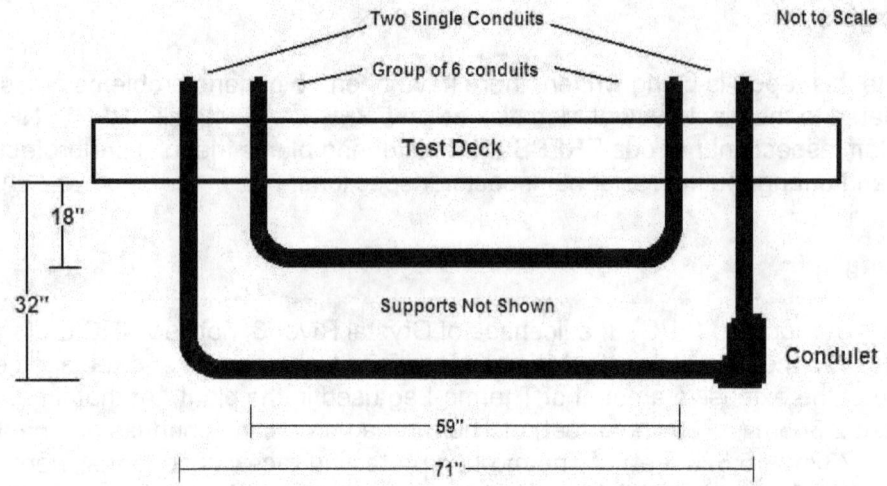

**Figure 5-15. Test Assembly of 1- and 3-hour Thermo-Lag/Mecatiss Test (Author, 2009)**

The ERFBS configurations used for these tests varied, as documented in Table 5-26 and Table 5-27, a Thermo-Lag stand alone barrier and a Thermo-Lag with Mecatiss overlay barrier were tested. For the 1-hour test, all conduits were protected with a base Thermo-Lag ERFBS consisting of pre-shaped conduit sections with the stress skin on the interior face. For the group of six conduits, pre-shaped conduit sections were secured to the outside of the two exterior conduits while the standard nominal 1.3 cm (0.5 in) Thermo-Lag panels (with ribs and stress skin) were fit to the top and bottom of the conduit group and butt jointed to the exterior pre-shaped conduit sections. Stainless steel band straps and 18 gauge stainless steel tie wire were used to secure the panels to the raceway, with trowel grade Thermo-Lag 330-1 used to pre-butter and post-butter the joints during construction. For the 3-hour test assembly, the Thermo-Lag base layer was the same as above, with the exceptions that the prefabricated panels and conduit sections being 2.858 cm to 3.175 cm (1.125 to 1.250 in) thick with a 0.858 cm (0.125 in) square coated mesh stress skin on the inside and outside of each section.

The Thermo-Lag materials were obtained from storage at FPC, except for the material placed on the single 1.9 cm (0.75 in) conduit, which was removed from an actual plant installation. Installation of the Thermo-Lag was installed by certified installers at FPC, witnessed by UL laboratory staff and FPC's Quality Control organization.

The test articles were then shipped to France where the Mecatiss overlay was installed. The 1-hour Mecatiss system consisted of (in order from conduit to exterior):

- A woven glass cloth coated on one surface with a silicone sealant. The coated glass cloth was identified as Silco.
- An adhesive identified as No. 75A
- A nominal 2.54 cm (1.0-in) thick refractory mineral wool insulation identified as MPF60
- Nylon Twine
- An adhesive identified as F-active

For the 1-hour test assembly, the Mecatiss barrier was installed on top of the Thermo-Lag base layer by performing the following:

- applying approximately 0.64 cm (0.25 in) diameter beads of Type 75A adhesive to the

- coated side of the Silco
- wrapping the Silco (inner layer) around the Thermo-Lag with the coated side of the Silco in contact with the Thermo-Lag
- applying type F active adhesive to the uncoated side of the Silco and on one side of the MPF60 insulation
- applying the MPF60 insulation to the silco and securing the insulation in place with nylon twine
- applying the Type F active adhesive to the exposed surface of the MPF60 insulation and on an uncoated side of Silco
- wrapping the MPF60 insulation with a layer of Silco (outer layer) with the uncoated side of the Silco being in contact with the MPF60 insulation
- applying bead of type 75A adhesive to the exposed Silco joints.

The installation of the Mecatiss barrier was performed by employees of Mecatiss and witnessed by a member of UL engineering staff and representative of FPC and Babcock and Wilcox (B&W) Nuclear technologies.

For the 3-hour test, the Mecatiss overlay was as described above with the exception of the mineral wool being different, namely a nominal 2.54 cm (1.0 in) thick refractory mineral wool insulation identified as MPF180A and another 2.54 cm (1.0 in) thick refractory mineral wool identified as MPF180B was used instead of MPF60. Again, installations of the Mecatiss barriers were witnessed by UL staff and FPC staff.

Furnace thermocouples included four (4) used by Mecatiss to control furnace, four (4) were used as input to the data acquisition system and four (4) used by UL as long time constant thermocouples (ASTM E-199). Furnace dimensions were 2.6m by 1.1m by 1.4m(8.5 feet by 3.5 feet by 4.7 feet) and a pressure differential throughout this test was slightly negative at - 40 Pascal (-0.8 pounds per square foot).

Following each tests, a water hose spray test was administered. The hose spray test consisted of the following parameters:

- 3.8 cm (1.5 in) fog nozzle,
- 30° discharge angle,
- pump pressure between 596397 and 695681 Pascal (86.5 and 100.9 lbs per in$^2$), and
- minimum discharge of 423 liters per minute (93 gallons per minute (gpm))

The hose spray was applied to all exposed surfaces for 5.5 minutes for a nominal distance of 1.5 m (5.0 ft) from test article that was rotated at approximately 6 revolutions per minute. Hose stream testing began approximately 4 minutes after the completion of the thermal endurance test, and no conduits were identified as being exposed at the end of the hose stream test.

Post test observations identified the following: Mecatiss appeared homogenous, without distortion, but with several blow holes seen, as the result of the off gassing of the inner TL layer. There was complete burn-through of the aluminum conduit on the 90° bend of conduit A (Thermo-Lag only article). Molten aluminum had formed pools at the bottom of the test furnace. In addition, to the bend having melted, it was also observed that the horizontal portion of Conduit A had completely melted, leaving only the vertical portion. Table 5-26 and Table 5-27 provide descriptions of each test article along with the actual endurance rating achieved during the test.

Table 5-26. 1-hour FPC Mecatiss Testing in France

| Article | Barrier | Failure time when $\Delta T_{avg} = 131°C$ (250°F) | Failure time when $\Delta T_{max} = 181°C$ (325°F) |
|---|---|---|---|
| Group of Six Conduits | Thermo-Lag | 44 minutes | 45 minutes |
| ¾" Conduit | Thermo-Lag | 26 minutes | 29 minutes |
| ¾" Conduit | Thermo-Lag & Mecatiss Overlay | Did not exceed | Did not exceed |

Table 5-27. 3-hour Mecatiss Testing in France

| Article | Barrier | Failure time $\Delta T_{avg} = 139°C$ (250°F) | Failure time $\Delta T_{max} = 181°C$ (325°F) | Max. Avg. Temp. (°C)* | Max. Single Temp. (°C)* |
|---|---|---|---|---|---|
| Group of Six Conduits ¾" | Thermo-Lag & Mecatiss Overlay | N/A | N/A | 270 | 238 |
| ¾" Conduit | Thermo-Lag | 1 hr. 9 min | 1 hr. 13 min. | 598 | 1148 |
| ¾" Conduit | Thermo-Lag & Mecatiss Overlay | 2 hr. 56 min | 3 hr. 1 min. | 146 | 239 |

* Temperature measurement at 3 hours and 11 minutes

These results clearly demonstrate that the use of Mecatiss as an overlay to existing Thermo-Lag 330-1 ERFBS provides additional fire endurance rating to the barrier system. For instance, the test results indicate that the Thermo-Lag only conduit covering exceeds the temperature rise criteria at approximately 73 minutes, while the additional Mecatiss overlay provided approximately 108 additional minutes of protection.

**NRC review of French Mecatiss Tests**

By letter dated April 7, 1995, NRC provided FPC with comments on the tests conducted in Morestel, France on the Thermo-Lag and Mecatiss ERFBS. The intent of this letter was to provide comment for the licensees' consideration when developing a plan for future testing. NRC comments pointed out the following concerns with the French Mecatiss Testing:

- Ambient temperature at start of test was outside the range identified in ASTM E-119 and not representative of conditions expected at CR-3. This deviation is expected to affect the test results for assemblies that meet the temperature criteria with little to no margin.
- The negative furnace pressure in the 1-hour test deviated from the 3-hour test by 16.00 Pascal (0.06 in of water).
- Furnace and burner specifics were not reported
- Only 4 thermocouples were used to record the furnace temperature, while ASTM E-119 requires a minimum of nine thermocouples symmetrically distributed.
- The furnace temperature, as measured by the UL thermocouples, was below the E-119

temperature and the area under the actual temperature curve was not within the 10% as required by E-119 for several exposure periods in both tests.
- The 1.90 cm (0.75 in) conduit protected with Thermo-Lag and Mecatiss failed to meet the 3-hour rating. (If these results are to be used, the licensee was directed to request a deviation from the acceptance criteria based on an engineering evaluation acceptable to the staff, such as demonstration of cable functionality.)

The licensee responded to NRC comments in its letter dated May 17, 1995. As explained in that letter, much of the deviations were a result of the test furnace size and instrumentation available to conduct the testing. As discussed previously, the licensee conducted the French Mecatiss testing to identify the viability of this barrier system to resolve Thermo-Lag issues and planned to conduct qualification testing in the USA, if favorable results were obtained.

### 5.5.3.2 UL Testing of Mecatiss for Florida Power Corporation

Following FPCs viability testing of Mecatiss at Morestal, France, the Mecatiss qualification tests were conducted at UL. Five fire tests were conducted at UL to qualify a 1- and 3-hour standalone Mecatiss barrier (MTS-1 & MTS-3), along with a 1- and 3-hour upgraded overlay Mecatiss barrier system (MPF-60 & MPF-180). The test plan was provided to NRC staff for comment.

The proprietary UL fire endurance test reports were submitted to NRC on March 30, 1996, while the non-proprietary versions were submitted on July 31, 1996. NRC staff reviewed these test reports and concluded in an SE dated January 29, 1997, that the Mecatiss fire barrier system, when designed and installed in accordance with the techniques utilized for the test specimens, meets the acceptance criteria specified in Supplement 1 to GL 86-10, and is, therefore, acceptable for use as a fire barrier systems relied upon by the licensee to meet NRC fire protection requirements. The following description of the test and results are based on public information, including NRC staff trip reports, NRC safety evaluations and a public FPC test results summery letter. A summary table of the results and details of this testing is presented in Table 5-28.

Test Deck No. 1 was subjected to a 1-hr test while test decks 2 and 6 were subjected to a 3-hour test, all in the UL column furnace, as specified by GL 86-10 Supplement 1. Test Deck No. 3 and 4 were tested in the UL "Floor Furnace," while Test Deck No. 5 contained three test articles were tested in the UL "wall furnace."

Table 5-28. Results of UL Mecatiss Testing

| Test Deck No. | Article No. | Raceway | ERFBS | Fire Test Rating (minutes) |
|---|---|---|---|---|
| 1 | 1 | ¾" conduit | MTS-1 | 60 |
|   | 2 | 24" cable tray | ½ length TL[1] + MPF-60 ½ length MTS-1 | 60 |
|   | 3 | ¾" conduit | 1-hr TL + MPF-60 | 60 |
| 2 | 4 | ¾" conduit | MTS-3 | 180 |
|   | 5 | 24" tray | 3-hr TL + MPF-180 | 180 |
|   | 6 | ¾" conduit | 3-hr TL + MPF-180 | 180 |
| 3 | 7 | 24" cable tray | MTS-1 | 60 |
|   | 8 | ¾ & 4" conduit | 1-hr TL + MPF-60 | 60 |
|   | 9 | two ¾" conduit | 1-hr TL + MPF-60 | 60 |
|   | 10 | 24" cable tray tee with four conduit stubs | 3-hr TL + MPF-180 | 60 |
| 4 | 11 | 24" cable tray | MTS-3 | 80 |
|   | 12 | ¾" & 4" conduit | 3-hr TL + MPF-180 | 80 |
|   | 13 | two ¾" conduit | 3-hr TL + MPF-180 | 80 |
|   | 14 | 24" cable tray tee with four conduit stubs | 3-hr TL + MPF-180 | 80 |
| 5 | 15 | 24" cable tray | 1-hr TL + MPF-60 | 60 |
|   | 16 | ¾" conduit & 12"x14"x6" junction box | 1-hr TL + MPF-60 | 60 |
|   | 17 | ¾" & 4" conduit | 1-hr TL + MPF-60 | 60 |
| 6 | 18 | 6" cable tray | 1-hr TL + MPF-60 | 93 |
|   | 19 | 6" cable tray | MTS-1 | 102 |
|   | 20 | 6" cable tray | 3-hr TL + MPF-180 | 115 |
|   | 21 | 6" cable tray | MTS-3 | 180 |
|   | 22 | 6"x6" cable wire way | ½ length TL[1] + MPF-60 ½ length MTS-1 | 110 |

[1] TL is an abbreviation for Thermo-Lag 330

It should be noted that for the results reported for Test 6 in Table 5-28, are the times when the barrier first exceeded the temperature rise criteria of GL 86-10 Supplement 1. All test articles of test deck No. 6 were subjected to a 3-hour E-119 fire endurance test, although not all were designed to be 3-hour barriers.

*5.5.3.3 Ampacity Derating Tests*

UL also performed several ampacity derating tests for FPC for the Mecatiss and Thermo-Lag/Mecatiss ERFBS used at CR-3. The testing was performed in accordance with IEEE P848/D15 "Procedure for the Determination of the Ampacity Derating of Fire Protected Cables," dated January 1, 1995. The results of this testing were submitted to NRC for review, but were determined to be proprietary to FPC and will not be presented in this repot. However, it can be said that the ampacity derating values of the Mecatiss ERFBS were consistent with derating values of other barrier types.

## 5.5.4 Resolution and Staff Conclusion

The Mecatiss ERFBS was used as a replacement barrier to Thermo-Lag or other ERFBS that were found incapable of performing there design function. There have been no generic problems identified with the use of Mecatiss to provide the required 1- or 3-hour protection of equipment important to safe shutdown. Therefore, the staff concluded that the use of Mecatiss as an ERFBS in accordance with applicable testing results will provide adequate assurance that the structures, systems, and components will be protected by the use of Mecatiss in configurations bounded by test results.

## 5.6 Kaowool and FP-60 (Insulative ERFBS)

Kaowool is manufactured by Thermal Ceramics. The Kaowool blanket material is produced from kaolin, a naturally occurring alumina-silica fire clay. It is a noncombustible, flexible, ceramic-fiber blanket, composed primarily of silica and alumina compounds ($SiO_2$ and $Al_2O_3$) and has a melting point of 1760°C (3200°F). FP-60 is basically an upgraded version of Kaowool with a 2.00 mm (0.08-in) aluminum foil skin laminated to both sides of a ceramic-fiber blanket. Kaowool and FP-60 are used to construct barriers intended to have a 1-hour fire resistance rating, but neither are rated for 3-hour use. In addition to their use as a fire barrier, some licensees used these barriers for separation of certain electrical systems in accordance with the guidance of RG 1.75, "Physical Independence of Electrical Systems," or to limit combustible sources within a fire area by wrapping it with either of these materials. San Onofre Nuclear Generating Station, Units 2 and 3 uses a product called Cerablanket which is a flexible, ceramic-fiber blanket similar to Kaowool and is shown in .

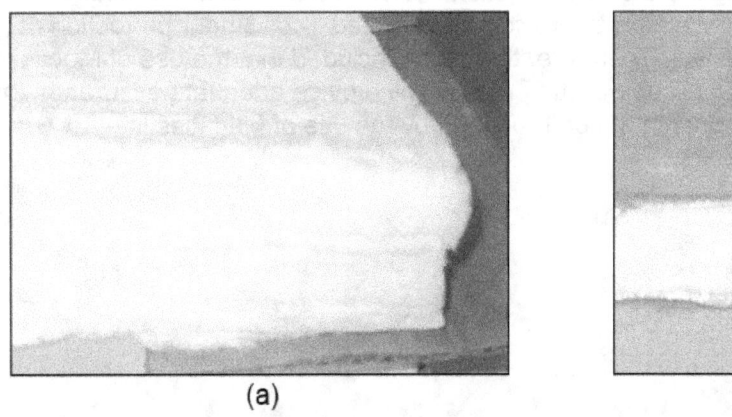
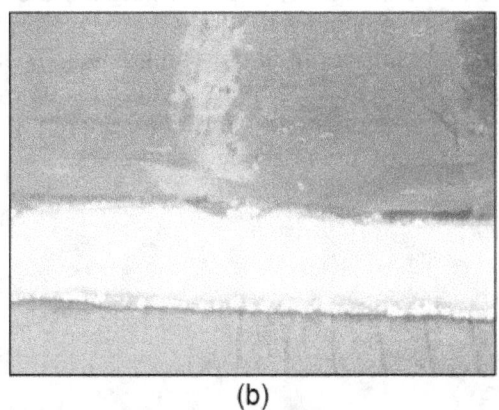

(a)          (b)

**Figure 5-16. Picture of (a) Kaowool and (b) FP-60 Material (Author, 2009)**

The installation of the Kaowool and FP-60 ERFBS are similar. For cable trays, the trays are first prepped with steel banding wrapped around the tray and spaced approximately 35.6 cm (14.0 in) on center (OC), then a wire mesh cut to the width to the cable tray is secured to the top of the tray with filament tape. The cable trays are then wrapped with two to six layers of the ceramic fiber blanket. Each wrap consists of a nominal width blanket 61.0 to 122 cm (24 to 48 in) overlapping on the top side of the raceway. During installation, each layer is held in place with filament tape. The adjacent wrap overlaps the preceding wrap by a minimum of 2.54 cm (1.0 in). Successive layer of the ceramic wrap are installed in the same manner with the overlap locations for succeeding layer offset a minimum of 30.5 cm (12 in) from the overlap of the preceding layer. As a final step, a nominal 1.9 cm (0.75 in) wide by 0.038 cm (0.015 in) thick steel band straps are wrapped around the cable tray system and secured with 2.54 cm (1.0 in) long channel-shaped crimp clips. The steel bandings are spaced at a maximum of 35.6 cm (14 in) on center and a maximum of 10.1 cm (4 in) from any joint on the outer blanket. After the installation of the ceramic fiber blanket is completed, an optional silica or glass fiber cloth can be wrapped around the outer layer. This will help protect the base Kaowool material but is not required. At wall and floor interfaces loose ceramic fiber is firmly packed round the periphery of the cable raceway.

Kaowool and FP-60 ERFBS can be installed on conduits and air drop configurations in the same manner as done with cable trays. , provides a diagram showing the various layer and configurations used to construct a Kaowool or FP-60 ERFBS.

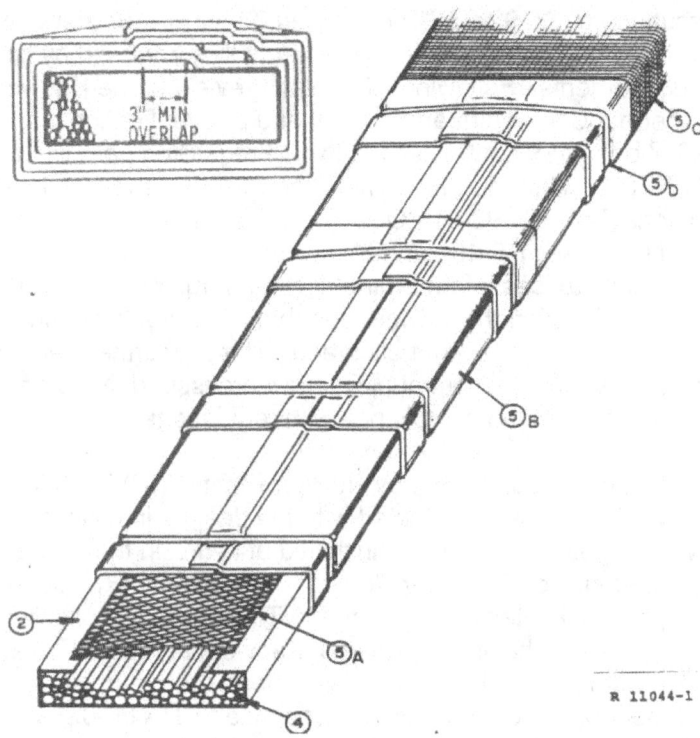

Figure 5-17. Sketch of Kaowool / FP-50 Installation (Fire Test, 1978)

Kaowool is available in numerous thicknesses and licensees used various configurations (number of layers versus thickness of material) to construct their ERFBS. When the FP-60 material came into use, the 1.3 cm (0.5-in) thick material provided somewhat better fire resistance rating. This is a result of multiple layers of thinner material will outperform the fewer layers of thicker material, because trapped air in space between the layers acts as an insulator. The consequences of a joint failure are also lessened because of the multiple joints and staggering of the joints.

### 5.6.1 History

Kaowool was originally developed by Babcock and Wilcox to be an asbestos replacement material and is commonly used to insulate high temperature furnaces, forges, and kilns. Its low density of 64.1, 96.1, or 128.1 kg per m$^3$ (4.0, 6.0, or 8.0 lbs per ft$^3$), very low thermal conductivity and ease of handling and cutting resulted in Kaowool being used in numerous commercial applications and configurations.

As a result of its successful use in other industrial applications, Kaowool insulation was one of the first materials to be used in protecting electrical raceways containing FSSD circuits. This is primarily due to the fact that, during the late 1970's and early 1980's there were concerns that there were no material commercially available that clearly met a 1-hour fire barrier requirement and because some licensees had experience with Kaowool installation to meet BTP APCSB

9.5.1 guidance, Kaowool was chosen as a viable solution to Appendix R fire barrier applications. Although the use of Kaowool and FP-60 material in the nuclear industry was a result of Appendix R requirements, the material was in existence well before the need for its employment as an ERFBS.

Subsequent to the issuance of GL 92-08, "Thermo-Lag 330-1 Fire Barriers," NRC began evaluation other known fire barrier materials and systems that are used by licensees to fulfill NRC fire protection requirements. Following NRC staffs review of the Kaowool and FP-60 test reports provided by Thermal Ceramic (manufacture) and a reverification inspection at Salem Nuclear Generating Station in 1993, NRC issued two IN regarding potential problems with Kaowool and FP-60 ERFBS. IN 93-40, "Fire Endurance Test Results for Thermal Ceramics FP-60 Fire Barrier Material," and IN 93-41, "One Hour Fire Endurance Test Results for Thermal Ceramics Kaowool, 3M Company FS-195 and 3M Company Interam E-50 Fire Barrier Systems." Both Information Notices informed the nuclear industry of deficiencies found in the test reports concerning qualification the barriers and that NRC would continue its review of the barriers ability to perform its fire resistive function and will issue further generic communications, if needed. Although no other generic communications were issued, NRC continued to interface with industry to determine its use and compliance with regulations.

During an NRC inspection of Joseph M. Farley Nuclear Plant (FNP) in 1996, the NRC inspectors identified technical issues associated with the design, installation, and fire-resistive performance of Kaowool raceway fire barriers installed at FNP. In the later part of 1996, NRC Region II offices requested Office of Nuclear Reactor Regulation (NRR) support in reviewing the identified issues through a Task Interface Agreement (TIA) 96-023. Following a detailed review of the performance and licensing basis of the use of Kaowool at FNP, NRC staff determined that the fire rating of the Kaowool installed at FNP was indeterminate, but less than the 1-hour needed to meet the Appendix R requirements. A response to TIA 96-023 was provided to the Region II offices on June 18, 1999. The response concluded that the licensee FNP did not have a sound technical basis for concluding that the Kaowool ERFBS installed at FNP meet the regulatory requirements or provided an adequate level of fire protection for the post-fire safe-shutdown capability. In SECY-99-204, NRC staff informed the Commission of its review of this matter and provided its TIA 96-023 response as an attachment.

When SECY 99-204 was issued on August 4, 1999, there were 15 unites (9 plants) that used Kaowool or FP-60 to meet the Appendix R regulatory requirements, they were:

- Farley 1 & 2
- Grand Gulf
- Prairie Island 1 & 2
- Sequoyah 1 & 2
- Susquehanna 1 & 2
- Fitzpatrick
- Hatch 1 & 2
- Salem 1 & 2
- Summer

Of the 15 units identified, all licensees except FNP, James A. Fitzpatrick Nuclear Power Plant, Grand Gulf Nuclear Station, and Virgil C. Summer Nuclear Station elected to voluntarily eliminate the use of Kaowool to meet the regulatory requirements. Farley Nuclear Plant used the largest quantity of Kaowool approximately 1,920 m (6,300 linear feet), Summer used 304.8 m (1,000 linear feet) (3 layers, each 2.54 cm (1-in) thick), Fitzpatrick used 18 m (60 linear feet) (six layers, each layer 1.3 cm (0.5 in) thick) and Grand Gulf used nearly 457 m (1,500 linear feet) (two layers, each 2.54 cm (1-in) thick).

### 5.6.2 Problems

As presented in the testing section that followings, the pure insulation properties of the ceramic fiber was not capable of reducing the heat transfer from the extreme heat of the test furnace (ASTM E-119 curve) to the protected cable raceways. In fact, there were only two test specimens (see Virgil C Summer Nuclear Station testing) which provide the rated 1-hour protection as specified in Supplement 1. All other testing of Kaowool and FP-60 ERFBS failed the Supplement 1 criteria or the test report was not sufficient in detail to determine if the test specimen would pass the GL 86-10 Supplement 1 guidance.

The construction of Kaowool resulted in a major problem with maintaining the barrier, namely any physical contact with the ERFBS had the potential to damage the barrier. Unlike other barriers that were a rigid or semi-rigid system, the loose Kaowool ceramic fiber could easily be pulled away with little to no effort. As a result, anytime a barrier was bumped into, had a tool dropped on, or was improperly used as a step or human support during maintenance activities, the barrier would typically become damaged and likely have a reduction in its thermal insulation capacity. As a result, several inspection findings and violations have been issued for damaged or missing Kaowool barriers. This is one of the reasons some licensees replaced or upgrade the Kaowool ERFBS with the FP-60 material which has the 2-mil aluminum foil skin on the exterior layer to protect the internal Kaowool fiber. As discussed in the Section 5.4, Hemyc and MT ERFBS were another solution which provided protection to the base Kaowool fiber.

A phenomena referred to as "wicking," also had the potential to degrade the barrier and cause additional fire hazards within the plant. Wicking is when a flammable liquid, like diesel fuel or flammable solvents (liquid hydrocarbon), are spilled or leak onto the noncombustible Kaowool fabric, causing the barrier to aid combustion if a fire were to occur. Think of an old kerosene lantern, same idea. Not only will wicking aid a fire, but when the Kaowool fabric becomes wet, the moisture and added weight has been known to cause layers of the material to fall away, thus reducing the quality of the barrier.

As discussed above, plants licensed to operate prior to incorporating Appendix R in the regulations, are required to satisfy Appendix R requirements. However, GL 86-10 states that licensees need not replace Kaowool materials that were installed before Appendix R became effective and that were accepted by NRC as 1-hour fire-rated barriers. According to SECY 99-204, NRC staff and the licensees have interpreted the GL 86-10 guidance to mean that Kaowool raceway fire barriers installed before Appendix R became effective and that were accepted by NRC as a 1-hour fire-rated barriers are "grandfathered" and that exemptions are not needed even though the barriers may not meet the technical requirements of Appendix R. However, GL 86-10 guidance does not relieve the licensee from establishing and maintaining the design bases for the fire barriers it has installed to satisfy NRC's fire-protection requirements. NRC staff review of FNP, determined that the licensee did not establish an acceptable design basis for the Kaowool ERFBS installed to satisfy Appendix R.

### 5.6.3 Testing

During NRC review and response to TIA 96-023, the licensee of FNP submitted 16 Kaowool fire test reports. In reviewing the testing documentation submitted by the licensee, fundamental generic testing deficiencies were discovered in most of the tests reviewed. These generic deficiencies included (1) non-standard full-scale test furnaces, (2) non-standard furnace instrumentation, (3) non-standard fire exposures, and (4) no hose stream testing.

The following test summaries are those applicable to testing ERFBS, although only one was performed in accordance with Supplement 1, these results do indicate the inability of Kaowool to provide the required fire protection of safe shutdown equipment.

*5.6.3.1 Sandia National Laboratories (SNL) Testing*

On September 15, 1978, a full-scale fire test was conducted at UL to demonstrate the effectiveness of a ceramic fiber blanket and automatic fire suppression system to protect cables in a vertical cable tray configuration. An open pool fire fueled by liquid hydrocarbon was used in the test.

The test was carried out in a corner-ceiling assembly approximately 6.1 m long by 6.1 m wide by 4.6 m high (20 ft x 20 ft x 15 ft). The walls of the assembly consisted of steel framing and 1.3 cm (0.5 in) thick Marinite boards covering the steel. Five open ladder cable trays were installed in the test assembly 45.7 cm (18 in) wide with rungs installed at 22.9 cm (9 in) intervals. Three conductor cables were run through the cable trays. The cables were 1.2 cm (0.468 in) diameter with 600 V rating made of #12 AWG stranded copper. A conductor insulation of 0.7 mm (0.027 in) covered the 9.4 m (31 ft) long cables which were bundled in groups of eight.

Kaowool fire barrier was installed along the entire length of each cable tray with a thickness of 2.54 cm (1 in) and fastened to the cable trays with 1.9 cm (0.75 in) thick steel bands and band clips. Additional pieces of Kaowool covered 1) the entire front surface of the cable bundle along the vertical overlapping joint of the outer layer, 2) the back surface of the cable bundle and tray at horizontal butt joints of the outer layer, 3) the horizontal butt joints of the outer layer, and 4) the ceiling and floor butt joints.

Seven Type K thermocouples were located in each tray and additionally between cable trays over the fire pan and near each open head sprinkler. The test began by pouring two gallons of n-heptane into a pan below the test assembly and igniting by torch. Within ten seconds of ignition, the flames had reached a maximum of 1.2 m (4 ft) and covered the entire fuel pool. All flaming ceased after 40 minutes. Cable trays clad with flame-engulfed Kaowool are shown in .

**Figure 5-18. Fire Engulfing Cable Trays Clad with Kaowool. (NUREG/CR-0596)**

Post test observations indicated that the Kaowool was blackened on the base of each tray upward about 0.3 m (1 ft) but remained largely unaffected on the inner surface except for light brown areas in the bases of trays 1, 2, 3, and 4. Thermal damage of cables was noted in all trays except for tray five, approximately 3 to 6 inches above the fire pan. Cable material in all four trays were melted and charred with the greatest damage occurring in cable tray 3. Maximum temperatures recorded for cable trays one through five were 56.9°C, 58.8°C, 60.6°C, 54.7°C, and 35.2°C (134.5°F, 137.9°F, 141.1°F, 130.4°F, and 95.3°F) respectively.

During testing, cables were energized with low voltage and conductors with the same color code in each tray were connected in parallel to provide three circuits per tray. Each circuit had low current flow during the test and was monitored continuously for shorts between conductors or between conductors and trays. A short circuit between conductors 1 and 3 in cable tray 3 was indicated at 3 minutes 13 seconds into the test. Three minutes 55 seconds into the testing, erratic measurements were recorded in tray 1 indicating intermittent short circuits. The complete detail of this testing can be found in NUREG/CR-0596, "A Preliminary Report on Fire Protection Research Program Fire Barriers and Suppression (September 15, 1978, Test)."

*5.6.3.2 NIST Small Scale Testing*

NIST performed small-scale testing of a 1-hour FP-60 barrier system consisting of four layers of a ceramic fiber blanket nominally 12.7 mm (0.5-in) thick with 0.051 mm (0.002-in) aluminum foil laminated on both surfaces. The test assembly was subjected to a 1-hour ASTM E-119 fire exposure. The results indicated that the average and maximum single point temperature rise criterion were reached during the 3.5 hour exposure.

*5.6.3.3 Vendor Testing*

5.6.3.3.1 Babcock and Wilcox Testing

This test series evaluated that protection provided by varying layers of Kaowool insulation installed on cable trays and conduits. The cable trays and conduits tested were loaded with IEEE-383 "Qualifying Class 1E Electric Cables and Field Splices for Nuclear Power Generating Stations," qualified and non-qualified cables with 5 to10 thermocouples placed on the exterior of the cables, within the cable tray. The natural gas fired 91.4 cm by 91.4 cm (36.0-in by 36.0-in) furnace was controlled to the ASTM E-119 time-temperature curve for all tests. For each test, the tray raceway was located such that the flames from the burners would be along the side and bottom of the cable tray. The conduit raceway was suspended above the cable tray assembly. The cables were connected to an incandescent display board capable of monitoring 20 circuits and powered by either 440 or 110 Volts AC power through a circuit breaker. This circuit configuration allowed for the detection of cable-to-cable and cable-to-raceway shorts, as well as open circuit faults.

Test 1 evaluated the performance of cables when no Kaowool protection, which resulted in circuit failure of the non-qualified cable within eight minutes. The configurations and results of all tests are shown in .

## Table 5-29. Kaowool ERFBS Test Results (10/24/1978)

| Test # | Raceway Configuration | Barrier Configuration | Min. Cable Failure Time (minutes) |
|---|---|---|---|
| 1 | Tray: Solid bottom, Steel Galvanized<br>Conduit: Steel | Tray: None<br>Conduit: None | 8[a] |
| 2 | Same as Test #1 | Tray: 1" Kaowool on top of cables, 2 la with 3" overlaps | 51 |
| | | Conduit: 2 layers of 1" Kaowool | 65 |
| 3A | Tray: Aluminum open ladder back<br>Conduit: Aluminum | Same as Test #2 | 11[a] |
| 3B | Same as Test #3A | Same as Test #3A except brackets used to hold exterior of blanket near butt joints were relocated to 3" on each side of butt joint, instead of 2" as done in Test #3A. | 61[a] |
| 4 | Same as Test #2 | Tray: 1" Kaowool on top of cables, 1 layer of 1" Kaowool wrapped with 3" overlaps. | |
| | | Conduit: 1 layer of 1" Kaowool | 40[a] |
| | | For both raceways, a 4" wide strip of Kaowool was wrapped around the butt joints and held in place by steel banding. | |

[a] Report did not specify failure location (tray or conduit)

Post test evaluation of the barrier with comparison of the thermocouple data indicated that butt joint failed resulting in the direct heat exposure to the cables and early failure times. Test 3B incorporated design changes to make the butt joints more thermally robust by adding a Kaowool collar surrounding all exterior butt joints. As shown in , the collar design change significantly improved the performance of the Kaowool barrier. The test report concluded with the following insights.

- Unprotected cables fail early (eight minutes) in complete engulfment fires
- Wrapping solid bottom and open ladder trays and conduit with 5.1 cm (2.0-in) of Kaowool blanket (with all butt joints tight) provides approximately 50 minutes of protection in complete engulfment fires.
- Wrapping solid bottom trays with 2.54 cm (1.0 inch) of Kaowool blanket (10.1 cm (4.0-in) collar over butt joints) provides approximately 40 minutes of protection in complete engulfment fires.
- Loose and open butt joints in insulation may lead to early cable failure in engulfment fires.

## 5.6.3.3.2 Underwriters Laboratories Testing

UL performed fire endurance and hose stream testing on a FP-60 floor protected assembly (File R11044-1 Project 84NK9356). The testing followed UL Subject 1724 (dated May 1984) and was performed by UL on September 26, 1984. The materials used in the test assembly were readily installed by qualified workers with tools and methods commonly used for construction work of this nature.

The ERFBS was installed, with four layers of ceramic fiber used on cable trays and the cable air drop configuration was protected with six layer of ceramic fiber. All steel supports were protected with the ceramic blanket wrapped with approximately 2½ layers; however, the test report does not specify the distance from the raceway which the supports are protected.

Every conductor was energized and monitored for circuit integrity throughout the fire endurance and hose stream testing. The test assembly consisted of eight configurations, four 91.4 cm (36.0 in) wide cable trays (half open ladder back and half solid back), two nominally 12.7 cm (5.0 in) diameter rigid steel conduit, one 30.5 cm by 30.5 cm by 15.2 cm (12.0-in by 12.0-in by 6.0 in) junction box, and one air drop cable. Cable tray cable fill used 1/C 300MCM power, 7/C # 12 AWG control, and 2/C # 16 AWG instrumentation cables. Actual percent cable fill varied from 18.5 percent to 87.6 percent for the cable trays and from 6.9 to 100 percent fill for the conduits raceways.

Ten days following assembly of the ERFBS, the endurance test was conducted in accordance with UL Subject 1724 and terminated at 61 minutes; the assembly was raised and subjected to a 206.8 Pascal (30 lbs per sq. in), 2.604 cm (1.025 in) diameter nozzle hose stream 525.8 cm (17.0 ft 3 in) away, for duration of 2.5 minutes. It was noted at 20 minutes into the fire test that the blanket wrap on the bottom surface of System No. 2 had slipped from beneath the steel banding strap and was bowing downward such that a maximum 1.3 cm (0.5-in) wide vertical opening was present at the center of the sheet edge. At the completion of the test, this opening had grown to be approximately 8.9 cm (3.5-in). The test report concluded that, "the 1-hour fire rating of the FP-60 ERFBS was established by evaluating the performance of the system with respect to maintaining the integrity of the electrical circuits under fire exposure conditions and during a hose stream test following the fire exposure." Based on this rating criterion UL determined that the FP-60 barrier tested provided the 1-hour rated protection.

Although the FP-60 barrier passed the test, per the UL Subject 1724 testing criteria, it was noted in the test report the following visual damage, as summarized in . As you can see from the results, although the circuit integrity was never lost during the fire endurance and hose stream portions of the test, the cable jackets and FP-60 ERFBS did experience a finite amount of damage.

### Table 5-30. Summary of UL FP-60 Fire Resistance Test Results

| Test Article | Description |
|---|---|
| #1 : 36" tray | ...approximately 50 percent of the two conductor No. 16 AWG located on the north bend of the tray were fused together. |
| #2 : 36" tray | ...approximately ½ of the blanket was eroded by the water hose stream such that the bottom surface of the cable tray was exposed. It was also noted that cables were fused together at numerous points. |
| #3 : 36" tray | ...approximately ½ of the insulation on the bottom surface was (missing following hose stream test). ...cable jacket of the seven conductor cables were fused together. |
| #4 : 36" tray | ...approximately ¾ of the fiberglass cloth was consumed. Steel banding straps were missing. Bottom of cable wrap was eroded. Blistering was present on the cable jacket, ranging in size from ¼" to ½" in diameter. |
| #5 : 5" Conduit | ...1/2 of blanket on bottom of conduit was consumed. At one point, all three cables fused together. |
| #6 : 5" Conduit | Side outer two layers were consumed. Outer three layers of bottom were consumed. All three cable types were undamaged. |
| #7 : Junction Box | All four Layers were consumed. Jackets of cables within junction box were fused together. |
| #8 : Air Drop | No information was provided on the air drop configuration. |

If you were to use these test results and the Supplement 1 acceptance criteria, it is easy to see that all configurations would not have been found acceptable (with maybe the exception of #8) to use this barrier as a qualified 1-hour ERFBS.

#### 5.6.3.3.3 Southwest Research Institute Testing

SwRI performed testing of the FP-60 ERFBS (Project No. 01-8305-053). The testing was performed in accordance with the ANI/MAERP standard. The test assembly slab was 2.1 m by 2.1 m (7.0 ft by 7.0 ft) square and 3 configurations were installed, namely a 2.54 cm (1.0 in) diameter rigid conduit adjacent to concrete slab (less than 1.3 cm (0.5 inches) from concrete slab) containing one control and one instrument cable, a cable air drop consisting of one power, one control, and one instrument cable, and a 3.7 m by 3.7 m by 15.2 cm (12.0 ft by 12.0 ft by 6.0 in) junction boxed mounted in a pendulum configuration using a 5.1 cm (2 in) diameter rigid conduit. The junction box contained the three types of cables, all instrumented with thermocouples (thermocouple spacing was not specified). The testing used a small scale horizontal exposure furnace with an expansion collar to fit the test deck assembly. The test furnace followed the standard time/temperature curve and at 60 minutes it was 5.2 percent above the corresponding area under the standard curve (ASTM E-119 allows for ± 10%). All cables were monitored for circuit integrity before the start of the test, at 50 minutes into the test, and after the hose stream test. No circuit integrity was lost during these periodic checks. The hose stream test consisted of a 6.4 cm (2.5-in) National Standard playpipe equipped with a 2.858 cm (1.125 in) tip, at a nozzle pressure of 206842.7 Pascal (30 lbs per sq. in) from a distance of 6.1 m (20.0 ft) for 2-1/2 minutes.

### 5.6.3.3.4 Southwest Research Institute Ampacity Derating Tests

SwRI performed ampacity derating testing of the FP-60 ERFBS. The test derived ampacity derating values for cable trays and conduits, both filled with 100 percent visual fill of a 3/C # 6 AWG XLPE insulated and CSPE (Hypalon) jacketed power cable. The cables were approximately 3.7 m (12.0 feet) in length with 0.3 to 0.6 m (1 to 2 ft) extending outside of each wrap. Type K (Chromel-Alumel) thermocouples were used to measure the temperature of the copper conductor within the cable. In the cable tray test, 39 thermocouples were used, while 15 were used when testing conduits. All cables were connected in series and supplies with 60Hz single phase AC power sufficient to reach a steady-state temperature of 90°C (194°F) at the hottest single point monitored. The test was conducted in the summer of 1986 and resulted in an ampacity derating factor of 62.2 percent for cable trays and 38.75 percent for conduits protected with the FP-60 ERFBS.

### 5.6.3.4 Industry Testing

#### 5.6.3.4.1 Virgil C. Summer Nuclear Station, Unit 1 Testing

On December 28, 1999, NRC staff witnessed testing conducted by the licensee of Virgil C. Summer Nuclear Station (VCSNS) for several Summer-specific Kaowool configurations. LER 1999-014 informed NRC that engineering personnel determined that some as installed applications may not meet the current regulatory requirements for one train free of fire damage for one hour.

NRC staff review the test results provided by the licensee and concluded that the fire endurance rating of the Kaowool ERFBS is highly configuration dependent. As shown in , the rating varies depending on the particular configuration tested. NRC staff also noted that the testing did not perform continuous megger testing, as suggested in Supplement 1 to GL 86-10 for evaluation of cable performance.

#### Table 5-31. VCSNS Kaowool Testing Results

| Item No. | Size | Configuration | Cable Weight (lb/ft) | Raceway Weight (lb/ft) | Fire Resistance Rating (minutes) |
|---|---|---|---|---|---|
| **Rigid Steel Conduit** | | | | | |
| 4 | 1-inch | Free Air | 0.4 | 1.5 | 43 |
| 7 | 1-inch | Free Air | 0.4 | 1.5 | 44 |
| 2 | 1-1/4 & 4-inch | Wall/Ceiling Mount | 5.1 | 9.8 | 56 |
| 1 | 4-inch | Free Air | 6.7 | 9.8 | 60 |
| 6 | 1-1/4-inch | Wall/Ceiling Mount | 0.3 | 2.0 | 60 |

Table 5-31. VCSNS Kaowool Testing Results (Continued)

| Item No. | Size | Configuration | Cable Weight (lb/ft) | Raceway Weight (lb/ft) | Fire Resistance Rating (minutes) |
|---|---|---|---|---|---|
| Steel Ladder Back Cable Tray | | | | | |
| 3 | 6x6-inch | Free Air | 5.3 | 8.0 | 46 |
| 5 | 6x36-inch | Free Air | 18.5 | 15.0 | 58 |
| Air Drop | | | | | |
| 10 | Air Drop | Free Air | 0.34 | 0 | 31 |

### 5.6.4 Resolution and Staff Conclusion

Plant specific resolution of Kaowool is provided in Section 6. Currently, the only plant that uses the Kaowool as an ERFBS is VCSNS, and the only plant using FP-60 ERFBS is James A. Fitzpatrick (JAF).

In some instances VCSNS uses Kaowool to meet the RG 1.75 criteria for safety related cable separation not related to Appendix R separation and/or safe shutdown. Where Kaowool is relied upon for Appendix R separation and/or safe shutdown, the licensee South Carolina Electric and Gas Company (SCE&G) informed NRC that,

> "In some applications Kaowool was found to be acceptable, after modification, in configurations determined by station supported testing of the product. The fire endurance testing for Kaowool was conducted in accordance with GL 86-10, Supplement 1."

Kaowool barriers that did not meet the fire endurance rating for the configurations that they are installed at VCSNS are being replaced with 3M product.

Initially the licensee of JFP determined that the FP-60 barrier used is operable and as such did not initiate compensatory measures. It also believed that its barrier is of a different construction than at other plants and did not join any industry initiative to resolve the issue. However, following issuance of SECY-99-204 questioning the ability of the FP-60 to withstand a 1-hour fire rating, FitzPatrick applied for and received an exemption from the requirement of Appendix R for its use of the FP-60 ERFBS, on May 29, 2001. Based on fire barrier testing, the licensee determined that the FP-60 ERFBS exceeded test acceptance criteria at 30 minutes. NRC staffs safety review concluded that an adequate level of fire safety such that there is reasonable assurance that at least one means of achieving and maintaining safe shutdown conditions will remain available during and after any postulated fire in the plant, and therefore, the underlying purpose of the rule is met.

Prairie Island Nuclear Generating Plant, Units, 1 and 2, Sequoyah Nuclear Plant, Units, 1 and 2, Susquehanna Steam Electric Station, Units 1 and 2, Edwin I. Hatch Units 1 and 2, and Salem Nuclear Generating Station, Units 1 and 2 all removed Kaowool from their plants as a method of meeting Section III.G of Appendix R. In addition, Farley Unit 1 and 2 have removed the use of

Kaowool from protection of safe-shutdown components, and Grand Gulf Nuclear Station has replaced Kaowool with a qualified 3M product.

In limited cases, Kaowool and FP-60 have been determined to adequately perform its design function for a specific time frame in combination with other fire protection features. However, in general, the staff concluded that Kaowool and FP-60 are unacceptable ERFBS as rated 1-hr and 3-hr ERFBS without proper testing to show otherwise.

## 5.7 Promat (Hydrate)

Promat Fire Protection is a division of Eternit Inc. which manufactures PROMAT-H rigid calcium silicate cement boards. Typically, 2.54 cm (1.0 in) thick boards are used for construction of cable tray and conduits ERFBS. The properties of these two boards are shown in Table 5-32.

Table 5-32. PROMAT Properties

| Property | PROMAT-H | PROMAT-L |
| --- | --- | --- |
| Density | 870 kg/m$^3$ | 430 kg/m$^3$ |
| Thermal Conductivity | 0.175 W/mK | 0.083 W/mK |
| Flame Spread | 0 | 0 |
| Smoke Development | 0 | 0 |
| Combustibility | Non Combustible | Non Combustible |

Installation of Promat-H involves cutting support strips 7.6 cm (3.0 in) wide and as long as the cable tray width or conduit outside diameter. These support strips are attached to the top and bottom of the raceway with 1.90 cm (0.750 in) by 0.025 gauge steel banding, spaced 30.5 to 61.0 cm (12.0 to 24.0 in) apart, along the entire length of raceway to be protected. These support strips provide a base to attach the Promat panels to surround the raceway. Individual layers of Promat are placed around the raceway to form a box and secured to each other by self-drilling screws. The first layer is also secured to the support strips. A 1-hour barrier typically has 2 layers of 2.54 cm (1.0 inch) thick Promat-H, while a 3-hour barrier has 4 layers of 2.54 cm (1.0 in) thick Promat-H cement board. Any gaps and joints are filled with an approved fire resistant caulk. An installation diagram for cable tray and conduit applications are shown below in Figure 5-19 and Figure 5-20 respectively.

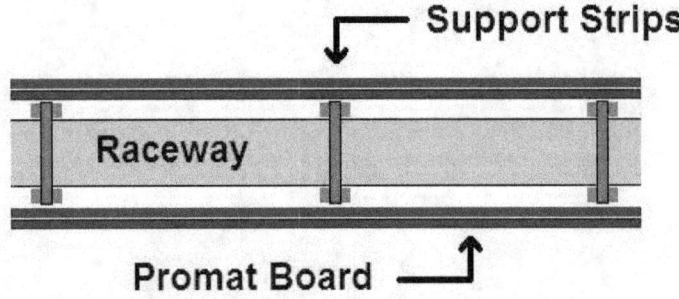

Figure 5-19. Promat-H Cable Tray Protection (Author, 2009)

Figure 5-20. Promat-H Conduit Protection (Author, 2009)

### 5.7.1 History

Promat fire protection products have a long history in providing passive fire protection products for various structural members (steel I-beams, walls, grease ducts, concrete, membrane ceilings, and roof decks). However, the use of Promat in the commercial U.S. NPP fleet is limited to Farley Nuclear Plant and Hatch Nuclear Plant.

### 5.7.2 Problems

No generic problems have been identified with the use of Promat-H or Promat-L rigid boards for ERFBS installations.

### 5.7.3 Testing

*5.7.3.1 NIST Testing*

Several samples of PROMAT-H were tested in the NIST small scale testing program. For a description of the NIST test program, please refer to Section 5.2.3.1. The first test (D1-1) consisted of two layers of 25 mm (0.98 in) thick ceramic board (i.e., a 1-hour barrier). The test assembly was subjected to an ASTM E-119 fire exposure. The results indicated that the average unexposed surface temperature rise criterion was met at 1 hour and 18 minutes, while the maximum unexposed single point surface temperature rise was no reached.

The second test (E3-1) consisted of two layers of 50 mm (1.97 in) thick ceramic board (i.e., a 3-hour barrier). The test assembly was subjected to the ASTM E-119 fire exposure. The results indicated that the average unexposed surface temperature rise criterion was met at 3 hours and 46 minutes, while the maximum unexposed single point surface temperature rise was no reached during the 3.92 hr test.

These small scale tests prove that homogeneous layers of Promat-H can provide the acceptable level of protection. However joints and bends are typically encountered in actual NPP ERFBS installations and workmanship of the installed barrier may affect its fire resistance. For this and other reasons discussed above, the applicability of these tests is limited. By comparison to other barriers tested in the NIST small scale program, the Promat-H performed well.

*5.7.3.2 Fire Endurance Testing*

Although no fire endurance test reports have been submitted on the plant docket for either plant that use these barriers, an NRC Inspection of the Promat-H barriers installed at Hatch Nuclear Plant resulted in no findings. NRC Inspection team reviewed the installed fire barriers in detail to verify that the as-built configurations met design requirements, licensee commitments, and standard industry practice and had been either properly evaluated or qualified by appropriate fire endurance tests. This review also included confirmation that the ERFBS were installed on the required circuits, fire barrier was of appropriate fire rating, and the barrier installations were consistent with the tested configurations.

Farley identified in its response to GL 06-03, that Promat was tested and qualified to ASTM E119-88 standard by Performance Contracting Inc. and conducted at OPL under Project No. 8806-90254 (Promat Report SR90-005). In addition to this report, the vendor identifies the following test reports related to Promat.

**Table 5-33. List of Promat Test Reports**

| Test Lab. | Description | Report Number |
|---|---|---|
| SwRI[a] | 1 Hr Cable Tray 0% Fill | 01-2299-001 |
| SwRI | 1 Hr Conduit | 01-2299-001 |
| SwRI | 1 Hr Cable Drop | 01-2299-001 |
| OPL[b] | 3 Hr Cable 80% Visual Fill | 8806-89017a |
| OPL | 3 Hr Cable Tray 0% Visual Fill | 8806-89006 |
| OPL | 3 Hr Conduit Against Concrete | 8806-89004 |
| OPL | 3 Hr Conduit Free Standing | 8806-89017b |
| OPL | 3 Hr Cable Drop | 8806-88053 |

[a] SwRI – Southwest Research Institute  [b] OPL – Omega Point Laboratories

*5.7.3.3 Ampacity Derating*

The PROMAT vendor manual provides the following ampacity derating values for a PROMAT-H ERFBS.

**Table 5-34. PROMAT-H Ampacity Derating**

| Configuration | 1 – Hour | 3 – Hour |
|---|---|---|
| Solid Bottom Tray | 8.7% | 27.2% |
| Ladder Bottom Tray | 31.8% | 45.3% |
| Conduit | 15.9% | 36.7% |

### 5.7.4 Resolution and Staff Conclusion

No generic issues or deficiencies have been identified and associated with this ERFBS, there has been no need for generic resolution related to Promat ERFBS. Therefore, the staff concluded that Promat when installed to bound as tested configurations will satisfactory perform its intended design function.

## 5.8 Pyrocrete (Hydrate)

Pyrocrete is a cementitious inorganic fireproofing material. It is supplied as a single powder component that is mixed with water before application. Typical installation includes mounting a galvanized metal lath around the structure to be protected. This lath will support the pyrocrete and help adhere it to the raceway until the pyrocrete cures. The pyrocrete powder is mixed with water and sprayed onto the area to be protected with special equipment. The pyrocrete mix can also be toweled onto the raceway. A trowel, roller or brush can typically be used to smooth the top of the pyrocrete applications to ensure a smooth and consistent thickness. Vendor testing of Pyrocrete 241 produced by Carboline® using ASTM E-84 resulted in zero (0) flame spread and zero (0) smoke development.

### 5.8.1 History

Pyrocrete is not widely used in the commercial NPP industry as an ERFBS. It is more commonly used to provide fire proofing to structural steel beams or ceiling members. As of this writing, there are only three sites that use Pyrocrete as an ERFBS, Surry Power Station, Units 1 and 2, Diablo Canyon Nuclear Power Plant Units 1 and 2, and Fort Calhoun Station, Unit 1.

### 5.8.2 Problems

No generic problems have been identified with the use of Pyrocrete as an ERFBS at US NPPs. However, Licensee Event Report 95-003-01 reports the licensee of Diablo Canyon Nuclear Power Plant identified untested configuration of Pyrocrete Fire Barriers. NRC staff inspections conducted in 1997 provided closure to the problems identified in the event report.

### 5.8.3 Testing

No fire endurance test reports were found to have been submitted to NRC for review; however the following summarizes NRC inspections of the Pyrocrete ERFBS installation in use at the three plants mentioned above. During these inspections, NRC staff reviewed fire tests and visually inspected the installed barriers.

NRC Inspection Reports 50-275/97-17 and 50-323/97-17, documents NRC inspection staffs review of the application of Pyrocrete as an ERFBS at Diablo Canyon Nuclear Power Plant Units 1 and 2. The report concludes that the barriers installed in the plant were acceptable. The inspectors based their conclusions on their review of pyrocrete fire tests completed to test the installed barriers and visually inspected pyrocrete fire barriers installed in the plant. The inspectors observed that the fire test for the pyrocrete passed the 3-hour test required by NRC and also observed that the configurations installed in the plant were in accordance with the configurations tested.

NRC Triennial Fire Protection Inspection Report 05000280/2006009 and 05000281/200609, dated April 11, 2006 documents NRC inspection staffs review of the Pyrocrete ERFBS used at Surry Power Station, Units 1 and 2. That report stated,

> The team inspected the material condition of accessible passive fire barriers surrounding and within the fire areas selected for review. Barriers in use included walls, ceilings, floors, mechanical and electrical penetration seals, doors,

dampers and cementitious fire resistive coatings. Construction details and fire endurance test data which established the ratings of fire barriers and fire resistive material were reviewed by the team. Engineering evaluations and relevant exemptions described in NRC safety evaluations related to fire barriers were reviewed. Where applicable, the team examined installed barriers to compare the configuration of the barrier to the rated configuration.

The report concludes that "no findings of significance were identified" related to the review of passive fire protection features.

According to a Surry Power Station Response to GL 06-03, Pyrocrete 241 has been qualified by Thermal Transmission Test (ref. Tech. Report EP-001 1) that uses the ASTM E-119 fire exposure and the failure criteria of an average temperature of 121°C (250°F) or single point temperature 162°C (325°F) above ambient backside temperature. The barrier was installed prior to the issuance of GL 86-10 Supplement 1.

The Fort Calhoun Station triennial fire protection inspection performed in 2008 concluded that no findings of significance were identified. The team observed the material condition and configuration of the installed barriers, seals, doors and cables. The team compared the installed configurations to the approved construction details and supporting fire tests. The team reviewed licensee documentation, such as NRC safety evaluation reports, exemptions from NRC regulations and deviations from the National Fire Protection Association codes, to verify that fire protection features met license commitments.

Niagra Mohawk, the licensee of Nine Mile Point Unit 1 submitted a fire endurance test reported completed by Industrial Testing Laboratories (ITL)[18] of St. Louis Missouri, dated September 6, 1979. In that test report, a planar 5.1 cm (2.0 in) thick pyrocrete barrier was exposed to that standard time-temperature curve exposure of E-119 and five thermocouples on the unexposed side of the barrier measured temperature rise. This testing was similar to the NIST small-scale testing. The ITL report indicated that the average temperature rise of $\Delta T_{avg} \geq 121°C$ (250°F) occurred at approximately 257 minutes and the maximum temperature rise ($\Delta T_{max} \geq 121°C$ (250°F)) did not occur prior to the shutdown of the furnace.

In addition to this testing, Niagra Mohawk also provided NRC staff with a report entitled, "Thermal Transmission of Pyrocrete 241 at Varying Thicknesses," completed by Johns-Manville. The purpose of this report was to experimentally determine the time the back side temperature of Pyrocrete 241 coated steel panels reached 121°C (250°F) above ambient, when applied at varying thicknesses and exposed to the ASTM E-119 time-temperature curve. The report documented the results, reproduced in :

---

[18] This is the same laboratory that plead guilty to falsifying records for Thermo-Lag testing.

Table 5-35. Results of Pyrocrete 241 Thermal Transmission

| Pyrocrete Thickness | Time to Reach $\Delta T \geq 250°F$ (minutes) |
|---|---|
| 0.635 cm (¼ in.) | 6 |
| 1.27 cm (½ in.) | 10 |
| 2.54 cm (1 in.) | 31 |
| 3.81 cm (1.5 in.) | 73 |
| 5.08 cm (2 in.) | 257 (ITL testing) |

These results indicate that the thickness of the Pyrocrete applied will affect the fire endurance rating. Although these results are not representative of a full-scale fire endurance test raceway assembly, the 257 minutes to reach the failure criteria indicates that relative to other materials tested in this manner, the 5.1 cm (2.0 in) Pyrocrete barrier can exhibit greater thermal resistance to the standard time-temperature curve exposures than most ERFBS.

## 5.8.4 Resolution and Staff Conclusion

No generic issues or deficiencies have been identified and associated with this ERFBS, there has been no need for generic resolution related to Pyrocrete ERFBS. Staff inspections have found the use of Pyrocrete adequately perform its intended design function. As such, the staff concluded that when installed to bound as tested configurations Pyrocrete will satisfactory perform its intended design function.

## 5.9 Versawrap (Hydrate/Insulative/Intumescent)

Versawrap was developed by Transco Products, Inc., as a stand-alone fire barrier and as a potential upgrade for existing raceway fire barriers, such as Thermo-Lag 330-1. Versawrap fire barriers are installed as individual layers of foil, water filled Mylar tubes, fiber blankets, foil and intumescent-coated fiberglass cloth. The numbers and arrangements of the specific barrier components are dependent on the type of items to be protected (raceway, cable, support, etc.) and the desired fire rating.   and  provide a graphical view of the barrier construction for a typical 1-hour barrier.

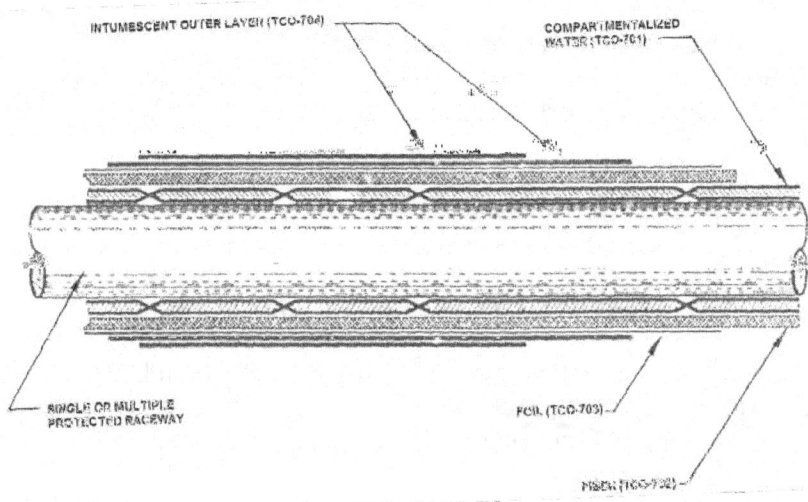

Figure 5-21.  Sketch of Layers Used in Versawarp ERFBS (NRC Trip Report 1997)

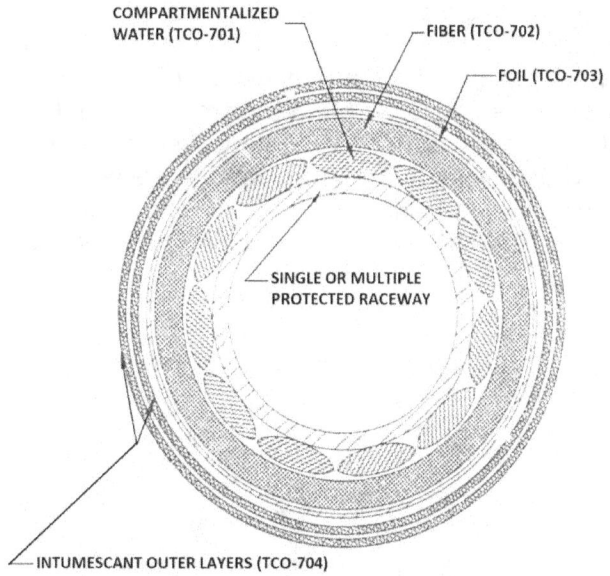

Figure 5-22.  Cut away of Versawrap ERFBS showing individual Layers (NRC Trip Report, 1997)

### 5.9.1 History

In 1997, NRC staff met with Entergy Operations, Inc., the licensee of Arkansas Nuclear One (ANO), to discuss the use of Versawrap as an upgrade to its Thermo-Lag 330-1 ERFBS and to observe full scale fire endurance testing of the product for 1- and 3-hour rating periods. Following the testing, NRC issued IN 97-59, "Fire Endurance Test Results of Versawrap Fire Barriers," to alert licensees of the preliminary results of the testing and the potential for Versawrap to not perform its design function for the rated period of time.

### 5.9.2 Problems

IN 97-59, "Fire Endurance Test Results of Versawrap Fire Barriers," dated August 1, 1997, identified several test assemblies that didn't pass NRC acceptance criteria, as specified in GL 86-10, Supplement 1. Section **Error! Reference source not found.** below discuss the results of this testing.

### 5.9.3 Testing

#### 5.9.3.1 UL Testing of Versawrap for Arkansas Nuclear One and Prairie Island

Transco Products, Inc. performed testing of its Versawrap ERFBS at UL on April 10, 1997. The purpose of this testing was to qualify barriers that are planned to be installed at Arkansas Nuclear One Unit 1 (1-hour and 3-hour) and at Prairie Island Nuclear Power Station (1-hour).

This single 3-hour fire exposure test followed by a solid-stream hose stream test that was intended to demonstrate the ability of both 1- and 3-hour Versawrap fire barrier systems (in the same test) to meet the acceptance criteria of GL 86-10 Supplement 1. The overall tests assembly consisted of a steel and concrete horizontal test deck from which 18 individual 1- and 3-hour test articles were suspended into UL's large floor furnace. The Versawrap ERFBS were installed on all articles in accordance with Transco instruction and procedures. The test articles were instrumented in accordance with GL 86-10 Supplement 1, with UL and Transco quality assurance personnel witnessing and documenting the test specimen construction. All testing used a single bare #8 AWG conductor to measure temperature within the raceway. The fire endurance hose stream tests were administered in accordance with GL 86-10 Supplement 1 guidance. , provides the results of the 1-hour and 3-hour test results for each raceway.

Table 5-36. Summary of UL Testing of Versawrap

| Article No. | Raceway Description | Desired Rating | Pass/Fail[1] |
|---|---|---|---|
| A | 2"x2" Aluminum Solid-Back Cable Tray/Tube Track | 1-hr | Pass |
| B | 30"x4" Steel Ladder-Back Cable Tray | 1-hr | Pass |
| C | ¾" diameter Rigid Steel Conduit | 1-hr | Pass |
| D | 4" diameter Rigid Steel Conduit | 1-hr | Pass |
| E | ¾" diameter Rigid Steel Conduit Near Concrete Barrier | 1-hr | Fail |
| F | Multiple intersecting Unistrut Hangers and ¾" dia. Rigid Steel Conduit | 3-hr | Pass |
| G | 2"x2" Aluminum Solid-Back Cable Tray/Tube Track | 3-hr | Pass |
| H | 30"x4" Steel Ladder-Back Cable Tray | 3-hr | Pass |
| I | ¾" diameter Rigid Steel Conduit | 3-hr | Pass |
| J | 4" diameter Rigid Steel Conduit | 3-hr | Fail |
| K | ¾" diameter Rigid Steel Conduit Near Concrete Barrier | 3-hr | Fail |
| L | Multiple intersecting Unistrut Hangers and ¾" dia. Rigid Steel Conduit | 1-hr | Pass |
| M | 1-½" diameter Steel Conduit Covered with Modified Versawrap (modified meaning that an unprotected Unistrut support will be used to support the horizontal section of the raceway. | 3-hr | Fail |
| N | 12"x12"x12"(16 gauge) Steel Box attached to ¾" dia. Steel Conduit | 3-hr | Fail |
| O | Two 6"x6"x6" (16 gauge) Steel Boxes attached to a ¾" diameter steel Conduit | 1-hr | Pass |
| P | Two 6"x6"x6" (16 gauge) Steel Boxes attached to a ¾" diameter Steel Conduit | 3-hr | Fail |
| Q | Air Drop - 3/C 500MCM Aluminum Armored Cable (this test was conducted to bound plant specific application) | 1-hr | Fail |

[1] The Pass/Fail results are based on NRC staff observation of preliminary data collected during observation of test.

*5.9.3.2 Omega Point Laboratories Testing of Versawrap for Transco Products, Inc.*

Transco Products, Inc. preformed testing of its Versawrap ERFBS at OPL on November 21, 1997. OPL Test Report Product No. 12000-101105 documents the test setup and procedure, along with the results. Seventeen test articles were tested in the single fire test. Conduit sizes from 1.9 cm (0.75 in) up to 12.7 cm (5.0 in) diameter were tested along with 61.0 cm (24.0 in) and 76.2 inches (30.0 in) cable trays and several junction boxes. A single #8 AWG bare copper conductor instrumented with thermocouples every 15.2 cm (6.0 in), along with thermocouples attached to the raceways (every six inches) were used to monitor the unexposed side temperature rise. GL 86-10, Supplement 1 acceptance criteria was used to evaluate the performance of the test articles. Although both 1-hour and 3-hour assemblies were contained in this one test, all articles were exposed to a 3-hour ASTM E 119 standard fire exposure, after which the entire assembly was subjected to the hose stream test (as stipulated in NRC guidance). Of the 17 test items, three failed to meet the temperature rise acceptance criteria. Item 1, a 1.9 cm (0.75 in) diameter rigid steel conduit failed at 170 minutes into a 180 minute test, post test examination revealed that a collar had slipped down, exposing the bare conduit. Item 15 a horizontal solid bottom steel cable tray 76.2 cm (30.0 in) wide by 10.1 cm (4.0 in) deep) failed at 28 minutes of a 1-hour rating, post test examination was not documented for any 1-hour test specimen. Item 15 was protected with a Thermo-lag upgrade (i.e. thermo-lag base layer enclosed by a 1-hour layer of Versa-Wrap). Item 17a was a 5.1 cm (2.0 in) diameter rigid steel conduit protected by a 3-hour Versa-Wrap design failed at 84 minutes of a 180 minute test. Post test examination of item 17a indicated that the Mylar tubes had melted and no water remained inside, discoloration was also noted on the cloth, foil, and blanket. provides a description of each test article along with the actual endurance rating achieved during the test.

Table 5-37. Summary of Omega Point Testing of Versawrap

| Article No. | Raceway Description | Desired Rating | Actual Rating (min) | |
|---|---|---|---|---|
| 1 | ¾" diameter rigid steel conduit | 3-hr | 170 | Fail |
| 1a | Mock Box Assembly protected with a 3-hr Darmatt ERFBS with two ¾" steel conduits protected with a 3-hr Versawrap ERFBS | 3-hr | 180 | |
| 2a | 3" diameter rigid steel conduit | 1-hr | 60 | |
| 2b | 3" diameter rigid steel conduit | 1-hr | 60 | |
| 3 | "L" shaped 4" diameter rigid steel conduit with cast iron LB fitting | 3-hr | 180 | |
| 4 | two vertical 2" diameter rigid steel conduits connected to common horizontal Unistrut support | 1-hr | 60 | |
| 5 | same as article 4 | 3-hr | 180 | |
| 6 | 6" high x 40" wide x 18" deep sheet steel junction box bolted in direct contact with concrete slab | 1-hr | 60 | |
| 7 | Pair of RSC (3/4" and 4") transitioning from vertical to horizontal thought large radius bend | 1-hr | 60 | |
| 8a | 6"x6"x6" sheet steel junction box | 1-hr | 60 | |
| 8b | ¾" diameter vertically oriented rigid steel conduit | 1-hr | 60 | |
| 8c | 6"x6"x6" sheet steel junction box in direct contact with concrete test slab | 1-hr | 60 | |
| 9 | 4" diameter horizontally oriented rigid steel conduit | 1-hr | 60 | |
| 10 | 3" diameter vertically oriented rigid steel conduit | 1-hr | 60 | |

Table 5-37. Summary of Omega Point Testing of Versawrap (Continued)

| Article No. | Raceway Description | Desired Rating | Actual Rating (min) | |
|---|---|---|---|---|
| 11 | 2"x2" aluminum wire way | 1-hr | 60 | |
| 12 | 24" wide by 4" deep solid bottom steel cable tray | 3-hr | 180 | |
| 13 | 12" diameter schedule 80 steel pipe penetration (Darmatt ERFBS) | 3-hr | 180 | |
| 14 | "L" shaped solid bottom steel cable tray (30"x4") | 1-hr | 60 | |
| 15 | 30"x4" solid bottom steel cable tray in horizontal orientation<br>1-hr Thermo-Lag base layer with Upgrade Versawrap | 1-hr | 28 | Fail |
| 16 | 24"x4" solid bottom steel cable tray in horizontal orientation<br>1-hr 3M-CS195 CT upgraded with Versawrap | 1-hr | 60 | |
| 17a | 2" diameter rigid steel conduit | 3-hr | 84 | Fail |
| 17b | 2" diameter rigid steel conduit | 1-hr | 60 | |
| 17c | 2" diameter rigid steel conduit | 1-hr | 60 | |
| 17d | 6"x6" sheet steel wire way horizontally oriented | 1-hr | 60 | |

*5.9.3.3 Omega Point Laboratories Testing of Versawrap for PP&L*

On April 16, 1998, a staff member from NRC witnessed fire endurance testing conducted at OPL for Pennsylvania Power and Light. The test assemblies consisted of 20 Thermo-Lag configurations, some of which were upgraded with a Versawrap overlay system. The testing followed the guidance specified in Supplement 1 to GL 86-10. Although the results are proprietary, approximately 2/3 of the test assemblies failed to meet the temperature rise acceptance criteria for 1-hour qualification.

**5.9.4 Resolution and Staff Conclusion**

No generic issues or deficiencies have been identified and associated with this ERFBS; there has been no need for generic resolution related to Versawrap ERFBS. Versawrap has been used as replacement materials to other deficient ERFBS.

As a result of its late entry into NPP applications and testing in accordance with GL 86-10 Supplement 1, the staff concluded that Versawrap when installed to bound as tested configurations will satisfactory perform its intended design function.

## 5.10 Pabco (Intumescent ERFBS)

### 5.10.1 History

Pabco was sold to Johns Manville in 1998, and the material was renamed "Thermo-12 Gold". Pabco Super Caltemp Gold Insulation contains about ½ to ¾ percent by weight of rayon fiber, ½ percent pulp, < ½ percent alkali resistant fiber glass, < ½ percent yellow iron oxide for color and the remainder is Tobermorite calcium silicate. In initial heat up, the rayon and pulp will decompose and give off water vapor and $CO_2$. Tobermorite is a crystalline calcium silicate mineral that is stable up to 677°C (1250°F).

Fort Calhoun Station (FCS) is the only US NPP which uses Pabco material as fire barrier protection for redundant trains of safe shutdown equipment located in the same fire area to satisfy 10 CFR 50, Appendix R Section III.G requirements. Licensee installation and inspection procedures have verified that these fire barrier materials were installed in a manner consistent with tested configurations. Deviations from tested and analyzed configurations were evaluated in accordance with GL 86-10, Supplement 1. These evaluations provide the necessary assurance that the installed fire barrier systems would possess the commensurate level of fire protection. OPPD inspects fire barrier configurations outside of containment on an 18 month frequency and inside containment on a refueling outage frequency.

### 5.10.2 Problems

No problems with the use of Pabco ERFBS as a redundant train fire separation found.

### 5.10.3 Testing

No public test reports for Pabco ERFBS found.

### 5.10.4 Resolution and Staff Conclusion

To date no generic issues related to Pabco ERFBS found. NRC staff inspections at Fort Calhoun Station have and will continue to allow for opportunities to evaluate this type of barrier.

## 5.11 Concrete (Insulative ERFBS)

### 5.11.1 History

Concrete is only used at one plant (Palisades Nuclear Plant) to provide a 1-hour barrier to separate redundant trains within the same fire area. The use of this concrete barrier was a result of resolving Thermo-Lag deficiencies in the plant. Palisades initially used approximately 44 m (144 linear ft) of 1-hour Thermo-Lag 330-1 and 9 m (30 linear ft) of a 3-hour Thermo-Lag 330-1 ERFBS. The concrete barrier was constructed by enclosing approximately a 1.22 m (4 ft) section of a 7.6 cm (3.0 in) diameter galvanized steel rigid conduit and pullbox within concrete. Resolution of all other applications of Thermo-Lag 330-1 was resolved by using a 1-hour fire rated cable or rerouting cables.

### 5.11.2 Problems

No problems with the use of Concete as an ERFBS for protection a redundant train fire separation have been identified.

### 5.11.3 Testing

No public records available to determine acceptable use of concrete as a rated 1-hour ERFBS. However, extensive experimental research has been done in the past to determine the required thickness of various types of structural concrete members to provide a specific fire rating. The Society of Fire Protection Engineering (SFPE) Handbook of Fire Protection Engineering has a section dedicated to analysis methods for determining fire resistance of concrete members. In addition, several experts knowledgeable with the effects of heat on concrete have developed models to determine the internal temperatures of concrete. Thus, there are several methods for licensees to determine the fire resistance rating of a concrete ERFBS with a specified minimum thickness.

### 5.11.4 Resolution and Staff Conclusion

Due to the limited use of concrete as an ERFBS material and its well understood heat transfer properties, there have not been any issues identified with the use of concrete as an ERFBS. Past and future NRC fire protection inspections will continue to provide the opportunities to evaluate this type of barrier.

# 6. Plant Specific Usage and Resolution of ERFBS Issues

This section provides information as to how each operating NPP resolved their compliance issues related to the respective ERFBS that they installed. It should be noted, that only operating plants are documented in this section. Plants in the decommissioning phase are not discussed; however, it is likely that these plants may have some type of ERFBS installed.

The information provided in this section is for reference purposes only and was obtained from numerous official communications between the NRC and its licensees. Although the intent of this section is to provide the most accurate information on plant usage of ERFBS, there may be instances where the documented application of ERFBS does not accurately represent what is actually used in plants. This information is also time sensitive, i.e., the information is current at the writing of the report. Licensees may, in the future, perform plant/licensing basis changes which will make this information obsolete. The reader is cautioned to verify the latest plant specific information. Each plants licensing basis documents the usage of ERFBS and exemptions from requirements or guidance. Therefore, the information provided is NOT intended to represent official NRC approval of fire barrier configurations. Only the plant specific licensing information can be relied upon for licensing basis judgments.

## 6.1 Arkansas Nuclear One, Units 1 and 2

Arkansas Nuclear One (ANO) initially used approximately 30 feet of a 3-hour rated Thermo-Lag 330-1 ERFBS in Arkansas Nuclear One Unit 2 to protect conduit raceway in the service water pump pits. The water pump pits are located in the intake structure and are pre-formed conduit shapes and panels on two 10.16 cm (4 in) conduits. The conduits contain power cables for two service water pumps. In a letter dated April 16, 1993, ANO stated that appropriate compensatory measures would be taken to fix the questionable Thermo-Lag installations. Following re-analysis the licensee determined that the barrier was no longer needed to meet 10CFR50.48 or Appendix R to Part 50, achieve physical independence of electrical systems, or meet a condition of the plant operating license.

By letter dated June 21, 1994, NRC concluded that the issues addressed in GL 92-08 were resolved and NRC review of the matter as tracked by TAC No. 85515 was complete.

Currently ANO uses Versawrap and Hemyc to meet the Appendix R requirements. In its response letter to GL 2006-03, the licensee stated that it uses Versawrap ERFBS to meet Appendix R separation requirements and all Versawrap applications have been qualified by various fire tests conducted by independent laboratories consistent with the guidance provided in GL 86-10 Supplement 1. In addition, the response stated that for this category of fire barrier, bases documentation will be maintained on site, and will be subjected to inspection during the normal NRC inspection process.

Approximately 1000 linear feet of Hemyc is use between the two units (802 linear feet Unit 1; 197 linear feet Unit 2). After reviewing NRC Hemyc test reports, the licensee determined that the Hemyc installed at ANO does not conform to the licensing basis and has been declared inoperable. Hourly or continuous fire watches have been established (depending on area fire detection operability) as a compensatory measure until the operability of Hemyc can be established. ANO has informed NRC of its plan to transition both units to NFPA 805, and plans to address the Hemyc issues during the implementation of NFPA 805. In the case that ANO does not transition to NFPA 805, then they will have to submit exemption requests to the NRC providing justification for the specific Hemyc configurations, or modify these configurations to come into compliance. A discussion of NFPA 805 is provided above in Section 3, "ERFBS Regulations."

## 6.2 Beaver Valley Power Station, Units 1 and 2

Beaver Valley Power Station (BVPS) uses Thermo-Lag, 3M Interam E-50 Series, and Darmatt KM-1 ERFBSs to provide separation of redundant trains locate in a single fire area. Following issuance of GL 92-08 and IN 93-41 all barriers were evaluated and found to be acceptable for the configurations that they are installed. In addition, tests and evaluations were preformed for Beaver Valley to determine that the Thermo-Lag material properties and attributes conform to NRC regulations. Tests and evaluations were also preformed on the 3M Interam E-50 Series and Darmatt KM-1 ERFBS which concluded that these barriers conform to NRC requirements.

BVPS Unit 1 initially used approximately 3 feet of a 1.5-hour rated Thermo-Lag 330-1 ERFBS to protect conduit penetration seals. In this application, the Thermo-Lag material is not used to protect the electrical cables, but to maintain the integrity of the fire area boundary of the cable mezzanine floor. Following re-analysis the licensee determined that the barrier was no longer needed.

BVPS Unit 1 used Hemyc ERFBS to protect electrical raceways associated with charging pump feeder cable CH-P-1B. The Hemyc fire barrier was replaced with Darmatt KM-1 material. Regional staff verified the Darmatt KM-1 fire barrier installation and configuration during site inspection and documented this review in NRC Inspection Report (IR) 50-344/02-04 and IR 50-412/02-04.

BVPS Unit 2 uses approximately 2900 linear feet of 1-hour and 1500 linear feet of a 3-hour Thermo-Lag barrier system to protected electrical raceway conduits. In addition, 1370 ft.$^2$ of 1-hour and 1770 ft.$^2$ of 3-hour Thermo-Lag barriers are used to protected supports, junction boxes, removable floor plugs, seismic gap seals, conduit sleeve extensions, and structural steel support plates for oversized penetration seals.

The actions required for review of GL 92-08 were communicated to be closed by NRC in letters dated September 23, 1994 and April 6, 1998 for Units 1 and 2, respectively. NRC letter dated October 1, 1998, identifies the completion and acceptance of its review of the ampacity derating methodology used at Beaver Valley.

## 6.3 Braidwood Station, Units 1 and 2

Braidwood Station Unit 1 initially relied on Thermo-Lag 330-1 barriers to meet 10 CFR 50.48 regulations and to provide separation between redundant electrical systems. Following issuance of GL 92-08 and the results from the NUMARC testing program, the licensee decided to abandon the Thermo-Lag barriers in place. The licensee then reduced the number of cables requiring protection by re-evaluating its safe shutdown analysis, re-routing cables and began installing a rated Darmatt KM-1 barrier. However, initial attempts to install the Darmatt KM-1 barrier identified that cable rerouting of all required cables would be more efficient and cost effective and therefore the use of Darmatt as a solution was abandoned. NRC closed out its review of Thermo-Lag actions for Braidwood station in its letter dated May 13, 1997.

Braidwood Station Unit 2 uses 3M Interam ERFBS in two locations (i.e., Unit 2 Cable Tunnel and Unit 2 Lower Cable Spreading Room) to ensure separation of redundant trains in the same fire zone. The 3M Interam ERFBS were installed during the original plant construction in accordance with Braidwood specifications. Following IN 95-52 and IN 95-52 Supplement 1, the licensee evaluated the 3M Interam installed configurations and determined that the barriers were capable of providing the necessary level of protection. Specifically, these evaluations determined that those ERFBS are considered to have at least a 49 minute fire rating when exposed to ASTM E-119 fire test and following the acceptance criteria of GL 86-10, Supplement 1. The licensee based the acceptance of the 49 minute fire rating on the basis that fire loads normally present in these two rooms are only capable of producing fire duration of 30 minutes if an ASTM E-119 type fire was postulated to occur.

By letter and SE dated November 2, 1999 NRC determined that all ampacity related concerns have been resolved and the licensee has provided adequate technical basis to ensure that all of ERFBS enclosed cables are operating within acceptable ampacity limits.

### 6.4   Browns Ferry Nuclear Plant, Units 1, 2 and 3, Sequoyah Nuclear Plant, Units 1 &2

TVA relies upon Thermo-Lag fire barrier material to protect fire safe shutdown circuits at BFN Units 1, 2 and 3, Sequoyah Nuclear Plant (SQN) Units 1 and 2, and WBN Unit 1 as approved by NRC Staff. Configurations installed at TVA facilities are in accordance with the tested configurations or have been evaluated by persons knowledgeable in fire barrier design and installation. The results of both the testing and engineering evaluations have been documented consistent with accepted engineering and industry standards. These configurations, both those specifically tested and unique configurations, are documented in facility design basis documentation that are controlled and maintained in accordance with TVA's Design Control and Quality Assurance Programs.

Initially Browns Ferry Unit 2 was the only unit to originally use Thermo-Lag 330-1. Approximately 200 linear feet of conduit barrier was used in Unit 2, which was subsequently upgraded per TVA tested configurations. TVA has Thermo-Lag installed in the BFN1 Intake Pumping Station to provide a 1-hour fire barrier. Most of the Thermo-Lag material in Browns Ferry Unit 1 was found to be unnecessary to comply with 10CFR50 and abandoned in place. Instead of costly amounts of fire barrier material, the plant chose to reroute electrical cables essential to the plant's safe shutdown. Abandoned Thermo-Lag which was accessible and cost effective to remove was discarded completely by June 20, 1996.

Ampacity derating issues at Browns Ferry Nuclear Plant Units 1-3 with regards to GL 92-08 were considered complete by NRC in a letter dated July 16, 1999.

### 6.5   Brunswick Steam Electric Plant, Units 1 and 2

Brunswick Steam Electric Plant (BSEP) has used Thermo-Lag 330-1 to satisfy 10 CFR 50.48 separation requirements, licensing commitments, and conditions associated with its Fire Protection Program. Only 12m (40 linear feet) of a 1-hour rated Thermo-Lag 330-1 barrier were used to protect two 6 m (20 foot) sections of cable trays. The majority of Thermo-Lag used was applied to conduits, which included 320m (1050 linear feet) of a 1-hour barrier and 430m (1410 linear feet) of a 3-hour barrier. In addition, 113 $m^2$ (1220 $ft^2$) of a 3-hour and 12 $m^2$ (130 $ft^2$) of a 1-hour barrier were used to protect junction boxes, equipment enclosures, door transoms, and penetration seals.

BSEP uses Kaowool as part of an approved Appendix A Fire Protection Program and Appendix R exemption. Kaowool provides additional protection defense-in-depth, but it is not credited as a 1-hour or 3-hour barrier as required by Appendix R.

BSEP uses 3M Interam E50A and E54A ERFBS materials for Appendix R purposes. The materials used in the 3M barriers have been installed to manufacturer's instructions and has been independently tested in accordance with national standards.

## 6.6   Byron Station, Units 1 and 2

Byron Station used about 954 m (3129 linear feet) of Thermo-Lag 330-1 ERFBS to protect redundant cable trains within the same fire area. Both 1- and 3-hour Thermo-Lag configurations were used in both units. By letter dated January 17, 1997, three methods of resolution were utilized, which included; 1) re-analysis of Safe Shutdown Analyses to eliminate the need for the fire barrier, 2) re-routing of cables such that redundant safe shutdown trains are not located in the same fire zone, or 3) replacement of the Thermo-Lag 330-1 with a qualified fire barrier (see Section 5.3 above on Darmatt KM-1). The licensee informed NRC that all planned modifications had been completed, as a result of GL 92-08. These modifications included, removing Thermo-Lag 330-1 ERFBS from several safe shutdown cables located in Unit 1 and protecting them with Darmatt KM-1 ERFBS, rerouting the remaining cables and their redundant counterparts and associated support equipment cables such that they are not located in the same zone. Circuits that no longer required protection have the Thermo-Lag 330-1 ERFBS abandoned in place.

Byron Station currently uses Darmatt KM-1 fire barrier on all electrical raceways where a fire barrier is required to ensure separation of redundant trains in the same fire zone. Darmatt KM-1 material was installed as a qualified replacement of the Thermo-Lag 330-1 fire barrier as part of Byron Station's corrective actions in response to GL 92-08.

By letter and SE dated November 2, 1999 NRC determined that all ampacity related concerns have been resolved and the licensee has provided adequate technical basis to ensure that all of ERFBS enclosed cables are operating within acceptable ampacity limits.

## 6.7   Callaway Plant, Unit 1

The Callaway Plant initially used limited quantities of Thermo-Lag 330-1 for raceway protection. The plant utilized approximately 31 linear feet of 3-hour cable tray barriers, 132 linear feet of 1-hour conduit barriers, and 614 linear feet of 3-hour conduit barriers. In order to satisfy the requirements of 10 CFR 50, Appendix R, the licensee made several modifications to the use of Thermo-Lag ERFBS. An Appendix R, Section III.G.2 reanalysis was performed which documented the technical basis for removal of Thermo-Lag from cables that were not needed for safe shutdown. A conduit was re-routed to meet the 20 foot separation criteria and local manual controls were added to the "B" and "C" steam dump valves in order to eliminate the need for the pre-existing fire barrier and Thermo-Lag on the raceways. In areas of the plant where a barrier remained necessary for compliance with Appendix R, Thermo-lag was removed and Darmatt KM-1 was installed. The Callaway Plant notified NRC that the issues identified in GL 92-08 had been completed as of December 31, 1996.

As noted above, the Callaway Plant use Darmatt KM-1 ERFBS for protection of redundant trains located in the same fire area that satisfies 10 CFR 50, Appendix R, III.G requirements. The fire barriers were tested and installed according to the guidance provided in GL 86-10, Supplement

1 with any deviations from the tested configurations evaluated against GL 86-10. NUREG 0830, supplement 3 provides NRC determination that the use of a 1-hour rated barrier at Callaway was found acceptable. Ampacity derating testing was provided to NRC on December 11, 1996. Darmatt KM-1 was added to additional raceways requiring protection where Thermo-Lag was not present.

## 6.8 Calvert Cliffs Nuclear Power Plant, Units 1 and 2

There are no ERFBS in use at Calvert Cliffs Nuclear Power Plant, Units 1 and 2.

## 6.9 Catawba Nuclear Station (Catawba), Units 1 and 2

Catawba Nuclear Station (Catawba) uses Hemyc ERFBS as a 1-hour rated barrier to provide compliance with Appendix R. Hemyc is used in both Units Auxiliary Feedwater (AFW) Pump rooms, with each pump room containing approximately 91 m (183 m total) (300 linear feet (600 linear feet total)). Following IN 2005-07, the licensee evaluated their use of Hemyc fire barrier and determined that the Hemyc ERFBS does not meet the required 1-hour fire rating. As compensatory measure, the licensee has implemented additional control on transient combustible/flammable materials entering these affected areas and established continuous fire watches under certain circumstances. On February 28, 2006, the licensee submitted its intent to voluntarily transition the Catawba Fire Protection Licensing Basis to NFPA 805 in accordance with 10 CFR 50.48(c). The NFPA 805 transition process is expected to bring the Hemyc concerns to resolution. However, if the licensee decides to not transition, then they will have to submit exemption requests providing justification for the specific Hemyc configurations, or modify these configurations to come into compliance. A discussion of NFPA 805 is provided above in Section 3, "ERFBS Regulations."

## 6.10 Clinton Power Station, Unit 1

Clinton Power Station (CPS) currently uses 3M Interam ERFBS where required to ensure separation of redundant trains in the same fire zone. Clinton Power Station originally utilized 167 m (547 linear feet) of 1-hour Thermo-Lag 330-1 cable tray fire barriers and 45 m (149 linear feet) of 3-hour fire barriers at 10 different locations throughout the plant. In addition, CPS uses 34 m (112 linear feet) of 1-hour conduit fire barriers and 31 m (103 linear feet) of 3-hour conduit fire barriers. Following issuance of GL 92-08, CPS implemented a Thermo-Lag corrective actions program to document the station's engineering evaluations to ensure the Thermo-Lag ERFBS provide the necessary level of protection. A letter dated June 26, 1998 from NRC declared that CPS should proceed with corrective actions in accordance with the plant's proposed schedule of completion by December 31, 1998. However, Clinton ultimately eliminated dependence on Thermo-Lag as a credited fire barrier through several different methods, including: modification of the existing design to provide divisional separation through rerouting of cables/conduits, installation other barrier designs, or development of deviations.

By letter dated September 29, 1998, NRC recognized that CPS Thermo-Lag 330-1 corrective actions and requested information in accordance with GL 92-08 were complete.

The 3M Interam ERFBS used at CPS were installed in the late 1990's. Engineering evaluations were conducted for these installations, including a review of the fire barrier design, materials, and installation configurations to ensure the ERFBS capability to provide the needed level of protection. In addition, CPS had Promatec Technologies Inc. provide test reports documenting acceptability of the 3M Interam E-54C system installed at CPS, in accordance with Appendix R

Section III.G.2.b and GL 86-10, Supplement 1. NRC Inspection Report 50-461/98026(DRS) documented that the licensees combustibility assessment adequately demonstrated that the 3M material used inside containment (i.e., Interam E-54C) was noncombustible per NRC guidance and testing conducted by OPL (January 17, 1995) and UL (September 26, 1983).

### 6.11 Columbia Generating Station

Columbia Generating Station (CGS) credits approximately 600 feet of Darmatt KM-1 to ensure the necessary level of protection of redundant trains located in a single fire area. The Darmatt KM-1 ERFBS are installed in 1- and 3-hr rated designs. CGS documented its Darmatt KM-1 qualification in Fire Protection File 1.2.3, Item 2, which concludes that the Darmatt configurations are installed and bounded by the fire testing of Darmatt KM-1 performed in accordance with GL 86-10, Supplement 1.  See Section 5.3 above.

CGS credits approximately 250 feet of 3M Interam 3-hour barrier to ensure the necessary level of protection of redundant trains located in a single fire area R-1. The 3M Interam ERFBS are installed in a 3-hr rated design. CGS documented 3M Interam qualification in Fire Protection File 1.2.2, Item 1, which concludes that the 3M Interam configurations used at CGS are installed and bounded by the fire testing performed in accordance with GL 86-10.

CGS credits approximately 2700 feet of Whittaker fire-rated mineral insulated (MI) cable (company now called Meggitt) to ensure the necessary level of protection of redundant trains located in a single fire area. The Whittaker MI cable ERFBS are 3-hr rated, but located in both 1- and 3-hr fire areas. In 3-hr fire areas, Whittaker MI cable is typically routed on the ceiling to prevent fire-induced debris impacts. CGS documented Whittaker MI cable qualification in Fire Protection File 1.11, Item 2, which concludes that the CGS Whittaker MI cable configurations are installed and bounded by the fire testing performed in accordance with ASTM E-119 and unique functionality testing. A modification was performed to address Information Notice (IN) 2006-02.

CGS originally credited approximately 5,500 feet of Thermo-lag 330-1 to comply with Appendix R commitments and Regulatory Guide 1.75. The Thermo-Lag 330-1 ERFBS were installed in 1- and 3-hr rated designs. Following issuance of GL 92-08, CGS abandoned reliance on Thermo-Lag 330-1 to meet Appendix R. Thermo-Lag 330-1 is still credited as a RG 1.75 electrical separation barrier and for structural steel fireproofing coating on instrument tube supports, and one cable tray support. CGS documented Thermo-Lag 330-1 qualification in Fire Protection File 1.12, Item 2. Most Thermo-Lag ERFBS were abandoned in place which created concerns with combustible loading (IN 92-82), seismic & hanger support loading (IN 95-49 Supplement 1), and ampacity derating (IN 92-46 & 94-22). Each of these issues was resolved. Ampacity derating values for CGS are well within prescribed safety margins and no cable rerouting or resizing is necessary.

### 6.12 Comanche Peak Steam Electric Station, Units 1 and 2

Comanche Peat Steam Electric Station (CPSES) uses Thermo-Lag 330-1 ERFBS to protect redundant trains located within a single fire area to satisfy fire protection program licensing commitments. CPSES had approximately 5500 linear feet of a 1-hour Thermo-Lag 330-1 ERFBS installed for Unit 1 and approximately 4700 linear feet of 1-hour ERFBS for Unit 2. The licensee directly qualified most ERFBS configurations by conducting a series of fire endurance tests that were performed to a site-specific test methodology and acceptance criteria approved by the NRC staff. In addition, all ERFBS configurations, including those not directly tested, have been evaluated in accordance with GL 86-10 guidance to ensure they are capable of providing the necessary fire endurance capability. The licensee also performed site-specific testing to determine appropriate electrical cable ampacity derating factors.

The NRC accepted closure of all Thermo-Lag ERFBS GL 92-08 technical issues (i.e., fire endurance capability, cable-ampacity derating factors, and seismic considerations) installed in CPSES Unit 2 via NUREG-0797, Supplements 26 and 27 dated February and April of 1993, respectively. Subsequently, the licensee performed additional fire endurance testing and implemented design upgrade modifications to address qualification of Thermo-Lag ERFBS configurations for Unit 1. The licensee received NRC acceptance for closure of all Thermo-Lag ERFBS GL 92-08 technical issues for Unit 1 via NRC correspondence dated May 14, 1999. In addition, an NRC safety evaluation dated June 14, 1995, concluded that there are no significant safety hazards at CPSES resulting from the ampacity derating concerns associated with the use of Thermo-Lag fire barriers to enclosed cables.

CPSES uses approximately 800 linear feet of Hemyc as a RES inside each unit's containment (1600 linear feet for both units). There are no applications at CPSES where Hemyc is used to provide a 1- or 3-hour fire barrier for separation of redundant post fire safe shutdown circuits. All other uses of Hemyc at CPSES relate to its licensing basis. In a letter dated December 20, 2007, the licensee informed NRC that after evaluating new information regarding NRC's testing of Hemyc; it concluded that the CPSES's use of Hemyc as a radiant energy shield continues to meet its licensing basis. Regional NRC staff verified that the licensee had appropriately dispositioned the issue. (IR 05000445/2008006 and 05000446/208006, July 3, 2008)

### 6.13 Cooper Nuclear Station

Cooper originally used a 1-hour rated Thermo-Lag 330-1 ERFBS to protect approximately 15.2 m (50 feet) of conduit. The Thermo-Lag was used in three separate locations; two in the cable spreading room and one in the cable expansion room. As of February 3, 1994, Cooper does not rely on Thermo-Lag for fire protection purposes. Two non-rated radiant energy shields located in the cable spreading room and one radiant energy shield in the cable expansion room were modified and the Thermo-Lag 330-1 removed. The Thermo-Lag ERFBS were removed and replaced with Promat-H material from Eternit or the cables that required protection were re-routed to comply with Appendix R.

In a letter from NRC dated May 30, 1995, NRC determined that the CNS response to GL 92-08 was acceptable. CNS stated that all three installations of Thermo-Lag 330-1 were sufficiently spaced to be considered as open-air installations per National Electric Code and did not require ampacity derating testing beyond the conduit itself. The conduit installations themselves are not further derated by the proximity of the Thermo-Lag material.

### 6.14 Crystal River Unit 3 Nuclear Generating Station

Crystal River Unit 3 Nuclear Generating Station (CR-3) uses Thermo-Lag and Mecatiss ERFBS to provide separation and/or safe shutdown purpose in accordance with the CR-3 licensing basis. The estimated total of Thermo-Lag 330-1 used to cover cable trays is 651 m (2,135 feet) and the estimated total used to cover conduit is 1,711 m (5,615 feet). This equates to approximately 44 m$^2$ (471 ft$^2$) of Thermo-Lag.

To resolve Thermo-Lag deficiencies identified in the mid 1990's, the licensee installed Mecatiss ERFBS on 71 circuits required for Appendix R safe shutdown, re-routed cables, installation of additional sprinkler systems to protect existing Thermo-Lag installations (exemption) or re-evaluation of Safe Shutdown Analysis credited alternative equipment and shutdown procedures. NRC approved an exemption from the regulations related to the use of enhanced sprinkler protection instead of Thermo-Lag barrier upgrade in the CR-3 auxiliary building. The use of Mecatiss ERFBS at CR-3 was approved by NRC in letter and SE dated January 29, 1997.

By letter dated November 7, 1997 NRC determined that all ampacity related concerns have been resolved and the licensee has provided adequate technical basis to ensure that all of ERFBS enclosed cables are operating within acceptable ampacity limits.

## 6.15  Davis Besse Nuclear Power Station, Unit 1 and Perry Nuclear Power Plant Unit 1

Davis Besse Nuclear Power Station (DBNPS) originally relied on Thermo-Lag to meet requirements of 10CFR50 Appendix R, to support commitments made in several exemptions in Appendix R, and to satisfy the plants' licensing commitments. Approximately 740 linear feet of 1-hour Thermo-Lag was used to cover conduits and 1000 $ft^2$ covered boxes. Approximately 170 linear feet of 3-hour Thermo-Lag was used to cover conduits and 860 $ft^2$ to cover boxes. Approximately 170 $ft^2$ was used in radiant energy shield applications, and 2200 $ft^2$ of Thermo-Lag was applied to structural steel for fire-proofing.

On February 20, 1996, Toledo Edison informed NRC of its decision to replace installed Thermo-Lag with alternate materials. Engineering analysis showed that at least four applications of Thermo-Lag were allowed exemptions due to modified control circuitry in the Davis Besse Heat Exchanger Pump Room, determination that fire barriers were not actually required on structural columns in a pump room, relocation of a circuit through penetration P2P5F, and depowering of electrical pumps that are no longer needed for plant safety. All Thermo-Lag was either removed or abandoned in place, and final modifications were completed by December 22, 1998.

DBNPS and Perry Nuclear Power Plant Unit 1 used 3M Interam E50 Flexible mat to replace Thermo-Lag 330-1 for circuits, Mandoval Fendolite to replace structural steel, and Promat-H cement board or 3M Interam E50 Flexible mat to replace Thermo-Lag for fire dampers.

## 6.16  Diablo Canyon Nuclear Power Plant, Units 1 and 2

Diablo Canyon Nuclear Power Plant (DCNPP) uses 3M and Pyrocrete as raceway fire barrier protection for redundant trains located in the same fire area to satisfy Appendix R requirements. The licensee verified that the installed ERFBS configurations were bounded by the tested configurations. Any deviations from the tested configuration were evaluated and those configurations not bounded by the test specimen were independently tested.

In the past, DCNPP Units 1 and 2 had used approximately 550 linear feet of a 1-hour and 65 m (212 linear feet) of a 3-hour Thermo-Lag ERFBS to protect conduits. In addition, there was 8 $m^2$ (87 $ft^2$) of Thermo-lag used to protect miscellaneous components. The licensee re-analyzed the need for the Thermo-Lag and determined that 40 m (130 feet) of conduits and 2 $m^2$ (24 $ft^2$) were needed. The remainder of the Thermo-Lag was replaced by 3M or Pyrocete materials.

## 6.17 Donald C. Cook Nuclear Power Plant, Units 1 and 2

Donald C. Cook (DC Cook) uses Thermo-Lag, Mecatiss, and Darmatt fire barriers to provide protection to various components within the plant. Prior to issuance of GL 92-08, DC Cook used Thermo-Lag 330-1 ERFBS exclusively to provide separation of redundant equipment. Following issuance of GL 92-08, the licensee took several approaches to resolve the Thermo-Lag barrier issues at DC Cook, including; re-assessment of the need for fire barrier protection, the replacement of Thermo-Lag with alternate fire barrier materials Darmatt KM-1 (see Section 5.3 above), and the continued limited use of Thermo-Lag as a fire area boundary.

These efforts resulted in Thermo-Lag no longer relied upon as a fire barrier to provided separation of redundant equipment within the same fire area per the requirements of section III.G.2 of 10 CFR 50, Appendix R. Thermo-lag is only relied on to provide separation between fire areas in the Unit 1 and Unit 2 hot shutdown panel enclosures. This application of Thermo-Lag was qualified by field testing. In addition, Thermo-Lag is used at DC Cook in the Unit 1 CD Diesel Generator Room and the Unit 2 AB Diesel Generator Room, in conjunction with the existing fire detection and fire suppression system. This barrier configuration has also been evaluated for compliance with 10 CFR 50, Appendix R, Section III.G.2(c). By letter and SE dated July 14, 1999, NRC concluded that the application of ampacity derating methodology used at CNP involve no significant safety hazards.

DC Cook Units 1 and 2 use Mecatiss® ERFBS as a cable tray 1-hour fire wrap in one fire zone to achieve compliance with 10 CFR, Appendix R, Section III.G.2.(b). Mecatiss® was installed at D.C. Cook in accordance with a Brand Fire Protection Service Manual which identified the fire test reports supporting the DC Cook applications. Brand Fire Protection Services provided support during installation of the Mecatiss® barrier at DC Cook.

Darmatt KM-1 is used at DC Cook and was installed as a replacement to most of the Thermo-Lag barriers previously used at the plant to support compliance with Appendix R train separation. All Darmatt barriers were installed by Transco Products, Inc. using an approved plant procedure based on vendor recommended procedures for installation of Darmatt KM-1. Fire test reports supporting the DC Cook Darmatt applications are identified in the procedure.

By SE dated July 14, 1999, NRC staff found no significant safety hazards associated with the application of the ampacity derating methodology used for DC Cook.

## 6.18 Dresden Nuclear Power Station, Units 1 and 2

Dresden Nuclear Power Station uses 3M Interam ERFBS on all electrical raceways where a fire barrier is required to ensure separation of redundant trains in the same fire zone. There are seven locations within the power station which utilize approximately 5.3 m$^2$ (57.2 ft$^2$) of 3M Interam on cable trays. Four of the locations contain two or three cable trays wrapped and bundled together. Dresden Nuclear Power Station commissioned an independent review of the 3M ERFBS installation and fire tests in 1994. Following the review, those configurations installed at the plant that didn't met the acceptance guidance provided in GL 92-08, were documented.

### 6.19 Duane Arnold Energy Center

Thermo-Lag ERFBS was initially installed at Duane Arnold Energy Center (DAEC) as ERFBS, steel coating and fireproofing and miscellaneous fire barrier installations such as penetration seals to meet 10 CFR 50.48 or Appendix R requirements. Approximately 64 m (211 linear feet) of a 1-hour and 264 m (865 linear feet) of a 3-hour Thermo-Lag ERFBS were installed at DEAC. The licensee resolved Thermo-Lag issues by performing safe shutdown reanalysis, rerouted cables, removed and replaced Thermo-Lag with Darmatt KM-1 ERFBS. In addition to using Thermo-Lag as an ERFBS, DAEC also used Thermo-Lag to construct structural steel fireproofing/barriers, penetration seals and door jam fire proofing. These non-ERFBS application have been resolved by the licensee; in some cases by performing independent fire testing.

DAEC uses Darmatt KM-1 ERFBS installed on certain electrical raceways in Fire Zone 2A.

DAEC does have an approved exemption for the use of an untested flexible conduit wrapping material (Hemyc) as documented in Letter dated October 14, 1987. DAEC is no longer utilizing this exemption and no longer has Hemyc installed.

NRC accepted DAEC ampacity derating methodology in its letter and SE dated January 26, 1999.

Industry chemical testing program confirmed the consistency of Thermo-lag material used at DAEC and supports the use of generic fire tests and other data developed by NEI and others.

### 6.20 Joseph M. Farley Nuclear Plant

Joseph M. Farley Nuclear Plant (FNP) requested a Kaowool related Appendix R exemption which was approved by NRC in 1985 & 1986. The 1986 exemption relied on Kaowool for protection of several cables in the Service Water Intake Structure. In light of the issues raised with the performance of the Kaowool EFRBS, the licensee re-evaluated the Appendix R program and determined that it no longer required the use of Kaowool as an ERFBS. To comply with Appendix R, without using Kaowool, the licensee performed approximately 35 modifications to eliminate its reliance on Kaowool. Some of these modifications included the use of Meggitt fire-rated cable, which was approved by NRC in SER dated February 13, 2006 and March 22, 2006.

FNP uses Promat-H board to construct cable tray enclosures to provide separation of redundant trains located in a single fire area. Promat-H material was tested and qualified to ASTM E-119-88 full scale wall assembly and small scale ceiling assembly by Performance Contracting Inc., under Omega Point Project No. 8806-90254 (Promat Report SR90-005). The acceptance criteria used to qualify this barrier assembly was ASTM E-119-88, which the licensee stated met the acceptance criteria of GL 86-10 Supplement 1.

FNP also uses 3M Interam E-50 Series wrap material and/or 45B formulated silicone elastomer at penetrations for the stairwell cable tray enclosures. 3M Interam E-50 wrap used at FNP has been tested in accordance with the acceptance criteria specified in GL 86-10 Supplement 1, for the configurations found within the plant. The 45B formulated silicone elastomer was tested in accordance with test conditions prescribed in ASTM E-119.

### 6.21 Fermi-2

In the past, Fermi-2 had used approximately 40 linear feet of a 3-hour Thermo-Lag 330-1 ERFBS to protect cable trays and approximately 1000 ft$^2$ for miscellaneous barriers. The licensee removed or replaced the Thermo-Lag with 3M material.

Approximately 450 linear feet of 3M Interam E-54A material is installed at Fermi-2 to provide a 1-hour rated barrier. A majority of the 3M material is located in the auxiliary building basement and on the plant's fifth floor to protect redundant cable trays and conduits within the same fire area per Appendix R requirements. Additional 1-hour 3M E-54A was installed on cable trays and cable tray supports and on a conduit in the Auxiliary building first floor mezzanine and cable tray areas. The additional 3M material was installed by December 2006. 3M Interam E-54A ERFBS qualification tests were performed using the requirements of GL 86-10 Supplement 1.

Older 3M M-20 and CS-195 material from the 1980s was removed as well from various cable trays and support barriers. The two 3M materials were installed in the auxiliary building in the 1980s in order to create a fire-safe portion of the building and replaced with 3M Interam E-54A fire barrier material.

### 6.22 James A. Fitzpatrick Nuclear Power Plant

James A. Fitzpatrick Nuclear Power Plan (JAFNPP) relies on Hemyc ERFBS to separate required safe shutdown equipment located in the same fire area. On September 27, 2006, NRC issued a SE approving JAFNPP's request for an exemption from 10 CFR 50.48. NRC regional integrated inspection report 05000333/200605, January 19, 2007, ADAMS Accession No. ML073390309 documents NRC inspectors' verification of the Hemyc fire barrier issue for JAFNPP.

In addition to the use of Hemyc at JAFNPP, FP-60 is also used to separate redundant trains located in a single fire area. For this application of FP-60, the licensee requested and received an exemption from NRC related to Appendix R requirements. The exemption was granted because the isolated area containing FP-60 fire wrap met NRCs three defense-in-depth conditions of preventing fires from starting, detecting, controlling, and extinguishing those that do occur, and to preserve safe shut-down of the plant from fire and fire extinguishing methods. JAFNPP received an exemption from NRC because there are no ignition sources in the area, there is a smoke detection and sprinkler system in-place, and fire wrap is used around the electrical cables providing 30 minutes of fire damage protection and approximately 52 minutes of functionality protection.

### 6.23 Fort Calhoun Station (FCS), Unit 1

FCS uses several ERFBS to protect redundant trains of safe shutdown equipment in accordance with Appendix R requirements. These ERFBS include: 3M Interam E50A, Pyrocrete, and Pabco barriers. The licensee stated in its response letter go GL 2006-03, that installation and inspection procedures have verified that these ERFBS are installed in a manner consistent with tested configurations and any deviations were evaluated with the guidance provided in GL 86-10, supplement 1.

## 6.24 R.E. Ginna Nuclear Power Plant

R.E. Ginna Nuclear Power Plant (Ginna) credited the Hemyc fire barrier system as a one hour fire rated barrier for Appendix R compliance purposes. Although Ginna's Appendix R analysis does not require any three hour fire rated Hemyc or MT configurations, there is one barrier location where the MT system is used to meet a one hour requirement (located on a source range nuclear instrument preamplifier to provide additional protection). The Hemyc system is also used in the Containment Building as a radiant energy shield, which is considered operable since the Hemyc system is non-combustible.

Ginna has approximately 130 m (425 feet) of Hemyc conduit wrap installed on conduits ranging in sizes from 1.3 cm to 6.4 cm (0.5 to 2.5 in) in diameter. In addition, there is approximately 30 m (100 ft) of Hemyc wrap installed on a single 61 cm (24 in) cable tray. Conduit wrap is directly applied, while the tray wrap is installed using the standard vendor design consisting of Hemyc wrap installed on a frame assembly that provides an air space between the wrap assembly and the raceway.

One three hour MT barrier (approximately four feet) is used to provide additional protection for a source range nuclear instrument preamplifier in the event of a fire. The Appendix R and fire analysis requirement for the location only requires a one hour fire barrier, which is the rating for other barriers in this area of the plant.

Hemyc configurations are located in the following plant areas: Battery Room, Intermediate Building Clean Side Basement, and Auxiliary Building Intermediate Floor and Basement level. The barrier material provides protection for the following systems/circuits: AC and DC Power Distribution, Steam Generator pressure indication, Source Range Nuclear Instrumentation, and Chemical and Volume Control System charging pump power.

Based on a review of NRC Information Notice 2005-07, all Hemyc configurations that are required for rated protection of circuits were determined to not be conforming to Ginna's current licensing basis and declared inoperable on April 6, 2005.

All Hemyc configurations that are required for rated protection of circuits have been declared inoperable. Hourly fire watch tours have been in place since April 6, 2005 and were implemented in accordance with site procedures. Additional compensatory measures, or changes to the current measures, may be considered based on NRC IN 97-48, Regulatory Issue Summary 2005-07, and future industry guidance.

Ginna has voluntarily committed to transition to NFPA 805 as stated in the letter of intent to NRC dated December 19, 2005, with an enforcement discretion period of three years. A project plan implementation schedule to transition to NFPA 805 has been developed. Based on the, outcome of the fire probabilistic risk assessment and modeling analysis, the Hemyc wrap will be replaced with an approved fire rated material, left as is, or eliminated through the use of change evaluations. In the event that Ginna does not transition, then they will have to submit exemption requests to the NRC providing justification for the specific Hemyc configurations, or modify these configurations to come into compliance. A discussion of NFPA 805 is provided above in Section 3, "ERFBS Regulations."

## 6.25 Grand Gulf Nuclear Station

Grand Gulf Nuclear Station (GGNS) uses two types of ERFBS to provide Appendix R protection of redundant trains required to safely shutdown the plant, they are; Thermo-Lag and 3M Interam. Thermo-Lag 330 is currently used in the Control and Auxiliary buildings and by letter dated April 21, 1997, NRC concluded that the GGNS program plan is acceptable to resolve the Thermo-Lag 330 issues identified in GL 92-08. Both Thermo-Lag and 3M barriers used at GGNS are qualified by fire tests in accordance with guidance provided in GL 86-10 Supplement 1.

GGNS in the past used Kaowool in several Appendix R applications. Following the issuance of IN 93-41 and subsequent meeting with NRC, the licensee of GGNS began evaluation possible solutions for the issues with Kaowool. GGNS initially intended to re-qualify the fire resistance rating and determine the overall acceptability of the Kaowool fire wrap system. Field walk downs and destructive examinations revealed additional installation deficiencies. It then became apparent to the licensee that Kaowool wrap would have to be completely reworked for any hope of qualification. The licensee then chose to replace and re-analysis (with risk-insights) the use of Kaowool and its replacement where still required with a 3M Interam ERFBS. By letter and SE dated September 29, 2006, NRC determined that the replacement of Kaowool ERFBS with 3M Interam material was acceptable for resolution of the Kaowool issues.

## 6.26 Shearon Harris Nuclear Plant

Shearon Harris Nuclear Plant (HNP) has used Thermo-Lag 330-1 to satisfy 10 CFR 50.48 separation requirements, licensing commitments, and conditions associated with its Fire Protection Program. Only 32 linear feet of a 1-hour rated Thermo-Lag 330-1 barrier were used to protect conduit raceways. In addition, 3270 sq. ft. of Thermo-Lag 330-1 barrier were used to protect area enclosures, partial height walls and door fireproofing.

HNP uses approximately 6,500 linear feet of Hemyc and 1200 linear feet of MT ERFBSs to satisfy Appendix R requirements. HNP has voluntarily committed to 10 CFR 50.48(c) to transition to NFPA 805, which it believes will resolve the Hemyc and MT issues by application of technical evaluations that consider potential adverse effects, risk, defense-in-depth, and safety margins as an acceptable alternative. In the event that HNP does not transition to NFPA 805, then they will have to submit exemption requests to the NRC providing justification for the specific Hemyc configurations, or modify these configurations to come into compliance.
A discussion of NFPA 805 is provided above in Section 3, "ERFBS Regulations."

In addition to the use of Hemyc and MT, HNP also uses Thermo-Lag 330-1 and 3M Interam E-54A barriers to provide Appendix R related equipment `protection. Both of these materials were tested per the guidance of GL 86-10 supplement 1 for specific applications used at HNP. Vendor testing was used for the 3M material and proprietary HNP fire testing was performed to qualify the Thermo-Lag installations.

## 6.27 Edwin I. Hatch Nuclear Plant

Edwin I. Hatch Nuclear Plant (Hatch) uses Promat-H material to construct cable tray ERFBS required for separation of redundant trains located within a single fire area. The licensee (Sothern Company) stated that "Promat-H is a material tested in accordance with UL standard 263, "Fire Tests of Building Construction and Materials," which references ASTM E-119 and NFPA 251 tests." Promat-H configurations used at Hatch were tested and qualified to ASTM E-

119-83 by Performance Contracting, Inc. under Omega Point Project No. 8806-90254 (Promat Report SR90-005).

The test acceptance criteria used were that of ASTM E119-83 Section 16 "Conditions of Acceptance" which meets the acceptance criteria of GL 86-10, Supplement 1. This criterion allows a maximum temperature rise of 250 degrees Fahrenheit above the initial temperature. (ADAMS Accession No. ML072060088) Testing on Promat-H included time-temperature tests, full scale fire testing on the wall, and small scale fire testing on the ceiling, all of which is documented in Promat Report SR90-005. The initial ambient temperature used during experimentation was 75 degrees Fahrenheit.

Hatch used approximately 1,250 linear feet of FP-60 in its river intake structure, and procured approximately 4,000 linear feet for installation in its Control Building and Reactor Building. Hatch installed its Kaowool barriers in 1984. On April 18, 1984, NRC granted Hatch an exemption for the use of Kaowool in the river intake structure with the area-wide automatic fire suppression system not required for the entire river intake structure. On January 2, 1987, another exemption was granted to the extent that a 20-foot separation was not required for cable in conduit and cable in trays wrapped with Kaowool blankets. The Kaowool ERFBS used at Hatch were subsequently replaced with FP-60 material in the 1992-1993 timeframe because of wear and degradation of the Kaowool material. The addition of the 2-mil aluminum skin covering provides protection to the Kaowool ceramic material. Hatch also uses Kaowool to provide physical separation RG 1.75 and to reduce combustible loading in a given fire area for compliance with Appendix A to Branch Technical Position APCSB 9.5-1. The licensee has submitted an evaluation to Region II staff that Kaowool is not used as a 1-hour fire barrier. The Regional staff's review indicates that this application is acceptable.

### 6.28 Hope Creek Generating Station

Hope Creek Generating Station does not use any ERFBS.

### 6.29 Indian Point Nuclear Generating Units 2 and 3 (IP-2 and IP-3)

Hemyc ERFBS is used at Indian Point Nuclear Generating Units 2 and 3 (IP-2 and IP-3) to provide separation and/or safe shutdown protection for compliance with Appendix R requirements. Approximately 31 linear meters (102 linear ft) of Hemyc is used in IP-2 and approximately 90 linear meters (295 linear ft) is used in IP-3.

Exemptions from the requirements of Appendix R have been granted for each case where the Hemyc ERFBS is used and credited to provide a fire resistance rating of 30 minutes or one hour. The exemptions were granted based on minimal fire challenge and other mitigating defense-in-depth factors.

IP-3 also uses Hemyc as a RES inside containment. As part of closeout actions associated with GL 2006-03, NRC inspection staff verified that appropriate corrective actions were taken while Hemyc ERFBS were considered inoperable, and that Hemyc ERFBS upgrade modifications, where required, had been completed.

IP-2 also uses 3M Interam E-54 for Appendix R purposes, configured as a 3-hour rated ERFBS outside the reactor containment, and as a RES inside the reactor containment. The 3M Interam E-54A installations were evaluated to ensure the capability to provide the necessary level of protection at the time the barriers were installed, and were reevaluated to confirm their adequacy after issuance of IN 95-52.

## 6.30 Kewaunee Power Station

Kewaunee Power Station uses 3M Interam E-50A endothermic mat to meet 3-hour rated configurations on conduits. These conduit ERFBS were installed in accordance with UL design listing (UL Electrical Circuit Protective System (FHIT) No. 7).

Although not used to provide Appendix R protection, Kewaunee Power Station also uses a 3-hour fire-rated Marinite board/Kaowool/Flameastic electrical circuit large pull box protective enclosure.

## 6.31 LaSalle County Station, Units 1 and 2

LaSalle County Station (LCS) initially relied on Thermo-Lag 330-1 barriers to meet 10 CFR 50.48 regulations and to provide separation between redundant electrical systems. Darmatt KM-1 material was installed as a qualified replacement for the 112 linear feet of Thermo-Lag 330-1 fire barrier as part of LCS response to GL 92-08. Therefore, LCS now uses Darmatt KM-1 fire barriers in areas where a fire barrier is required to ensure separation of redundant trains in the same fire zone. By letter dated January 17, 1997, the licensee of LCS informed NRC that all Thermo-Lag 330-1 ERFBS material had been replaced with Darmatt KM-1 material. By letter and SE dated December 22, 1999, NRC staff determination that all ampacity derating concerns were resolved for LCS Units 1 and 2, and the licensee provided an adequate technical basis to ensure that all of the fire barrier enclosed cables are operating within acceptable limits.

LCS also uses a limited amount of Kaowool fire barrier in one reactor building to augment the approximately 12 m (40-foot) spatial separation between cabling of redundant trains, and extends protection out to 15 m (50 feet) from the redundant cable. NRC has approved use of Kaowool in this limited application due to lack of automatic fire suppression in the area. The Kaowool used has a performance rating of 90 minutes and is layered approximately 7.6 cm (3.0 in) thick along the length of fire-protected area.

## 6.32 Limerick Generating Station Units 1 and 2 and Peach Bottom Atomic Power Station, Units 2 and 3

Limerick Generating Station (LGS) and Peach Bottom Atomic Power Station (PBAPS) use Darmatt KM-1 and Thermo-Lag where a fire barrier is required to ensure separation of redundant trains the same fire area. As a result of GL 92-08, both sites implemented a Thermo-Lag corrective actions plan that documented the analysis, testing, and modifications to ensure ERFBS relied upon to provide separation of redundant safe shutdown trains within the same fire area provide the necessary level of protection.

LGS and PBAPS both use Thermo-Lag 330-1 ERFBS to comply with their fire protection plans. The licensee uses this barrier to protect electrical power and control cables for systems and components used for achieving and maintaining safe shutdown conditions. Thermo-Lag was also used to provide physical independence (RG 1.75) and in some instances, both physical independence and safe shutdown capability. One- and three-hour Thermo-lag installations are used at both sites accounting for approximately 1450 m (4,750 feet) of Thermo-Lag at LGS and 1341 m (4,400 feet) of Thermo-Lag at PBAPS. In response to GL 92-08, the licensee identified that it had not performed plant specific fire endurance tests of Thermo-Lag 330-1 material, but relied on the manufactures (TSI) and other licensee tests to qualify the licensees' installations.

Safe shutdown re-analysis was completed to minimize reliance on Thermo-Lag 330-1 by use of operation actions and economically justifiable plant modifications and identification of cables that require protection by some type of ERFBS. Destructive examination of a sample of Thermo-Lag installations was performed at LGS to ensure that the Thermo-Lag installation was assembled with materials of acceptable quality (void of cracks, voids, and deformations).

A NRC inspection team reviewed the design and qualification testing for the Darmatt KM-1 electrical raceway fire barriers, and performed a walk down of installed barriers for the selected areas. This review was performed to verify that the selected items of the fire barrier system met their design and licensing bases. No findings of significance were identified. (ADAMS Accession No. ML020080162)

By letter dated September 21, 1998, NRC informed the licensee that all information requested in GL 92-08 had been received and all actions related to Thermo-Lag 330-1 ERFBS used at LGS, except ampacity derating, had been closed out. LGS completed all of its Thermo-lag related corrective actions by September 1999 and PBAPS actions were completed by October 1999. Safety Evaluation dated January 12, 2000, documents NRC staff evaluation of the Thermo-Lag ampacity derating issues at the PBAPS and LGS. The staff found that the ampacity derating analysis results are acceptable and there are no significant safety hazards associated with the application of the licensee ampacity derating methodology.

### 6.33 McGuire Nuclear Station, Unit 1 and 2

Thermo-Lag was initially used at McGuire Nuclear Station (McGuire) but cable that used Thermo-Lag ERFBS to provide the required protection have been replaced with a fire resistive electrical cable manufactured by Meggitt Safety Systems (previously known as Whittaker Electronic Systems) for several "A" train cables that are not separated by greater than 20ft from redundant "B" train cables. This electrical cable is a type of mineral insulated cable and the use of this cable at McGuire has been approved by NRC SE dated January 13, 2003.

McGuire uses approximately 20 linear feet of Hemyc ERFBS in Unit 1 and 44 linear feet in Unit 2, as a 1-hour rated barrier for compliance with Appendix R requirements. In response NRC and industry testing results, the licensee determined that their use of Hemyc does not meet the 1-hour fire rating to comply with McGuire licensing basis. As compensatory measure, the licensee has implemented additional control on the types of materials introduced into areas containing Hemyc and performs routine fire watches in the affected areas. On April 18, 2006, McGuire licensee submitted it intent to transition to NFPA 805 in accordance with 10 CFR 50.48(c). The licensee expects to resolve all issues related to the Hemyc ERFBS during the NFPA 805 transition process. In the event that McGuire does not transition to NFPA 805, then they will have to submit exemption requests to the NRC providing justification for the specific Hemyc configurations, or modify these configurations to come into compliance. A discussion of NFPA 805 is provided above in Section 3, "ERFBS Regulations."

### 6.34 Millstone Power Station, Units 2 and 3

Millstone Power Station Unit 3 (Millstone 3) used a 1-hour rated Thermo-Lag 330-1 ERFBS to protect approximately 40 ft conduits containing Appendix R required cables. The licensee replaced the Appendix R required cables with a 1-hour fire rated cable, to eliminate its reliance on Thermo-Lag materials. Currently, Millstone Unit 3 does not use any ERFBS to separate redundant trains located within the same fire area. Compliance with Appendix R is achieved for Unit 3 by physical separation, water curtains or other approved deviations.

Millstone Power Station Unit 2 (Millstone 2) uses Thermo-Lag 330 and 770 ERFBS to protect conduit, cable trays and junction boxes. All Thermo-Lag used in Unit 2 were installed in accordance with the Omega Point Laboratories and NEI tests and didn't deviate from tested configurations.

In a letter dated May 9, 2002, Millstone 2 received a letter from the NRR stating that the ampacity derating issues were resolved due to the fact that the Millstone 2 had provided all necessary information mentioned in GL 92-08. (ML020700197)

Millstone 2 uses Thermo-Lag 330 and 770 ERFBS to protect conduit, cable trays and junction boxes. All Thermo-Lag used in Millstone 2 were installed in accordance with the Omega Point Laboratories and NEI tests and didn't deviate from tested configurations.

### 6.35 Monticello Nuclear Generating Plant, Unit 1

Monticello used approximately 21 m (70 ft) of a 3-hour rated Thermo-Lag 330-1 ERFBS design to protect conduits. The licensee rerouted two sections of these cables and also removed Thermo-Lag 330-1 in the plant in order to eliminate its reliance on the material. All material was removed and cables rerouted during the 1993 plant refueling. As a result, Monticello no longer utilizes any ERFBS to provide separation of redundant trains located in the same fire area. NRC replied by letter on May 27, 1993 that the issue is considered closed and Monticello NGP provided necessary information with regards to GL 92-08 and 10 CFR 50, Appendix R.

### 6.36 Nine Mile Point Nuclear Station, Units 1 and 2

Nine Mile Point Nuclear Station (NMP) previously used a small quantity of Thermo-Lag 330-1 to ensure safe shutdown capability and to meet the requirements of 10CFR50.48. The Thermo-Lag being used provided 3-hour fire protection in three locations in both NMP Units 1 and 2. There are nine conduit enclosed cables (3.8 cm (1.5 in) diameter) that utilize Thermo-Lag 330-1 in the missile barrier enclosure within Diesel Generator 103 Room which are intended to provide separation between Diesel Generator 102 and 103 control cables located within missile shield. There is also 3-hour, 2.5 cm (1.0 in) thick Thermo-Lag applied to HVAC duct work in the Turbine Building running from the Control Room to the Auxiliary Control Room. There are also very small quantities of conduit barriers installed in NMP1.

NMP Unit 2, a composite conduit box enclosure made of tube steel and angled framing had a coating of 1 inch thick, 3-hour Thermo-Lag fire barrier applied. 3-hour Thermo-Lag 330-1 was also used to provide separation for a safe shutdown area from an adjacent transformer yard in lieu of a fire damper in the outside wall. In addition, Thermo-Lag was installed in Unit 2 to provide separation of diesel day tank rooms from the diesel generator room. This installation is used in lieu of fire dampers on each end of an HVAC duct at the room's boundaries.

In a letter from Niagara Mohawk Power Corporation dated January 30, 1996, NMPC stated that no Thermo-Lag 330-1 fire barriers performing a safety function of separating redundant safe shutdown trains of equipment remained at Nine Mile Point. One application of Thermo-Lag was abandoned in place and the remaining applications were removed completely.

### 6.37 North Anna Power Station, Units 1 and 2

North Anna Power Station (NAPS) Units 1 and 2 use ERFBS made of 3M Interam E-53A series mat to provide a 1-hour rated fire barrier to protect power cables to a charging and component cooling water pump. These applications of 3M use an additional layer of E-53A series mat than the tested configuration, to improve fire resistance and ensured the 1-hour rating would be achieved.

Approximately 20 m (65 linear feet) and 5 m$^2$ (50 ft$^2$) of Thermo-Lag was replaced by 3M material and with gypsum board. An additional 12 m$^2$ (128 ft$^2$) of Thermo-Lag was addressed by engineering evaluations.

### 6.38 Oconee Nuclear Station Units 1, 2, and 3

Oconee Nuclear Station (Oconee) does not use any ERFBS for compliance with NRC regulations. However, Oconee does use Hemyc blanket pads to cover three wall penetration seals which provide additional thermal margins. The actual penetration seals do not rely on the application of the Hemyc pads to meet their 3-hour fire endurance rating. The three penetrations have the Hemyc covering total approximately 3 m$^2$ (35 ft$^2$).

### 6.39 Oyster Creek Nuclear Generating Station

Oyster Creek Nuclear Generating Station (OCGS) used to rely on approximately 322 m (1055 ft) of 1- and 3-hour Thermo-Lag fire barriers on various conduits and boxes in seven different areas of the plant.

OCGS originally intended to declare exemptions from GL 92-08, noting that Thermo-Lag applications did not protect electrical cables regarding safe shutdown of the nuclear plant. However, after examining cost of a 100 percent upgrade of the fire barrier system, the exemptions were withdrawn and a complete new barrier system was planned. Bounding tests were performed utilizing guidance provide in GL 86-10, Supplement 1 and any deviations from the fire tests were documented, evaluated, and addressed. Materials considered for the new fire barrier wraps were Darmatt, Versawrap, and Mecatiss. OCGS now uses a combination Thermo-Lag/Mecatiss ERFBS, along with a standalone Mecatiss barrier and a standalone 3M Interam ERFBS. Approximately 321 m (1055 linear feet) of Thermo-Lag located on 1-hour conduit and 1-hour boxes was replaced along with approximately 6 m (20 linear feet) of 3-hour conduit.

On January 30, 2001, OCGS provided NRC with a letter stating that all corrective actions required by GL 92-08 regarding Thermo-Lag deficiencies were completed by December 31, 2000.

### 6.40 Palisades Nuclear Plant

Palisades Nuclear Plant (PNP) uses concrete as a rated 1-hour fire barrier to separate redundant trains within the same fire area. The barrier was constructed by enclosing a conduit and pull box with concrete.

Palisades initially used a 1- and 3-hour Thermo-Lag barrier to protect 53 linear meters (174 linear ft) of conduit and two junction boxes. The licensee replaced 43 linear meters (140 linear ft) with a 1-hour rated fire cable and embedded 4 ft and associated junction boxes in concrete. The remaining 30 linear feet was rerouted to provide the required separation. PNP completed a written response to NRC on January 6, 1997 stating that the proposed corrective actions for the plant's Thermo-Lag installations were complete.

### 6.41  Palo Verde Nuclear Generating Station Units 1, 2, and 3

Palo Verde Nuclear Generating Station (PVNGS) uses Thermo-Lag ERFBS to provide protection as required by Appendix R. PVNGS does not utilize any 1-hour or 3-hour fire barriers required for cable trays. A large amount of fire barrier material for cable raceways was found unnecessary and nearly 80 percent of the Thermo-Lag at the nuclear station was removed. PVNGS has about 215 m (705 linear feet) of Thermo-Lag 330-1 installed for 1-hour conduit barriers. Palo Verde also has approximately 76 $m^2$ (820 $ft^2$) of Thermo-Lag 330-1 utilized for 3-hour HVAC and cable tray support barriers. About 58 $m^2$ (625 $ft^2$) of Thermo-Lag is being used for 1-hour junction box barriers, and there is also about 43 $m^2$ (460 $ft^2$) of radiant energy heat shield Thermo-Lag installed for conduits.

In a letter dated June 11, 1998, NRC stated that all corrective actions for the resolution of the Thermo-Lag issues identified in GL 92-08 had been implemented and the actions tracked by TAC Nos. M85583, M85584, and M85585 were complete.

Palo Verde conducted ampacity evaluations for specific Thermo-Lag 330-1 enclosed raceways. Testing performed on cables in Palo Verde Unit 1 was declared sufficient for Units 2 and 3 as well due to replicated configurations and materials. As of December 24, 1997, all ampacity issues associated with Thermo-Lag fire barriers were resolved for all three nuclear station units.

### 6.42  Peach Bottom Atomic Power Station, Units 2 and 3

See Section 6.32 Limerick Generating Station Units 1 and 2 and Peach Bottom Atomic Power Station above.

### 6.43  Perry Nuclear Power Plant Unit 1

See Section 6.15 Davis Besse Nuclear Power Station, Unit 1 and Perry Nuclear Power Plant Unit 1.

### 6.44  Pilgrim Nuclear Power Station

Mecatiss and 3M Interam ERFBS are used at Pilgrim Nuclear Power Station to provide protection of electrical and instrumentation cables associated with equipment that provides safe shutdown capability in accordance with Appendix R. These barriers were installed subsequent to GL 86-10 supplement 1 and IN 95-52, were evaluated against these criteria and addressed the indentified concerns.

### 6.45  Point Beach Nuclear Plant, Units 1 and 2

Point Beach Nuclear Plant Units 1 and 2 utilize 3M Interam E-50 Series ERFBS on selected raceways to provide separation between redundant shutdown trains to meet the separation of criteria of Section III.G.2.c of 10 CFR 50, Appendix R. The configurations were installed at PBNP before guidance provided in GL 86-10 supplement 1 was available, were qualified in accordance with ASTM E-119 and ANI/MAERP fire test specifications, available at the time of the installation. All 3M barriers installed post GL 86-10 Supplement 1 were qualified to the acceptance criteria provided in that guidance.

Some of the items tended to in order to comply with 10 CFR 50 include upgrading control building walls which could be ignited by turbine fires to a 3-hour rating, including fire dampers, fire doors, and penetration seals. The walls of the cable spreading room and diesel generator room were also upgraded to a 3-hour fire rating. The viewing window was also upgraded to a two-hour fire rating, as well as the walls of the control room were upgraded to a two-hour fire rating that separates the service building and general auxiliary building ventilation exhaust filters from the remainder of the auxiliary building.

### 6.46 Prairie Island Nuclear Generating Plant Units 1 and 2

Prairie Island Nuclear Generating Plant (PINGP), Units 1 and 2 use Darmatt KM-1 and 3M Interam E50 series ERFBS to provide Appendix R safe shutdown circuit protection. These two barriers were qualified for a 1-hour fire rating in accordance with GL 86-10 Supplement 1 guidance. These two ERFBS are not used inside containment; instead Marinite is used to provide radiant energy shields inside containment.

Applications of Kaowool at the nuclear plant were removed due to the fact that they are no longer required by 10CFR50, Appendix R. All applications of Kaowool fire barrier were analyzed before removal to make sure they are necessary to comply with 10CFR50. Prairie Island determined that several installations of the fire barrier were unnecessary, and also that several locations were in need of a fire barrier. Field testing showed that some barriers met the required 20 foot separation criteria, allowing Kaowool to be removed and not replaced with upgraded fire barriers. Cables were also rerouted during a scheduled power outage to meet separation criteria and reduce the plant's reliance on fire barrier applications.

Through NRC approval and by letter dated December 27, 2000, Prairie Island Nuclear Generating Plant declared its completion date for Kaowool replacement and cable rerouting as February 28, 2001.

### 6.47 Quad Cities Nuclear Power Station, Units 1 and 2

Quad Cities Nuclear Power Station (QCNPS) uses several different types of ERFBS to ensure separation of redundant trains in the same fire zone, including, Darmatt KM-1, Versa Wrap, and 3M Interam.

In 1994, QCNPS commissioned an independent review of the 3M Interam ERFBS installation and fire tests. The review evaluated installation configurations and test data the meet industry standards and GL 86-10 sup 1 acceptance criteria with the instillations at QCNPS. Where plant configurations that didn't bound the acceptable test configurations, a plant modification was performed to ensure the installed fire barrier was bounded by a tested configuration.

QCNPS uses Darmatt KM-1 and Versa Wrap 1-hour rated ERFBS. These barriers were installed in the late 1990's and an engineering evaluation for the modifications included a review of the fire barrier endurance testing to ensure the capability of these two ERFBS.

## 6.48 River Bend Station, Unit 1

River Bend Station (RBS) relies on Thermo-Lag 330 to provide the protection to safe shutdown circuits required by Appendix R. Originally, one and three hour Thermo-Lag materials are used for safe shutdown purposes at RBS. There was approximately 923 linear feet of 1-hour Thermo-Lag material and approximately 366 feet of 3-hour Thermo-Lag installed on cable trays. For conduits, there were about 4282 feet of 1-hour material and 1429 feet of 3-hour material. In addition, Thermo-Lag was utilized to cover approximately 741 ft$^2$ of 1-hour and 277 ft$^2$ of 3-hour items including junction boxes, instruments, instrument racks, motor operated valves, a ceiling assembly, a steel beam, and one radiant energy shield.

All Thermo-Lag barriers at RBS were declared inoperable on October 26, 1989. In response to its failed 3-hour fire endurance testing of Thermo-Lag 330, and supplementary testing by NEI, RBS developed a new post-fire safe shutdown analysis to reduce the plant's dependence on Thermo-Lag. In addition new Thermo-Lag configurations replaced the previous ones using new materials based on successful NEI test results. By using a combination of rerouting cables and revising its safe shutdown analysis, RBS reduced the amount of Thermo-Lag to approximately 500 feet. All Thermo-Lag configurations at RBS are of a 1-hour application.

Implementation of the new materials was delayed due to ampacity derating issues. Fire barriers at RBS were installed in accordance with TSI Technical Note 20684, but after planning and testing new fire barrier configurations, RBS became concerned that insufficient experimental results would prevent RBS from implementing any upgrades. RBS removed the cable configurations from service in order to perform cable degradation tests and reduce overload on numerous cables in order to resolve electrical concerns from NRC Electrical Engineering Branch and SNL. NRC sent a letter to RBS dated November 15, 1999 stating that there are no remaining ampacity derating issues as identified in GL 92-08.

## 6.49 H.B. Robinson Steam Electric Plant, Unit 2

Hemyc ERFBS is used at H.B. Robinson Steam Electric Plant (Robinson) as a 1-hour fire rated barrier to protect conduits in accordance with Appendix R. The use of Hemyc at HBRSEP was granted by NRC in exemptions dated October 25, 1984 and October 17, 1990. Approximately 120 linear feet of Hemyc is installed in the Component Cooling Water Pump Room to protect the pump power cables. This application includes two 10.16 cm (4 in) and two 7.62 cm (3 in) conduits. Following notification of the Hemyc fire testing failures, the licensee considered these Hemyc barriers inoperable and undertook compensatory measures until such barriers could be determined operable. On June 10, 2005, the licensee notified that it intended to transition to NFPA 805 and would disposition any Hemyc related issues then. However as a proactive measure, the licensee intends to replace the Hemyc with an ERFBS that has been tested and qualified to the required rating per guidance provided in GL 86-10 supplement 1. By letter dated August 2, 2007, the licensee notified NRC that all Hemyc installations had been removed and replaced with a 1-hour fire rated 3M Interam E-54A ERFBS. NRC inspection staff verified the licensees' installation of the 3M ERFBS, documented in IR 05000261/2007007 and Exercise of Enforcement Discretion, dated December 20, 2007. In the event that H.B. Robinson does not transition, then they will have to submit exemption requests to the NRC providing justification for the specific Hemyc configurations, or modify these configurations to come into compliance. A discussion of NFPA 805 is provided above in Section 3, "ERFBS Regulations."

MT fire barrier material is used at Robinson to cover both sides of two penetration seals containing the steam generator blowdown lines. Therefore, this application of MT material is not used an ERFBS to protect cables, but to provide added thermal insulation for the expansion and contraction of the steam generator blowdown lines.

## 6.50 St. Lucie Plant, Units 1 and 2

Approximately 110 feet of Hemyc material is used at St. Lucie Plant as a noncombustible radiant energy shield inside Unit 2 containment to satisfy a license basis requirement for separation of safe shutdown cables in the event of a fire. In a safety evaluation dated March 27, 1984, NRC determined that installation of a 1-1/2 inch insulating blanket manufactured by B&B Insulation, Inc. for protection of cable tray configuration inside containment was acceptable. The noncombustible RESs are installed beneath the lowest redundant Division A & B cable trays at each elevation and all conduits inside Unit 2 Containment not separated by 20 feet are enclosed with a 1-hour fire-rated barrier. For the St. Lucie plant application, the noncombustible RES (Hemyc) are intended to deflect heat away from the protected cables so that it will dissipate into the voluminous containment atmosphere. As such, for cable tray applications, only the bottom and sides of the lowest cable tray are covered with the Hemyc material. The material is attached to the cable tray with staples such that the material will survive a design bases event (DBE). In some applications, the licensee uses Mecatiss instead of Hemyc to construct the RES.

B&B Insulation Inc. testing via ASTM E-119-80 and ANI/MAERP Bulletin No. 5 (79) rated the Hemyc barrier with a 1-hour fire rating.

The original application of Thermo-Lag 330-1 material at St. Lucie Plant consisted of 1-hour and 3-hour fire barrier conduit protection, 3-hour fire area boundary walls and ceilings, and containment radiant energy shields. The 1-hour and 3-hour protection installations for conduit consisted of one-half inch (minimum) thickness and 2.54 cm (1 in) (minimum) thickness of Thermo-Lag 330-1 preformed sections tie-wired or banded tot eh conduit, respectively. Thermo-Lag was used to achieve independence of electrical systems and for raceway fireproofing, along with a number of walls and wall sections to meet separation requirements.

St. Lucie Plant also uses Thermo-Lag to provide compliance with Appendix R requirements and to achieve 20ft separation of redundant circuits, Thermo-Lag is not an intervening combustible in any area where Appendix R, Section III.G.2(b) is credited for by design. The configurations used have been qualified by direct testing performed by Fire Protection Evaluation Records which documents the Thermo-Lag fire resistance pursuant to GL 86-10. In addition, the licensee has ensured that the quality of the Thermo-Lag installed was consistent with tested materials and that the critical properties and characteristics of the procured material is within acceptable limits. No cable trays are protected with Thermo-Lag material in either unit. Following issuance of Bulletin 92-01 the licensee declared all Thermo-lag barriers inoperable and implemented compensatory measures. To resolve the Thermo-lag inoperability, the licensee preformed the following measures:

- evaluate where Thermo-lag no longer needed
- re-route cables through separate fire area
- inside containment Thermo-lag applications were replaced with or encapsulated in stainless steel sheet metal to provide a RES
- verification of properly installed base layers,
- upgrades for the 1-hour barrier consisted of reinforcement of seams and joints through the addition of stress skins and tie wires and the addition of 1-quarter inch (minimum)

thickness Thermo-Lag 330-1 overlays for conduits smaller than three inches in diameter, and

- 3-hour barrier upgrades included reinforcement at conduit and support interfaces with the use of wire mesh or stress skin. An additional layers of Thermo-Lag 770-1 mat was applied over the base layer (number of layers depended on conduit size and cable fill) and an application of a Thermo-Lag 770-1 trowel grade top coat finish.

By letter dated June 23, 1998, the licensee notified NRC that all corrective actions associated with Thermo-Lag resolution had been completed. As identified in its June 23, 1998 letter, the licensee uses approximately 335 linear meters (1100 linear ft) of a 1-hour rated and 3 linear meters (10 linear ft) of a 3-hour rated Thermo-Lag barrier to protect conduits. No cable trays are protected with this barrier. In addition, $158m^2$ ($1700ft^2$) of Thermo-Lag material is used to provide 3-hour rated walls, floors and ceilings.

NRC SE dated March 26, 1999, approved the St. Lucie ampacity derating methods of analysis.

The licensee preformed combustibility loading calculations to document the acceptability of this increased combustible loading for each affected fire area and a 10 CFR 50.59 safety evaluation including an final safety analysis report change package was issued.

## 6.51 Salem Nuclear Generating Station, Units 1 and 2

In response to NRC GL 92-08, Salem Generating Station does not rely on Thermo-Lag 330-1 for fire protection purposes. SGS relies on approximately 1,981 m (6,500 feet) of fire barrier material including Kaowool, 3M FS-195, and 3M Interam™ E-50 Series ERFBS.

After NRC inspection of the Salem Generating Station between April 14 and 18, 1997, it was determined that SGS relied on the same fire wraps and configurations utilized since May of 1993. On June 6, 1997, SGS notified NRC that it had established a "Cable Raceway Fire Wrap Resolution Plan" to update and improve the plant's fire wrap system. By December 31, 2002, Salem reduced the amount of fire barrier materials used, including Kaowool, 3M FS-195, and 3M E-50, from 6,096 m (20,000 feet) of fire wrap to 1,981 m (6,500 feet) of fire wrap in order to improve their outdated fire barrier system and improve plant safety. The plant was updated and improved through the use of cross-ties, modified control circuits, cable re-routes, and removal of unnecessary Kaowool and 3M fire wrap.

## 6.52 San Onofre Nuclear Generating Station Units 2 and 3

San Onofre Nuclear Generating Station (SONGS) originally used Thermo-Lag 330-1 in four applications, two of which were only for Unit 1 reactor that has been permanently shutdown. The only application where Thermo-Lag was used for Appendix R purposes has been replaced. Additionally, the licensee used approximately $256ft^2$ of Thermo-Lag 330-1 as a non-cable protection related barrier; it has replaced that barrier with Pyrocrete 241.

Submittals and SER identify that the Cerablanket ERFBS did not meet the entire acceptance criteria delineated in the standard fire tests and only maintained circuit integrity for 49 minutes. NRC acceptance of the Cerablanket raceway fire barrier material was based on the existence of automatic fire detection and suppression systems in the areas in which the barrier material was installed, and the site fire department's ability to respond and initiate suppression activities. The SER concluded "that the deviations from BTP CMEB 9.5-1, associated with the barrier material, are not significant from a fire safety standpoint and are, therefore, acceptable."

Cerablanket has since been replaced by a 3M barrier. The criteria for acceptance of the 3M material was based on the cold-side temperature requirements of GL 86-10 and that the 3M ERFBS exceeded the endurance time of Cerablanket. In configurations where the 60 minute time rating could not be met, the licensee maintained the fire suppression system as described in the deviation.

## 6.53 Seabrook Station, Unit 1

Seabrook uses 3M Interam™ E-50 series fire wrap system as their sole ERFBS. All redundant train cables that required ERFBS protection are routed through conduit and protected with a 1-hour rated 3M Interam E-50 series ERFBS. The 1-hour fire endurance rating of the 3M ERFBS used at Seabrook were qualified by representative fire endurance tests, using ASTM E-119 and NRC GL 86-10 acceptance criteria. Following notification of 3M barrier fire endurance testing failures the licensee evaluated the impact on its utilization of the 3M ERFBS. It found 10 conduits with fill percentages ranging from 4% to 12% (the vendor identifying cable trays with less than 15% fill may result in a derated endurance rating). The supplemental testing results were extrapolated to evaluate the difference in the thickness of material used at Seabrook and the evaluation concluded that the installed 3M Interam E-50 series ERFBS for the identified 10 conduits were in fact capable of providing the rated 1-hour protection. All other conduits contained greater than 15% fill and therefore, met the new criteria. The smallest conduit protected with 3M material at Seabrook is 2.54 cm (1 in) in diameter and there are no cable trays which require protection by an ERFBS.

## 6.54 Sequoyah Nuclear Plant, Units 1 and 2

A general overview of TVA's resolution to Thermo-Lag usage is presented in Section 6.4 Browns Ferry Nuclear Plant, Units 1, 2 and 3, Sequoyah Nuclear Plant, Units 1 &2 above. As indicated in its June 30, 1999 letter to NRC, TVA completed its Thermo-Lag modifications to Sequoyah Nuclear Plant, which included; installation of an upgraded Thermo-Lag 330-1 barrier on conduits smaller than three inches, junction boxes, a cable tray and other applicable configurations identified via their Thermo-Lag testing program. In addition, TVA performed a Thermo-Lag reduction review which eliminated approximately 1300 linear feet of Thermo-Lag.

## 6.55 South Texas Project, Units 1 and 2

Thermo-Lag 330-1 fire barrier systems are used at South Texas Project (STP) to provide both 1-hour and 3-hour fire barrier separation of safe plant shutdown equipment outside containment. The material is also used to provide separation as a radiant energy shield and to achieve physical independence of electrical systems. STP installed Thermo-Lag 330-1 based on fire endurance tests from TSI which proved to be indeterminate. After further engineering evaluation of STP including fire endurance testing of both 1-and 3-hour configurations, STP determined that three different applications of Thermo-Lag needed to be removed.

There were 24 cable trays with Thermo-Lag installed at STP that required removal of the fire barrier material. The concern regarding the cable trays had to do with a potential impact on cable ampacity. There was also a configuration upgrade in Fire Area 07 of STP of one train of Qualified Display Processing System and sequencer control cables. Minor upgrades to the wall and box interface to meet the tested configuration have been implemented. The newly installed configurations now meet 1-hour fire rating standards consistent with 10CFR50, Appendix R. Thermo-Lag was also removed from inside the reactor containment buildings. A small amount of Thermo-Lag residue was left behind in order to prevent damage to various electrical and HVAC components.

Thermo-Lag is still utilized in several locations outside of the reactor building at South Texas Project. The material is used on supports, as a separation barrier to meet RG 1.75 separation requirements, and at other locations where there is not an impact on cable ampacity. The Thermo-Lag which remains at the plant is bounded by fire load analyses and is not considered a fire barrier.

In a letter to STP and the Houston Lighting and Power Company dated April 4, 1997, NRC informed the licensee that the necessary information and modifications regarding GL 92-08 had been provided and the actions tracked by TAC Numbers M85606 and M85607 were complete. By letter dated January 19, 1999, NRC acknowledged that any ampacity derating concerns at STP have been resolved in accordance with GL 92-08.

### 6.56 Virgil C. Summer Nuclear Station, Unit 1

Thermo-Lag 330 barrier material is used at five separate locations at Virgil C. Summer Nuclear Station (VC Summer). All five installations are installed to provide 1-hour fire barriers in order to meet separation requirements of 10CFR50, Appendix R. The fire barrier is used as panels to protect cable tray 3308, to protect a 7.62 cm (3 in) conduit that supplies DC power to the Main Control Board, to enclose two conduits which contained Nuclear Instrument signal cables, to protect Unistrut and threaded rods which suspend a M-board fire barrier, and to protect two conduits which provide "A" Train power to the "C" chiller within the plant.

Fire barrier configurations at VC Summer were installed in accordance with manufacturer's instructions and guidance and the plant itself had never performed any fire endurance testing. Ampacity testing in accordance with TSI in 1993 proved that there were significant concerns in regards to ampacity derating of the Thermo-Lag 330-1 installations. VC Summer was required to address installations of Thermo-Lag to resolve the ampacity derating issues.

One of the corrective actions VC Summer performed was to modify existing circuits to meet the requirements of 10CFR50 Appendix R without relying on Thermo-Lag. Cables were rerouted to meet separation criteria. Thermo-Lag was also removed from the VUL21A conduits and a 1-hour fire rated Gypsum board installed to maintain adequate circuit separation requirements for the HVAC system water chiller C transfer switch. In addition, a deviation request to Appendix R was submitted and approved by NRC for using Rockbestos Firezone R Cable to replace portions of safe shutdown circuits requiring protection to meet the requirements of 10CFR50, Appendix R. VC Summer also indicated that all installations of Thermo-Lag at the nuclear station were removed and replaced with an alternate fire barrier (e.g., 3M Interam).

In a letter dated May 5, 1998, NRC informed Virgil C. Summer Nuclear Station that all ampacity derating issues as well as Thermo-Lag corrections were resolved as identified by GL 92-08.

## 6.57 Surry Power Station, Units 1 and 2

Surry Power Station does not use any conventional ERFBS materials to protect redundant safe shutdown equipment. However, Pyrocrete manufactured by Carboline Company is used at Surry in conjunction with a Bio-K-10 mortar produced by Bio-Fire Protection, Inc on fiberglass piping.

Qualification of Pyrocrete 241 has been qualified by Thermal Transmission Test (ref. Tech. Report EP-001 1) that uses the ASTM E-119 fire exposure and the failure criteria of an average temperature of 250°F or single point temperature 325°F above ambient backside temperature. The barrier was installed prior to the issuance of GL 86-1 0 Supplement 1. Pyrocrete 241 has also been tested under UL designs N7I5, N716, N717, N7I8 and S706. The Bio K10 mortar has been tested in accordance with UL Design No. CAJ5006. Field installations of Pyrocete and Bio-K10 mortar do not deviate from tested configurations.

The Surry Power Station Units 1 and 2 use a combined total of 100 linear feet and 316 ft$^2$ of Thermo-Lag barriers for Appendix R compliance on electrical conduit only. Following the review related to GL 92-03, the licensee notified NRC on July 25, 1995, that Thermo-Lag 330-1 is no longer relied on for any 1-hour or 3-hour protection. PCI Promatec notified the NRC through public comment to this report that Surry also uses 3M Interam ERFBS.

## 6.58 Susquehanna Steam Electric Station (SSES), Units 1 and 2

Susquehanna Steam Electric Station (Susquehanna) originally used approximately 15,000 linear feet of Kaowool and Thermo-Lag ERFBS prior to identification of their deficiencies. The majority of Thermo-Lag was constructed using steel, 61 cm (24 inch) wide raceway and is intended for 1-hour applications. SSES relies on Thermo-Lag 330-1 to meet safe shutdown requirements of 10CFR50 Appendix R and to meet plant licensing commitments. Barrier elimination though analysis, modifications to the plant equipment and cabling, and modifications and upgrades to approximately 9,000 linear feet of ERFBS were completed by October 5, 1998. In addition, all Kaowool fire barriers were eliminated from the plant.

In a letter dated April 9, 1999, NRC provided response to Susquehanna that all ampacity derating issues dealing with cables wrapped in Thermo-Lag 330-1 were operating within reasonable limits and issues mentioned in GL 92-08 were resolved.

Susquehanna uses Darmatt material in only one of its fire barriers installed in the Unit 2 Reactor Building. Susquehanna has eliminated the use of Kaowool in its plant for compliance with Appendix R regulations.

## 6.59 Three Mile Island Nuclear Station, Unit 1

TMI-1 uses a combination of ERFBS to ensure separation of redundant trains in the same fire zones. Thermo-Lag was initially installed at the plant to comply with Appendix R requirements. Following implementation of a Thermo-Lag corrective action plan, the licensee upgraded or replaced Thermo-Lag installations with Mecatiss and in some applications requested exemptions for the continued use of Thermo-Lag. In addition, TMI was granted an exemption to Appendix R in several specific applications to allow the use of a Rockbestos fire resistant cable in lieu of a 1hour fire barrier.

NRC issued two Confirmatory Orders directing TMI-1 to complete final implementation of its corrective action program, which was confirmed to be completed by December of 1999. In January of 1999, NRC sent a letter to GPU Nuclear stating that all ampacity derating issues with regards to GL 92-08 were resolved at the Three Mile Island Nuclear Plant.

## 6.60 Turkey Point Nuclear Generating Units 3 and 4

Turkey Point Nuclear Generating Units 3 and 4 (Turkey Point) use Thermo-Lag ERFBS to provide raceway protection in accordance with Appendix R. The Thermo-Lag system designs are based on direct qualification testing or fire resistance equivalency evaluations performed using guidance from GL 86-10 and supplement 1. Approximately 14,608 linear feet and 1,287 sq ft of a 1-hour rated thermo-lag ERFBS and 602 linear feet and ≤38sq ft of a 3-hour rated thermo-lag ERFBS are used at Turkey Point.

Thermo-Lag 330-1 is used as an RES inside containments, approximately 700 feet in both units. To address the combustibility issues related to Thermo-Lag materials, the licensee replaced or encased the RES Thermo-lag barriers with stainless steel overlays. Outside containment, Turkey Point uses Thermo-Lag a 1-hour and 3-hour rated fire barriers. Approximately 600 feet in both units were upgraded with a Thermo-Lag 770-1 overlay of existing barriers.

Turkey Point requested exemption from upgrading Thermo-Lag fire barriers in 10 different locations within Units three and four. Exemptions were granted by NRC in a letter dated June 15, 1998 because either the fire zones and barriers met the 20 foot separation criteria or the cables located inside the raceways were not needed for the plant's safe shutdown. Fire Zone 106R was denied an exemption because there was only a 10 foot separation, falling short of the criteria from Appendix R. Turkey Point completed all planning and upgrades by December 31, 1999.

NRC SE dated March 26, 1999, approved the Turkey Point ampacity derating methods of analysis.

The samples provided to NEI for chemical composition testing indicated that the Turkey Point samples were consistent with other samples provide by industry and meet the acceptance criteria set by this testing program.

Exemption dated February 24, 1998 grants the use of a 25-minute fire rated ERFBS in lieu of a 1-hour fire barrier system as required by Section III.G.2 of Appendix R

Exemption dated October 8, 1998 grants the use of a 25-minute fire rated ERFBS in lieu of a 3-hour barrier as required by Section III.G.2 of Appendix R.

Exemption dated May 4, 1999, grants the use of a 25-minute fire rated ERFBS in lieu of a 3-hour barrier as required by Section III.G.2 of Appendix R, for fire zone 106R.

## 6.61 Vermont Yankee Nuclear Power Station

Vermont Yankee Nuclear Power Station (VYNPS) initially utilized Hemyc ERFBS to protect selected raceways to meet the separation requirements of 10 CFR 50, Appendix R. VYNPS completed its replacement of Hemyc ERFBS on systems credited in the VY Safe Shutdown Capability Analysis supporting compliance with Appendix R on July 28, 2005. Any Hemyc material that remains in the plant is not relied on for Appendix R compliance.

Thermo-Lag was used to a very limited extent at VYNPS (5 conduits two 3/4", one 3" and one 4" in a 3-hour configuration to achieve physical independence of electric power systems, approximately 48 linear feet). The two 3/4" conduits were removed as a result of equipment upgrades related to RG 1.97 suppression chamber water level and temperature upgrades. VYNPS did not perform its own independent testing of Thermo-Lag 330-1 ERFBS. VYNPS applied a 10% ADF based on ITL report No. 84-10-5 and a 50% ADF for all raceways that contain power cables. To eliminate reliance on Thermo-Lag 330-1 ERFBS at VYNPS, the licensee rerouted raceways to meet the 20 foot separation criteria (with detection and suppression). In other instances, the licensee replaced the existing Thermo-Lag materials with a qualified 3-hour fire wrap manufactured by 3M Company (3M Interam E-54A Flexible Fire Barrier Wrap). Based on derating factors provided by the vendor, VYNPS used an ADF of 30% in the design analysis for application of the 3M material. By letters dated June 1 and 28, 1993, the licensee informed NRC that it no longer relied on Thermo-Lag 330-1 material to achieve compliance with the requirements of 10 CFR 50, Appendix R. NRC notified VYNPS that actions required to address issues identified in GL 92-08 had been completed in its letter dated April 12, 1995.

Previous Hemyc or Thermo-Lag applications that required protection are now protected with 3M Interam® E-54A material used to construct ERFBS with a design rating of 1-hour. After receiving NRC IN 95-52 regarding 3M Interam test results, the licensee performed engineering evaluations on its use of 3M to determine that all 3M Interam® ERFBS installed at VY provide adequate fire barrier performance. In addition, the adequacy of the replacement barrier is also based on the vendors test results for conduit size and fire-wrap configurations that are determined to be bounded by acceptable test results summarized in IN 95-52. NRC inspection staff determined that there were no performance deficiencies associated with the licensees' replacement of the Hemyc material with the 3M Interam ERFBS.

### 6.62 Vogtle Electric Generating Plant, Units 1 and 2

Vogtle Electric Generating Plant (Vogtle) initially used approximately 247 linear meters (810 linear ft) and 102 m$^2$ (1100 ft$^2$) of 3-hour rated Thermo-Lag 330-1 ERFBS to protect raceways (conduits and cable trays) and junction boxes, respectively. The Thermo-lag was use to meet the requirements of Appendix R and to provide physical independence of electrical systems. Following the issuance of GL 92-08 the licensee notified NRC via letter dated May, 10, 1995, that upon further review and conversations with NRC staff, the licensee decided to remove Thermo-Lag 330-1 materials form its Vogtle plant. Vogtle decided to resolve the Thermo-Lag issue by rerouting circuits, re-evaluation of safe shutdown equipment requirements, and redefinition of existing fire areas.

3M Interam E53C, E54A, and E54C materials are used at Vogtle to construct ERFBS. These materials have been tested in accordance with GL 86-10, Supplement 1, as documented in test reports Omega Point Laboratory CTP-2005, 14540-99416; Vogtle Document AX3AJ08-00001.

Vogtle also protects some conduits and junction boxes with a cementitious spray-applied fire resistant coating per UL design Y707 and Y708. The masonry unit assemblies, composite assemblies of structural materials, and spray applied coatings were tested in accordance with UL Standard 263, "Fire Tests of Building Construction Material," which references ASTM E-119 and NFPR 251 tests.

### 6.63 Waterford Steam Electric Station, Unit 3

Waterford Steam Electric Station Unit 3 (Waterford 3) credits Hemyc ERFBS as a 1-hour fire rated barrier for Appendix R compliance along with using Hemyc in the containment building as a RES. Waterford 3 uses approximately 2000 feet of Hemyc conduit wrap installed on conduits ranging in sizes from ¾ to 5 inches in diameter. 1200 feet of Hemyc wrap installed on 24" wide cable trays, 7 electrical junction boxes and 5 containment electrical penetrations. Conduits, electrical junction boxes and containment penetration boxes are directly wrapped with Hemyc. Tray wrap is installed using the standard vendor design consisting of Hemyc wrap installed on a frame assembly that provides an air space between the wrap assembly and the raceway. Hemyc is credited in 19 fire areas/zones.

NRC approved Waterford 3 use of Hemyc wrap in SER NUREG-0787 Supplement 5 Section 9.5.1.4. That acceptance was based on testing performed by an independent laboratory using visual inspection and circuit integrity as acceptance criteria to the standard ASTM E-119 1-hour fire test exposure. However, based on NRC insights from its testing of Hemyc, the licensee determined that the Hemyc installed at Waterford 3 does not conform to the licensing basis and has been declared inoperable. While the Hemyc ERFBS remains inoperable the licensee has implemented the compensatory measures identified in its Technical Requirements Manual, which require a continuous fire watch on at least one side of the affected assembly, or verify the operability of the fire detectors on at least one side of the inoperable assembly and establish an hourly fire watch patrol. The licensee produced a Hemyc resolution plan, which includes qualification testing, resolution under NFPA 805, and partial replacement/upgrades. In the event that Waterford does not transition, then they will have to submit exemption requests to the NRC providing justification for the specific Hemyc configurations, or modify these configurations to come into compliance. A discussion of NFPA 805 is provided above in Section 3, "ERFBS Regulations."

Waterford 3 also uses the 3M Interam ERFBS in 1- and 3-hour rated configurations. The licensee stated that the 3M system is qualified by various fire tests conducted by independent testing laboratories consistent with the guidance provided in GL 86-10 supp 1. In addition, the licensee noted that, "3M Interam is the only ERFBS approved by Entergy for use in future installations at Waterford 3 Nuclear Station."

### 6.64 Watts Bar Nuclear Plant, Unit 1

Watts Bar Nuclear Plant Unit 1 (WBN) utilizes approximately 600 linear feet of Thermo-Lag 330-1 conduit fire barrier material to meet the requirements of 10CFR50 Appendix R, Section III G.2. The Thermo-Lag is applied to the Unit 1 Reactor Building Annulus.

WBN underwent a four phase testing program to analyze the condition of Thermo-Lag 330-1 utilized in the plant. Phase 1 involved testing and creating a set of acceptance criteria from the actual Thermo-Lag material used in fire, ampacity, and seismic testing. Phase 2 included chemical and physical material properties testing of previously procured Thermo-Lag 330-1 from circa 1985. Phase 3 involved additional procurement of Thermo-Lag during WBN fire barrier installation. Phase 4 included installation of 3-hour fire barrier material in the Watts Bar Nuclear Plant. Approximately 30 m (100 feet) of Thermo-Lag 770-1 was installed in the plant after being tested according to the criteria established in Phase 1.

Testing performed by the TVA showed that Thermo-Lag 330-1 is reliable for use in WBNP and could be used to protect electrical raceways in 1-hour fire protection applications. NRC gave notification on January 6, 1998 that the Watts Bar Nuclear Plant had sufficiently provided necessary information and performed the actions requested in GL 92-08 and action tracked by TAC number M85622.

Letter dated January 6, 1998 provides the SER related to the ampacity derating of Thermo-Lag 330-1 used in the WBN Unit 1. In that SER, the staff found that there are no significant safety hazards introduced with the use of ampacity test program results for cables enclosed by the subject Thermo-Lag fire barrier configurations at WBN Unit 1.

TVA also performed a detailed chemical analysis program of the Thermo-Lag material used in WBN Unit 1. Its four phase test program included Thermogravimetric analysis (TGA) (an empirical test method which develops a control decomposition curve for materials. This analysis provides verification that samples possess equivalent ratios of compounds.), Infrared (IR) spectroscopy (IR is used to identify organic and inorganic compounds in a material, this analysis demonstrates the distinct wave lengths as absorbed by specific compounds in the material), density (density testing is used to determine the weight and consistency of the material), Board Sear Strength (used to provide additional assurance of reliable mechanical properties for seismic qualification). Phase 1 verify consistency between various batches of material used in actual fire, ampacity, and seismic qualification testing. Phase II verify new old stock material is acceptable for use and made of consistent properties. Phase III testing of material prior to installations. Phase IV testing of thermo-lag 770-1.

## 6.65 Wolf Creek Generating Station, Units 1 and 2

Wolf Creek Generating Station (Wolf Creek) initially installed Thermo-Lag 330-1 type ERFBS to meet 10 CFR 50.48(a) requirements along with meeting the requirements of RG 1.75 physical independence of electrical system criteria. Approximately 675 linear feet of 1- and 3-hour Thermo-Lag 330-1 material was installed in cable trays and conduits in Wolf Creek to meet 50.48(e) and about 810 linear feet is used to meet RG 1.75. Following issuance of GL 92-08, detailed discussion with NRC staff and with other industry licensees, Wolf Creek chose to replace the Thermo-Lag barriers with an alternate material instead of justifying the continued use of Thermo-Lag. Wolf Creek resolved these thermo-lag issues by one for four solutions, (1) re-analysis of safe shutdown to demonstrate barrier is not needed, (2) modify the plant such that barrier is no longer needed for safe shutdown, (3) re-route wrapped conduits and raceways, or (4) replace Thermo-Lag 330-1 with Darmatt KM-1 material. Thermo-Lag 330-1 barriers not removed were left installed in the plant and evaluated and added to the plant combustibility loading for applicable fire areas. By letter dated June 23, 2006, NRC concluded that based on the licensees May 31, 2006 letter responding to GL 06-03, "the fire barrier systems at Wolf Creek, that separate redundant safe shutdown trains located within the same fire area, have been designed and installed in accordance with current NRC guidance."

NRC issued an SER to Wolf Creek on April 6, 1998, stating that based on the staffs review, the licensees ampacity derating analysis results are acceptable and there are no significant hazards associated with the licensees' ampacity derating methodology used at Wolf Creek.

Wolf Creek sent five (5) samples of Thermo-Lag material they use in the plant to NUCON International, Inc. for Pyrolysis Gas Chromatography testing. The test results showed that all five samples are consistent in terms of chemical composition.

## 7. Summary of Findings

(1) Use of ERFBS in NPPs is a direct result of the 1975 Brown Ferry Fire and the subsequent NRC fire protection regulations. The rush for NPPs to achieve compliance with the new regulation and wide use of ERFBS resulted in problems with proper testing, design, installation, maintenance, and ability of the barrier to perform its desired function.

(2) If ERFBS are properly designed, tested, configured, installed, inspected, and maintained, there is reasonable assurance that they will provide the fire resistance of the tested configuration.

(3) Plant specific deficiencies have been, and will continue to be found on occasion during routine licensee surveillances and NRC inspections. Fire protection defense in depth provides reasonable assurance that such deficiencies will not present an undue risk to the public health and safety.

(4) A large number of fire endurance tests have established the fire-resistive capabilities of the ERFBS material, designs, and constructions installed in NPPs. The test results support the conclusion that the regulatory requirements can be met by these fire barrier systems.

(5) Satisfactory NRC guidance on testing ERFBS, including performance, design, and acceptance criteria are available in Supplement 1 to GL 86-10. Availability of this guidance earlier would have eliminated many ERFBS problems identified in the past.

(6) The potential problems that were raised about ERFBS have been addressed. The staff did not find safety-significant plant-specific problems nor did it find problems with potential generic implications.

# 8. CONCLUSION

The implementation of Appendix R and acceptable use of ERFBS has taken considerable time to ensure the adequate protection of the systems needed to safely shutdown a plant in response to a fire. The promulgation of the rule, publishing of guidance to the licensees, confirmatory testing, and ongoing maintenance and inspection demonstrated that the ERFBS installed in nuclear power plants are sufficient to maintain defense-in-depth when combined with other defense-in-depth measures.

On the basis of the information found and assessed the staff concluded that the general condition of ERFBS used in industry is satisfactory. The staff did not find plant-specific problems of safety significance or concerns with generic implications. Even though the staff has concluded that the use of ERFBS in industry is satisfactory, it expects that plant-specific deficiencies will occasionally be found during future licensee surveillances and NRC inspections. However, likelihood of these occurrences is greatly reduced from the extensive historical testing and past NRC involvement with ERFBS issue resolution and clear NRC guidance.

# 9. DEFINITIONS

**Ablation** – the process of ablating, such as, surgical removal or by loss of part (as ice from a glacier or the outside of a nose cone) by melting or vaporization.[3]

**Air Drop** – lengths of electrical cable supported at each end with no use of continuous raceway support.[4]

**Ampacity Derating Factor** – A numeric value representing the fractional reduction from a base ampacity cable rating. Ampacity derating factors are associated with specific installation conditions.[2]

**Ampacity Correction Factor** – A numeric value equal to one minus the ampacity derating factor.[2]

**Cable Tray** – A raceway resembling a ladder (called ladder back) and usually constructed of steel or aluminum. Other styles of trays include solid-bottom and channel type.[2]

**Combustible Material** – Any material that will burn or sustain the combustion process when ignited or otherwise exposed to fire conditions.[1]

**Compensatory Measures** – interim step to restore operability or to otherwise enhance the capability of structures, systems, and components important to safety until the final corrective action is complete.[1]

**Electrical Raceway Fire Barrier System** – Non-load-bearing partition type envelope system installed around electrical components and cabling that are rated by test laboratories in hours of fire resistance and are used to maintain safe-shutdown functions free of fire damage.[1]

**Endothermic** – characterized by or formed with absorption of heat.[3]

**Exothermic** – characterized by or formed with evolution of heat.[3]

**Fire Barrier** – Those components of construction (walls, floors and their supports), including beams, joists, columns, penetration seals or closures, fire doors and fire dampers that are rated by approving laboratories in hours of resistance to fire and are used to prevent the spread of fire.[1]

**Fire Resistance** – The ability of an element of building construction, component, or structure to fulfill, for a stated period of time, the required load-bearing functions, integrity, thermal insulation, or other expected duty specified in a standard fire resistance test.[1]

**Fire-Resistance Rating** - The time that materials of a test assembly have withstood a standard ASTM E-119 fire exposure and have successfully met the established test acceptance criteria.[1]

**Free of Fire Damage** – the structure, system or component under consideration is capable of performing its intended function during and after the postulated fire, as needed.[1]

**Intumescence** – the property of a material to swell when heated: intumescent materials in bulk and sheet form are used as fire-proofing agents. [3]

**Noncombustible** – material which in the form in which it is used and under the conditions anticipated, will not ignite, burn, support combustion, or release flammable vapors when subjected to fire or heat.[1]

**Raceway** - An enclosed channel of metal or nonmetallic materials designed expressly for holding wires, cables, or busbars, with additional functions as permitted by code. Raceways include, but are not limited to, rigid metal conduit, rigid nonmetallic conduit, intermediate metal conduit, liquid-tight flexible conduit, flexible metallic tubing, flexible metal conduit, electrical nonmetallic tubing, electrical metallic tubing, underfloor raceways, cellular concrete floor raceways, cellular metal floor raceways, surface raceways, wireways, and busways.[1]

**Radiant Energy Shield (RES)** – is a shield designed to provide protection from redundant essential raceways or fire safe shutdown equipment against the radiant energy from an exposure fire. RES are typically installed within containment. [1]

**Standard Test Fire Exposure** – Fire exposure as specified in ASTM E-119 or NFPA 251.[7]

**Stress Skin** – A pretreated open weave carbon steel mesh used to provide a mechanical base for application of Thermo-Lag 330-1 bulk grade material during manufacture of prefabricated thermo-Lag materials It is also used as a mechanism to externally reinforced joints and seams between the prefabricated Thermo-Lag materials forming protective envelops. [6]

**Sublimation** – The process by which solids are transformed directly to the vapor state or vise versa without passing through the liquid phase.[6]

**Thermo-Lag** – A water-based, thermally-activated fire-resistant coating that operates on the principle of sublimation with partial intumescences. The performance of the product is based on the integrated effect of sublimation, heat blockage derived from endothermic reaction and decomposition and increased thermal resistance of a char layer developed through intumescences and the effect or reradiating.[6]

**Thermal-Short** – A path (typically metallic) where heat from an external source (fire) can be conducted into the ERFBS and cause failure of the protected electrical component.[4]

Definition References:
1. NRC Regulatory Guide 1.189, Rev. 1, "Fire Protection for NPPs," March 2007.
2. IEEE Standard 848-1996, "Standard Procedure for the Determination of the Ampacity Derating of Fire Protected Cables," The Institute of Electrical and Electronics Engineers, Inc., 345 East 47th Street, New York, NY 10017-2394, USA.
3. Merriam-Webster Dictionary, 2010
4. Author, 2009
5. NUREG-0800, SRP 9.5-1
6. TVA Thermo-Lag 330-1 Design Standard
7. GL 86-10 Supplement 1

# 10. REFERENCES

1. NRC NUREG-1742, Vol. 1, "Perspectives Gained from the Individual Plant Examinations of External Events (IPEEE) Program," April 2002.

2. NRC NUREG-1778, Draft Report for Comment, "Knowledge Base for Post-Fire Safe-Shutdown Analysis," January 2004

3. NRC NUREG/CR-5088, "Fire Risk Scoping Study: Investigation of Nuclear Power Plant Fire Risk, Including Previously Unaddressed Issues," January 1989.

4. NRC NUREG/CR-6042, Rev. 2, "Perspective on Reactor Safety," March 2002.

5. NUREG/CR-0596, "A Preliminary Report on Fire Protection Research Program Fire Barriers and Suppression (September 15, 1978, Test)."

6. NUREG-0800, "Standard Review Plan for the Review of Safety Analysis Reports for Nuclear Power Plants: LWR Edition," March 2009.

7. M.H. Salley, "An Examination of the Methods and Data Used to Determine Functionality of Electrical Cables When Exposed to Elevated Temperatures as a Results of a Fire in a Nuclear Power Plant," University of Maryland Masters of Science Thesis, 2000, U.S. NRC (ADAMS Accession No. ML051450082).

8. NRC Information Notice 84-09, "Lessons Learned From NRC Inspections of Fire Protection Safe Shutdown Systems (10 CFR 50, Appendix R)," February 13, 1984.

9. NRC Generic Letter 86-10, "Implementation of Fire Protection Requirements," April 24, 1986.

10. NRC Information Notice 91-47, "Failure of Thermo-Lag Fire Barrier Material to Pass Fire Endurance Test," August 6, 1991.

11. NRC Information Notice 91-79, "Deficiencies Found in Thermo-Lag Fire barrier Installation," December 6, 1991.

12. NRC Information Notice 92-18, "Potential for Loss of Remote Shutdown Capability During a Control Room Fire," February 28, 1992.

13. NRC Information Notice 92-46, "Thermo-Lag Fire Barrier Review Team Findings, Current Fire Endurance Tests, and Ampacity Calculation Errors," June 23, 1992.

14. NRC Information Notice 92-46, Attachment 1, "Final Report – Special Review Team for the Review of Thermo-Lag Fire Barrier Performance," April 21, 1992.

15. NRC Bulletin 92-01, "Failure of Thermo-Lag 330 Fire Barrier System to Maintain Cabling in Wide Cable Trays and Small Conduits Free from Damage," June 24, 1992.

16. NRC Information Notice 92-55, "Current Fire Endurance Test Results for Thermo-Lag Fire Barrier Material," July 27, 1992.

17. NRC Bulletin 92-01, Supplement 1, "Failure of Thermo-Lag 330 Fire barrier System to Perform its Specified Fire Endurance Function," August 28, 1992.

18. NRC Information Notice 92-82, "Results of Thermo-Lag 330-1 Combustibility Testing," December 15, 1992.

19. NRC Generic Letter 92-08, "Thermo-Lag 330-1 Fire Barriers," December 17, 1992.

20. NRC Information Notice 93-40, "Fire Endurance Test Results for Thermal Ceramics FP-60 Fire Barrier Material," May 26, 1993.

21. NRC Information Notice 93-41, "One Hour Fire Endurance Test Results for Thermal Ceramics Kaowool, 3M Company FR-195 and 3M Company Interam E-50 Fire Barrier Systems," May 28, 1993.

22. NRC Information Notice 94-22, "Fire Endurance and Ampacity Derating Test Results for 3-hour Fire Rated Thermo-Lag 330-1 Fire Barriers," March 16, 1994.

23. NRC Generic Letter 86-10, Supplement 1, "Fire Endurance Test Acceptance Criteria for Fire Barrier Systems Used to Separate Redundant Safe Shutdown Trains Within the Same Fire Area," March 25, 1994.

24. NRC Information Notice 94-22, "Thermo-Lag 330-660 Flexi-Blanket Ampacity Derating Concerns," May 13, 1994.

25. NRC Information Notice 91-79, Supplement 1, "Deficiencies Found in Thermo-Lag Fire barrier Installation," dated August 4, 1994.

26. NRC Information Notice 94-86, "Legal Actions Against Thermal Science, Inc., Manufacturer of Thermo-Lag," December 22, 1994.

27. NRC Information Notice 95-27, "NRC Review of Nuclear Energy Institute, Thermo-Lag 330-1 Combustibility Evaluation Methodology Plant Screening Guide," .May 31, 1995.

28. NRC Information Notice 95-32, "Thermo-Lag 330-1 Flame Spread Test Results," August 10, 1995.

29. NRC Information Notice 95-49, "Seismic Adequacy of Thermo-Lag Panels," October 27, 1995.

30. NRC Information Notice 95-52, "Fire Endurance Test Results for Electrical Raceway Fire Barrier Systems Constructed from 3M Interam Fire Barrier Materials," November 14, 1995

31. NRC Information Notice 94-86, Supplement 1, "Legal Actions Against Thermal Science, Inc., Manufacturer of Thermo-Lag," November 15, 1995.

32. NRC Information Notice 97-59, "Fire Endurance Test Results of Versawrap Fire Barriers," August 1, 1997.

33. NRC Information Notice 95-49, Supplement 1, Seismic Adequacy of Thermo-Lag Panels," December 10, 1997.

34. NRC Information Notice 95-52, Supplement 1, "Fire Endurance Test Results for Electrical Raceway Fire Barrier Systems Constructed from 3M Interam Fire Barrier Materials," March 17, 1998.

35. NRC Information Notice 2005-07, "Results of Hemyc Electrical Raceway Fire Barrier System Full Scale Fire Testing," dated April 1, 2005. (ADAMS Accession No. ML050890089)

36. NRC Generic Letter 2006-03, "Potentially Nonconforming Hemyc and MT Fire Barrier Configurations," April 10, 2006.

37. SECY 83-269, "Fire Protection Rules for Future Plants," July 5, 1983.

38. NRC SECY-99-204, "Kaowool and FP-60 Fire Barriers," August 4, 1999. (NUDOCS Accession No. 9909140100)

39. SECY-94-127, SECY-94-128, "Staff Requirements – Briefing on Status of Thermo-Lag (SECY-94-127, SECY 94-128), 10:00 A.M., Friday, May 20, 1994, Commissioners' Conference Room, One White Flint North, Rockville, Maryland (Open to Public Attendance)," dated May 26, 1994. (NUDOCS Accession No. 9406010141)

40. SECY-96-012, "Weekly Information Report – For the Weeks Ending January 5 and January 12, 1996," dated January 18, 1996.

41. OIG Inspection Report, "Inspection of NRC Staff's Acceptance and Review of Thermo-Lag 330-1 Fire Barrier Material," dated August 12, 1992, US NRC (NUDOCS Accession No. 9209250301)

42. NRC Regulatory Guide 1.189, Rev. 1, "Fire Protection for NPPs," March 2007.

43. NRC Appendix A to Branch Technical Position APCSB 9.5-1, "Guidelines for Fire Protection for NPPs Docketed Prior to July 1, 1976," August 23, 1976.

44. Sandia Report, SAND94-0146, "An Evaluation of the Fire Barrier System Thermo-Lag 330-1," Steven P. Nowlen, Steven Ross, September 1994.

45. NRC Technical Report, "Re-assessment of NRC Fire Protection Program," February 27, 1993. (NUDOCS Accession No. 9504190313)

46. SFPE Handbook of Fire Protection Engineering, $4^{th}$ ed., Society of Fire Protection Engineers, Quincy, Massachusetts, 02269, 2008.

47. American Nuclear Insurers, ANI Information Bulletin #5 (79), "ANI/MAERP Standard Fire Endurance Test Method To Qualify a Protective Envelope for Class 1E Electrical Circuits," July 1979.

48. American Society for Testing and Materials, Standard E-119, "Standard Test Methods for Fire Tests of Building Construction and Materials."

49. Underwriter Laboratories (UL) Standard Subject 1724, "Outline of Investigation for Fire Tests for Electrical Circuit Protection Systems."

50. American Society for Testing and Materials, Standard E-136, "Standard Test Method for Behavior of Materials in a Vertical Tube Furnace at 750 degree C."

51. American Society for Testing and Materials, Standard E-84, "Standard Test Method for Surface Burning Characteristics of Building Materials."

52. Letter from Robert M. Latta (U.S. NRC) to D.L. Farrar (Commonwealth Edison Company), "Safety Evaluation of 1-hour Fire-Rated Darmatt KM-1 Fire Barrier System Application at the LaSalle County Station," dated November 20, 1995. (NUDOCS Accession No. 9511270272)

53. Letter from J.C. Linville (U.S. NRC) to O.D. Kingsley (Exelon Generation Company, LLC), "NRC Triennial Fire Protection Inspection Report No. 50-352/01-14, 50-353/01-14," dated January 7, 2002. (ADAMS Accession No. ML020080162)

54. Letter from D.M. Skay (U.S. NRC) to O.D. Kingsley (Commonwealth Edison Company), "Closure of ampacity derating issues for fire barriers at Lasalle County Station, Units 1 and 2 (TAC NOS. MA3323 and MA 3324)" dated December 22, 1999. (ADAMS Accession No. ML993620265)

55. Letter from B.C. Buckley (U.S. NRC) to J.A. Hutton (PECO Energy Company), "Safety Evaluation Addressing Thermo-Lag Related Ampacity Derating Issues, Peach Bottom Atomic Power Station, Units 2 and 3, and Limerick Generating Station, Units 1 and 2 (TAC NOS. MA3404, MA3405, MA3872, and MA3873)" (ADAMS Accession No. ML003677255)

56. Letter from E.E. Fitzpatrick (Indiana Michigan Power) to U.S. NRC, "Donald C. Cook Nuclear Plant Units 1 and 2 Response to Generic Letter (GL) 92-08, Thermo-Lag 330-1 Fire Barriers," dated December 27, 1996. (NUDOCS Accession No. 9701060108)

57. Letter from M.A. Peifer (Indiana Michigan Power) to U.S. NRC, "Donald C. Cook Nuclear Plant Units 1 and 2 60 Day Response To Nuclear Regulatory Commission Generic Letter 2006-03: Potentially Nonconforming Hemyc And Mt Fire Barrier Configurations," dated June 1, 2006. (ADAMS Accession No. ML061600213)

58. Letter from J.F. Stang (U.S. NRC) to R.P. Powers (Indiana Michigan Power), "Completion of Licensing Action for Generic Letter 92-08, 'Thermo-Lag 330-1 Fire Barriers' for Donald C. Cook Nuclear Power Plant Units Nos. 1 and 2, TAC Nos. M85538 and M85539." (NUDOCS Accession No. 9808310265)

59. Letter from J.N. Hannon (U.S. NRC) to P. Gunter (Nuclear Information and Resource Service), "Completion Status for the 1998 Thermo-Lag 330-1 Confirmatory Orders (TAC No. MC2248)," dated April 23, 2004. (ADAMS Accession No. ML041120065)

60. Letter from A. Marion (NEI) to J.N. Hannon (NRC), "Promatec Hemyc 1-hour and MT 3-hour Fire Barrier Systems," dated April 25, 2001. (ADAMS Accession No. ML011220179)

61. NRC Inspection Report No. 50-400/99-13, "Fire Protection Inspection," dated February 3, 2000. (ADAMS Accession No. ML003685341)

62. Inspection Report No. 50-369/00-09 & 50-370/00-09, "McGuire Nuclear Station – NRC Inspection Report," dated December 15, 2000 (ADAMS Accession No. ML003778709)

63. Letter from A.Marion (NEI) to J.N. Hannon (NRC), "Promatec Hemyc 1-hour and MT 3-hour Fire Barrier Systems," dated April 25, 2001 (ADAMS Accession No. ML011220179)

64. Memorandum from D.C. Lew (NRC) to J. Hannon (NRC), "Preliminary Pass/Fail Test Results for Hemyc 1-hour Rated Electrical Raceway Fire Barrier Systems," Dated March 28, 2005 (ADAMS Accession No. ML050880176)

65. NRC Confirmatory Fire Performance Testing of Hemyc and MT ERFBS Test Reports and Related Documentation, dated April 29, 2005. (ADAMS Accession No. ML051190026)

66. Meeting Summary (US NRC), Summary of Meeting to Discuss Kaowool and FP-60 Fire Barriers, dated January 5, 2000 (ADAMS Accession No. ML003673753)

67. NRC Safety Evaluation (US NRC), "Safety Evaluation by The Office of Nuclear Reactor Regulation Related to Amendment No. 170," dated September 26, 2006.

68. Licensee Event Report 1999-014-03, "Kaowool Fire Barrier Outside 10 CFR 50 Appendix R Design Basis," dated May 17, 2006. (ADAMS Accession No. ML061420145)

69. Memorandum from E.W. Weiss (NRC) to R. Emch (NRC), "Review of Virgil C. Summer Nuclear Station Testing of Kaowool Fire Barrier Systems on December 28, 1999," dated November 6, 2000. (ADAMS Accession No. ML003766962)

70. NRC letter and SER, G.S. Vissing to M. Kansler (Entergy), James A. FitzPatrick Nuclear Power Plant, "Exemption from Certain Requirements of Section III.G.2.c of Appendix R to 10 CFR 50," dated May 29, 2001. (TAC No. MB0395) (ADAMS Accession No. ML010790125)

71. Test Report provided by Southern California Edison (SCE), "Test for fire protection for complete fire engulfment of cable trays and conduits containing grouped electrical conductors," report date October 24, 1978. (NUDOCS Accession No. 8403220024)

72. NRC letter to South Carolina Electric & Gas, "Review of Virgil C. Summer Nuclear Station Testing of Kaowool Fire Barrier Systems on December 28, 1999 (TAC No. MA9190)," dated February 20, 2001. (ADAMS Accession No. ML010510405)

73. NRC Letter to Florida Power Corporation, "Crystal River Nuclear Generating Plant Unit 3 – Review Of Report On Test Of Mecatiss Fire Barrier Material In Morestel, France, December 1994 (TAC No. M91772)," dated April 7, 1995. (NUDOCS Accession No. 9504140139)

74. Letter from to US NRC, from FPC, "Report on Test of Mecatiss Fire Barrier Material in Morestel France, December 1994," dated March 6, 1995. (NUDOCS Accession No. 9503130329)

75. Letter from G.L. Boldt (FPC) to US. NRC, "Response to NRC Staff Questions on Test of Mecatiss Fire Barrier Material in Morestel France, December 1994," dated May 17, 1995. (NUDOCS Accession No. 9505240353)

76. Letter from FPC, to US NRC, "Mecatiss Fire Barrier Endurance Test Results," dated March 30, 1996. (NUDOCS Accession No. 9604030021)

77. Safety Evaluation of Mecatiss Fire Barrier Test Program (TAC No. M91772), dated January 29, 1997 (NUDOCS Accession No. 9701310024)

78. Meeting Summary, "Summary Of Meeting on February 28, 1995, Regarding Thermo-Lag Resolution Issues," March 10, 1995. (NUDOCS Accession No. 9503160355)

79. Memorandum from E. Connell and P. Madden to K.S. West, "trip to underwriters laboratories (UL) – Florida Power Corporation Mecatiss Fire barrier fire endurance testing program (TAC No. M91772)," dated July 16, 1996. (NUDOCS Accession No. 9701140282)

80. NRC Inspection Report, "Fort Calhoun Station – NRC Triennial Fire Protection Inspection Report 05000285/2008009," dated September 24, 2008. (ADAMS Accession No. ML082690060)

81. NRC Inspection Report, "Surry Power Station- NRC Triennial Fire Protection Inspection Report 05000280/2006009 and 05000281/2006009," dated August 11, 2006. (ADAMS Accession No. ML062260007)

82. NRC Inspection Report, "NRC Inspection Report 50-275/97-17; 50-323/97-17 and Notice of Violation," dated October 29, 1997. (NUDOCS Accession No. 9711050064)

83. Letter from D.P. Dise (Niagra Mohawk Power Corporation) to T.A. Ippolito (NRC), "Submittal of ITL testing report of Pyrocrete," dated August 11, 1980. (NUDOCS Accession No. 8008180498)

84. Letter from D.P. Dise (Niagra Mohawk Power Corporation) to T.A. Ippolito (NRC), "Submittal of Vendor Report on Thermal Transmission of Pyrocrete 241 at Varying Thicknesses," dated April 17, 1979. (NUDOCS Accession No. 8005090346)

85. Fire Endurance Test, "Omega Point Fire Endurance Test of Versawrap Raceway Fire Barrier Systems for Conduits and Cable Trays," dated April 29, 1998. (http://www.nofire.com/approvals/TechManual_Aircraft_Nuclear_Utility.pdf)

86. Darchem Engineering, "Darchem Thermal Protection Brochure," Stillington, United Kingdom. http://www.esterline.com/Portals/8/Darchem/PDF/DTP_Brochure.pdf

87. SwRI Test Report 1208-001, "Nuclear Component Qualification Test Report for the Generic Seismic Qualification of 3M Interam E-50D 3-hour Fire Protection System," dated July 1986 (NUDOCS Accession No. 9308040186)

88. SwRI Test Report Project No. 01-7912(2), "Qualification Fire Test of a Protective Envelope System," dated June 1985. (NUDOCS Accession No. 9308040254)

89. IEEE Standard 848-1996, "Standard Procedure for the Determination of the Ampacity Derating of Fire Protected Cables," The Institute of Electrical and Electronics Engineers, Inc., 345 East 47$^h$ Street, New York, NY 10017-2394, USA.

90. Memorandum from T.E. Murley (Director NRR) to all 5 regional administrators, "Staff Review of Responses to NRC Bulletin 92-01 – Failure Of Thermo-Lag 330 Fire Barrier System," dated September 21, 1992. (NUDOCS Accession No. 9209300154)

91. Letter from John N. Hannon (U.S. NRR) to Paul Gunter (Reactor Watchdog Project), "Completion Status for the 1998 Thermo-Lag 330-1 Confirmatory Orders (TAC No. MC2248)," dated April 23, 2004. (ADAMS Accession No. ML041120065)

92. Letter from C.E. McCracken (U.S. NRC) to A.C. Thadani (U.S. NRC) "Results of Thermo-Lag 330 Combustibility Testing Performed by NIST, Report of Test FR 3989," dated October 8, 1992. (NUCOCS Accession No. 9210190176)

93. Letter from C.E. McCracken (U.S. NRC) to A. Marion (NEI), "Concludes that NRC Staff Will Not Accept use of NEI Application Guide to Justify use of Thermo-Lag Materials where Noncombustible Materials Specified by NRC Regulations," dated March 13, 1995. (NUDOCS Accession No. 9503200111)

94. Promatec Procedure IP-8400, "Fabrication of INSULCO/HEMYC Cable Protection System Components," dated March 7, 1985. (NUDOCS Accession No. 9308270288)

95. Meeting Summary, "Summary Of November 19, 1993, Meeting Between NRC Staff And NUMARC On Industry Thermo-Lag Fire Barrier Test Program," dated December 1, 1993. (NUDOCS Accession No. 9312130295)

96. Letter from R.P. Zimmerman (U.S. NRC) to C.K. McCoy (Georgia Power Company), "Follow-up To The Request For Additional Information Regarding Generic Letter 92-08, Issued Pursuant To 10 CFR 50.54(F)," dated December 30, 1994. (NUDOCS Accession No. 9501100141)

97. Memorandum from Commissioner I. Selin (U.S. NRC) to J.M. Taylor (U.S. NRC), "Inspector General's Inspection of the NRC Staff's Acceptance and Review of Thermo-Lag 330-1 Fire Barrier Material," dated August 17, 1992. (NUDOCS Accession No. 9209250291)

98. Letter from L.J. Callan (NRC) to T.E. Murley (NRC), "Request for additional information regarding generic letter 92-08, 'thermo-lag 330-1 fire barriers'," dated December 20, 1993 (NUDOCS Accession No. 9401050138)

99. Letter from L.J. Callan (NRC) to R.A. Stratman (Centerior Service Company), "Request For Additional Information Regarding Generic Letter 92-08, 'Thermo-Lag 330-1 Fire Barriers,' Pursuant to 10 CFR 50.54(f)," dated December 20, 1993. (NUDOCS Accession No. 9312300259)

100. Letter from Commonwealth Edison Co. to U.S. NRC, "Response to NRC Request for Additional Information Regarding Fire Testing of the Darmatt KM-1 Fire Protection System," dated June 2, 1995. (NUDOCS Accession No. 9506080643)

101. Memorandum from S. West (U.S. NRC) to L.B. March (U.S. NRC), "Trip Report – Trip to UL to observe three hour fire test of one and three hour Versawrap raceway fire barrier systems," dated May 9, 1997. (NUDOCS Accession No. 9705130324)

102. Transco Procedure No. TIOAP 9.20 for installation of KM-1 Fire Barrier Systems for Electrical Raceway Systems. (NUDOCS Accession No. 9501050344)

103. Letter from Commonwealth Edison Co. to U.S. NRC, "Corporate Quality Verification Dept Audit G-94-80, addressing Thermo-Lag Test Program," dated September 8, 1994. (NUDOCS Accession No. 9409130203)

104. Trip Report from D. Oudinot to K.S. West, "Trip Report Concerning Fire Endurance Testing of Darmatt KM-1 Fire Barrier Systems," dated February 14, 1996. (NUDOCS Accession No. 9602150357)

105. Letter from K.R. Cotton (U.S. NRR) to Steve Byrne (South Carolina Electric & Gas Co.), "Review of Virgil C. Summer Nuclear Station Testing of Kaowool Fire Barrier Systems on December 28, 1999 (TAC No. MA9190)," dated February 20, 2001. (ADAMS Accession No. ML010510405)

106. Letter from J.B. Archie (South Carolina Electric & Gas) to U.S. NRC, "Virgil C. Summer, 60-Day Response to NRC Generic Letter 2006-03: Potentially Non-Conforming Hemyc and MT Fire Barrier Configurations," dated June 5, 2006. (ADAMS Accession No. ML061590311)

107. Letter from J.B. Archie (South Carolina Electric & Gas) to U.S. NRC, "Virgil C. Summer, 60-Day Response to NRC Generic Letter 2006-03: Potentially Non-Conforming Hemyc and MT Fire Barrier Configurations – Request for Additional Information," dated August 4, 2006. (ADAMS Accession No. ML062220348)

108. U.S. NRC, "Edwin I. Hatch Nuclear Power Plant – NRC Triennial Fire Protection Inspection Report 05000321/2006006 and 05000366/2006006," dated May 31, 2006 (ADAMS Accession No. ML061520335)

109. Letter from C.E. McCracken (U.S. NRC) to A. Marion (NEI), "NEI Application Guide for Evaluation of Thermo-Lag 330 Fire Barrier Systems," dated October 16, 1995 (NUDOCS Accession No. 9510250078)

110. Letter from Union Electric to NRC, "Darmatt Qualification Tests," February 26, 1997 (NUDOCS Accession No. 9703100121)

111. Faverdale Technology Centre, "Test report for a 1 hour fire hose stream tests on Darmatt KM1 fire protection systems to ASTM E119 NRC GL 86/10 Supplement 1," dated January 9, 1994. (NUDOCS Accession No. 9801270098)

112. Letter From T.S.O'Neil (Exelon Generation Company, LLC) to U.S. NRC, "60-Day Response to Generic Letter 2006-03, 'Potentially Nonconforming Hemyc and MT Fire Barrier Configurations,'" dated June 6, 2006. (ADAMS Accession No. ML061640343)

113. Letter from J.E. Booker (Gulf States Utilities Company) to U.S. NRC, "NRC Bulletin 92-01, Supplement 1," dated April 14, 1993. (NUDOCS Accession No. 9304200036)

114. Letter and SE, "River Bend Station, Unit 1 – Thermo-Lag Related Ampacity Derating Issues," dated November 15, 1999. (ADAMS Accession No. ML993350552)

115. Letter from R.J. King (Entergy) to U.S. NRC, "NRC Generic Letter 2006-03, dated April 10, 2006, "Potentially Nonconforming Hemyc and MT Fire Barrier Configurations"," dated June 1, 2006. (ADAMS Accession No. ML061570394)

116. Letter from R.J. King (Entergy Operations, Inc.) to U.S. NRC, "Supplemental Response to Request for Additional Information (re: Ampacity Derating), Generic Letter 92-08," dated June 28, 1996. (NUDOCS Accession No. 9607080463)

117. Letter from Duquesne Light Company to U.S. NRC, "Response to Request for Additional Information for BVPS Unit No. 1, NRC Generic Letter 92-08, "Thermo-Lag 330-1 Fire Barriers," dated March 21, 1994. (NUDOCS Accession No. 9403280372)

118. Beaver Valley Power Station – NRC Inspection Report 50-344/02-04, 50-412/02-04, dated May 30, 2005. (ADAMS Accession No. ML021510069)

119. Letter from J.F. Lucas (Progress Energy) to U.S. NRC, "Response To NRC Generic Letter 2006-03, "Potentially Nonconforming Hemyc Ant) Mt Fire Barrier Configurations'," dated June 08, 2006. (ADAMS Accession No. ML061640136)

120. H.B. Robinson Steam Electric Plant - NRC Triennial Fire Protection Inspection Report 05000261/2007007 and Exercise of Enforcement Discretion, December 20, 2007. (ADAMS Accession No. ML073620541)

121. Letter from J.A. Stall (Florida Power and Light Company) to U.S. NRC, "Potentially Nonconforming Hemyc and MT Fire Barrier Configurations," dated June 9, 2006. (ADAMS Accession No. ML061640269)

122. Letter from J.A. Stall (Florida Power and Light Company) to U.S. NRC, "Generic Letter 92-08 Corrective Actions," dated June 23, 1998. (NUDOCS Accession No. 9806290310)

123. Letter from C.O. Thomas (U.S. NRC) to T.F. Plunkett (Florida Power and Light Company), "Closeout Report For The Ampacity Derating Issues Related To Generic Letter 92-08, 'Thermo-Lag 330-1 Fire Barriers' – St. Lucie Plants, Units 1 And 2, And Turkey Point, Units 3 And 4." Dated March 26, 1999. (NUDOCS Accession No. 9904020172)

124. Letter from A.E. Scherer (Southern California Edison) to U.S. NRC, "Response to Request for Additional Information on Response to Generic Letter 2006-03," dated June 18, 2007. (ADAMS Accession No. ML071710548)

125. Letter from G.St. Pierre (FPL Energy Seabrook, LLC) to U.S. NRC, "Response to Request for Additional Information Regarding Resolution of Generic Letter 2006-03, Potentially Nonconforming Hemyc and MT Fire Barrier Configurations," dated July 10, 2007. (ADAMS Accession No. ML071990101)

126. Letter from T.C. Feigenbaum (North Atlantic Energy Service Corporation) to U.S. NRC, "Response to Generic Letter 92-08 (TAC NO. 85603)," dated March 31, 1993. (NUDOCS Accession No. 9304130290)

127. Letter from P. Salas (TVA) to U.S, "Sequoyah Nuclear Plant (SQN) – Final Closeout Regarding Resolution of Thermo-Lag 330-1 Fire Barrier Upgrades," dated June 30, 1999. (NUDOCS Accession No. 9907080120)

128. Letter from S.E. Thomas (Houston Light & Power Company) to U.S. NRC, "Closure of Thermo-Lag Concerns at the South Texas Project," dated February 8, 1999. (NUDOCS Accession No. 9902120175)

129. Letter from U.S. NRC to W.T. Cottle (Houston Lighting & Power Company), "Completion of licensing action for generic letter 92-08, Thermo-Lag 330-1 fire barriers," dated April 4, 1997. (NUDOCS Accession No. 9704080234)

130. Letter from S.L. Rosen (Houston Light & Power Company) to U.S. NRC, "Thermo-Lag 330-1 Fire Barriers," dated April 16, 1993. (NUDOCS Accession No. 9304190275)

131. Letter from J.L. Skolds (SCE&G) to U.S. NRC, "Response to NRC Generic Letter 92-08 Thermo-Lag 330-1 Fire Barriers," dated February 17, 1993. (NUDOCS Accession No. 9302230141)

132. Letter from L.M. Padovan (U.S. NRC) to G.J. Taylor (Sothern Carolina Electric Company), "Thermo-Lag Related Ampacity Derating Issues and Completion Of Licensing Action For Generic Letter (GL) 92-08," dated May 5, 1998. (NUDOCS Accession No. 9805080208)

133. NRC letter and SE, "Thermo-Lag Related Ampacity Derating Issues and Completion of Licensing Action for Generic Letter (GL) 92-08, 'Thermo-Lag 330-1 Fire Barriers,' for Virgil C. Summer Nuclear Station," dated May 5, 1998. (NUDOCS Accession No. 9805080208)

134. Letter from G.T. Bischof (Virginia Electric and Power Co.) to U.S. NRC, "Kewaunee Unit 1, Millstone Units 2 & 3, North Anna Units 1 & 2, Surry Units 1 & 2, Generic Letter 2006-03, "Potentially Non-Conforming HEMYC and MT Fire Barrier Configurations, "Response to Request for Additional Information," dated May 31, 2007. (ADAMS Accession No. ML071520515)

135. Letter from U.S. NRC to J.P. O'Hanlon (Virginia Electric and Power Company), "Completion of Licensing Action for Generic Letter 92-08, 'Thermo-Lag 330-1 Fire Barriers,' for Surry Power Station Units 1 & 2," dated May 14, 1997. (NUDOCS Accession No. 9705190306)

136. Letter from R.E. Martin (U.S. NRC) to O.J. Zeringue (TVA), "Completion Of Licensing Action For Generic Letter 92-08, 'Thermo-Lag 330-1 Fire Barriers,'" And Supplemental Safety Evaluation Report On Ampacity Issues Related To Thermo-Lag Fire Barriers For Watts Bar Nuclear Plant, Unit 1. (NUDOCS Accession No. 9804090170)

137. Letter from D.V. Kehoe (TVA) to U.S. NRC, "Watts Bar Nuclear Plant, Request for Additional Information (RAI) Regarding GL 92-08, 'Thermo-Lag 330-1 Fire Barriers'," dated December 18, 1995 (ADAMS Accession No. ML072890454)

138. Letter from R.R. Baron (TVA) to U.S. NRC, "Response to RAI Regarding GL 92-08, 'Thermo-Lag Fire Barriers'," dated March 22, 1995. (NUDOCS Accession No. 9503310004)

139. Letter and SE, "Completion of Licensing Actions for Generic Letter 92-08 and Supplemental Safety Evaluation Report on Ampacity Issues Related to Thermo-Lag Fire

Barriers for Watts Bar Nuclear Plant, Unit 1," dated January 6, 1998. (ADAMS Accession No. ML073240205)

140. J.A. Fitzpatrick Nuclear Power Plant Exemption from the Requirements of 10 CFR Part 50, Appendix R, dated September 27, 2008. (ADAMS Accession No. ML062190377)

141. Letter from L.M. Stinson (Southern Company) to U.S. NRC, "NRC Generic Letter 2006-03 Response," dated June 9, 2006. (ADAMS Accession No. ML061600376)

142. Letter from M. Stevins (Luminant) to U.S. NRC, "Additional Information Provided Regarding NRC Generic Letter 2006-03, "Potentially Nonconforming Hemyc and MT Fire Barrier Configurations," dated December 20, 2007. (ADAMS Accession No. ML073620447)

143. Letter from D.W. Coleman (Energy Northwest) to U.S. NRC, "Completion of Thermo-Lag 330-1 Fire Barrier Corrective Actions," dated January 19, 2000. (ADAMS Accession No. ML003678400)

144. Letter from J.V. Parrish (WPPSS) to U.S. NRC, "Changes to Thermo-Lag 330-1 Resolution Plan," dated September 26, 1997. (ADAMS Accession No. ML041320336)

145. NRC Triennial Fire Protection Inspection Report 05000445/2008/006 and 05000446/2008006, Comanche Peak Steam Electric Station Units 1 and 2, dated July 3, 2008. (ADAMS Accession No. ML081890579)

146. Letter from K.D. Young (Ameren UE) to U.S. NRC, "Supplemental Information for the 60-day Response to NRC Generic Letter 2006-03," dated July 17, 2006. (ADAMS Accession No. ML062060383)

147. Letter from G.F. Dick (U.S. NRC) to O.D. Kingsley (Commonwealth Edison Group), "Thermo-Lag Related Ampacity Derating Issues, Byron Station, Units 1 and 2, and Braidwood Station, Units 1 and 2," dated November 2, 1999. (ADAMS Accession No. ML993200165)

148. Safety Evaluation Related to Generic Letter 92-08 Ampacity Derating Issues Byron Station Units 1 and 2, Braidwood Station Units 1 and 2, dated November 2, 1999. (ADAMS Accession No. ML993200165)

149. Letter for T.W. Alexion (U.S. NRC) to J.W. Yelverton (Entergy Operations, Inc.), "Response to NRC Bulletin 92-001, Supplement 1, 'Failure of Thermo-Lag 330 Fire Barrier System'", dated January 22, 1993. (NUCOCS Accession NO. 9301270011)

150. Letter from G. Kalman to U.S. NRC, "Response to GL 92-08, Regarding Configurations and Amounts of Thermo-Lag Fire Barriers Installed in Plant and Cable Loading," dated June 21, 1994. (NUDOCS Accession No. 9406270024)

151. Letter from T.A. Marlow (Entergy Operations, Inc.) to U.S. NRC, "Response to Generic Letter 2006-03, Potentially Nonconforming Hemyc and MT Fire Barrier Configurations," dated June 7, 2006. (ADAMS Accession No. ML061720459)

152. Letter from D. Holland (U.S. NRC) to J.S. Forbes (Entergy Operations, Inc.), "Evaluation of Arkansas Nuclear One, Units 1 and 2, Response to Generic Letter 2006-03, 'Potentially Nonconforming Hemyc and MT Fire Barrier Configurations'," dated September 29, 2006. (ADAMS Accession No. ML062620115)

153. Letter from J.D. Sieber (Duquesne Light Company) to U.S. NRC, "Provides Response to Request for Additional Info. Re. GL 92-08, 'Thermo-Lag 330-1 Fire Barriers,'" dated March 21, 1994. (NUDOCS Accession No. 9403280372)

154. Beaver Valley Power Station – NRC Inspection Report 50-344/02-04, 50-412/02-04, dated May 30, 2005. (ADAMS Accession No. ML021510069)

155. Letter from G.S. Thomas (Duquesne Light Company) to U.S. NRC, "Response to NRC Follow-up to Request for Additional Info. Re. GL 92-08, 'Thermo-Lag 330-1 Fire Barriers,' for BVPS Unit 2," dated December 22, 1994. (NUDOCS Accession No. 9412300237)

156. Letter from Duquesne Light Company to U.S. NRC, "Response to Request for Additional Information Regarding Generic Letter 92-08, 'Thermo-Lag 330-1 Fire Barrier'," dated December 22, 1994 (NUDOCS Accession No. 9412290093)

157. Letter from D.J. Chrzanowski (Commonwealth Edison) to T.E. Murley (U.S. NRC), "Forwards Response to GL 92-08, Thermo-Lag 330-1 Fire Barriers," dated April 16, 1993. (NUDOCS Accession No. 9304260035)

158. Letter from K.L. Kaup (Commonwealth Edison) to U.S. NRC, "Braidwood Station Status Update of Thermo-Lag (GL 92-08) Issues," dated September 25, 1995. (NUDOCS Accession No. 9510020266)

159. Letter from G.F. Dick, Jr. (U.S. NRR) to I.M. Johnson (Commonwealth Edison), "Completion of Licensing Action for GL 92-08 – Braidwood Station, Units 1 and 2," dated May 13, 1997. (NUDOCS Accession No. 9705150355)

160. Letter from D.M. Benyak (AmerGen Energy, Exelon Generation Company, LLC) to U.S. NRC, "Additional Info. Supporting the 60-Day Response to Generic Letter 2006-03, Potentially Nonconforming Hemyc and MT Fire Barrier Configurations," dated May 31, 2007. (ADAMS Accession No. ML071520085)

161. Letter from G.W. Morris (Tennessee Valley Authority) to U.S. NRC, "Browns Ferry, Sequoyah, Watts Bar, Potentially Non-Conforming Hemyc and MT Fire Barrier Configurations – 60 Day Response," dated June 7, 2006. (ADAMS Accession No. ML061600208)

162. Letter from D.E.Martin (Office of Nuclear Materials Safety and Safeguards) to S.A. Toelle (U.S. Enrichment Corporation), "Response: Request for Additional Information," dated May 31, 2006. (ADAMS Accession No. ML061500208)

163. Letter from R.D. Machon (Tennessee Valley Authority) to U.S. NRC, "Response: Request for Additional Information Regarding Generic Letter (GL) 92-08, Thermo-Lag 330-1 Fire Barriers," dated March 22, 1995. (NUDOCS Accession No. 9503290089)

164. Letter from W.O. Long (U.S. NRC) to J.A.Scalice (Tennessee Valley Authority), "Forwards SE Which Constitutes Staff Review and Approval of TVA Ampacity Derating Test and Analyses for Thermo-Lag Fire Barrier Configurations," dated July 16, 1999. (NUDOCS Accession No. 9907210115)

165. Letter from D.V. Kehoe (Tennessee Valley Authority) to U.S. NRC, "Results of Thermo-Lag Testing and Notified NRC that Work Associated with Thermo-Lag Completed for Facility, Per GL 92-08," dated December 18, 1995. (ADAMS Accession No. ML072890454)

166. Letter from R.R. Baron (Tennessee Valley Authority) to U.S. NRC, "WBN – Request for Additional Information Regarding Generic Letter (GL 92-08), Thermo-Lag 330-1 Fire Barriers," dated March 22, 1995. (NUDOCS Accession No. 9503310004)

167. Letter from R.E. Martin (U.S. NRC) to O.J. Zeringue (Tennessee Valley Authority), "Completion of Licensing Action for GL 92-08, Thermo-Lag 330-1 Fire Barriers and Supplemental Safety Evaluation Report on Ampacity Issues Related to Thermo-Lag Fire Barriers for Watts Bar Nuclear Plant Unit 1," dated January 6, 1998. (ADAMS Accession No. ML073240205)

168. Letter from Carolina Power and Light to U.S. NRC, "Request for Additional Information Regarding Generic Letter 92-08, 'Thermo-Lag 3301-1 Fire Barrier,' Pursuant to 10 CFR 50.54(F) – Brunswick Steam Electric Plant, Units 1 and 2, and Shearon Harris Nuclear Power Plant, Unit 1," dated February 15, 1994. (NUDOCS Accession No. 9402180274)

169. Letter from L.O. DelGeorge (Commonwealth Edison) to U.S. NRC, "Forwards Response to NRC Request for Additional Information re GL 92-08, 'Thermo-Lag 330-1 Fire Barriers,'" dated February 10, 1998. (NUDOCS Accession No. 9402250179)

170. Letter from J. Hosmer (Commonwealth Edison) to U.S. NRC, "Provides Response to Request for Additional Information re GL 92-08, 'Thermo-Lag 330-1 Fire Barriers' per 10CFR50.54(f)," dated December 16, 1994. (NUDOCS Accession No. 9412220241)

171. Letter from J.B. Hosmer (Commonwealth Edison) to U.S. NRC, "Provides Submitted Information as Update to NRC on Status of Actions to Address GL 92-08, 'Thermo-Lag 330-1 Fire Barriers,'" dated January 17, 1997. (NUDOCS Accession No. 9701270122)

172. Letter from T.S. O'Neill (AmerrGen Energy Co., LLC, Exelon Nuclear) to U.S. NRC, "60-Day Response to Generic Letter 2006-03, 'Potentially Nonconforming Hemyc and MT Fire Barrier Configurations,'" dated June 6, 2006. (ADAMS Accession No. ML061640343)

173. Letter from Donald F. Schnell (Union Electric) to U.S. NRC, "Closure Notification of Thermo-Lag Fire Barrier System Issues for Plant," dated January 31, 1997. (NUDOCS Accession No. 9702070238)

174. Letter from Donald F. Schnell (Union Electric) to U.S. NRC, "Forwards Response to NRC 931221 RAI re GL 92-08, "Thermo-Lag 330-1 Fire Barriers," dated February 10, 1994. (NUDOCS Accession No. 9402170325)

175. Letter from J.S. Perry (Illinois Power Company) to S.A. Varga (U.S. NRC), "Response to 931227 Request for Additional Information re GL 92-08, Thermo-Lag 330-1 Fire Barriers," dated February 9, 1994. (NUDOCS Accession No. 9403040225)

176. Letter from J.B. Hopkins (U.S. NRC) to J.V. Sipek (Clinton Power Station), "Completion of Licensing for Generic Letter 92-08, Thermo-Lag 330-1 Fire Barriers," dated September 29, 1998. (NUDOCS Accession No. 9810020229)

177. Letter from T.S. O'Neill (AmerGen Energy Company, LLC and Exelon Nuclear) to U.S. NRC, "60-Day Response to Generic Letter 2006-03, 'Potentially Nonconforming Hemyc and MT Fire Barrier Configurations,'" dated June 6, 2006. (ADAMS Accession No. ML061640343)

178. Letter from D.M. Benyak (AmerGen Energy Company, LLC and Exelon Nuclear) to U.S. NRC, "Braidwood Station, Units 1 and 2, and Clinton, Unit 1, Additional Information Supporting the 60-Day Response to Generic Letter 2006-03, 'Potentially Nonconforming Hemyc and MT Fire Barrier Configurations,'" dated May 31, 2007. (ADAMS Accession No. ML071520085)

179. Letter from W.S. Oxenford (Energy Northwest) to U.S. NRC, "Columbia, Response to Generic letter 2006-003, 'Potentially Nonconforming Hemyc and MT Fire Barrier Configurations,'" dated June 9, 2006. (ADAMS Accession No. ML061710470)

180. Letter from C.G. Sorenson (Washington Public Power Supply System) to U.S. NRC, "Columbia Response to Generic letter 92-08, 'Thermo-Lag 330-1 Fire Barriers,'" dated April 13, 1993. (ADAMS Accession No. ML041320331)

181. Letter from D.H. Jaffe (U.S. NRC) to C.L. Terry (Texas Utilities Electric Company), "Forwards Safety Evaluation Accepting Licensee Response to GL 92-08, 'Thermo-Lag 330-1 Fire Barriers,' dated 921217, for Comanche Peak Electric Station, Unit 1," dated May 14, 1999. (NUDOCS Accession No. 9905190038)

182. Letter from F.W. Madden (TXU Power) to U.S. NRC, "Comanche Peak – 60-Day Response to NRC Generic Letter 2006-03, 'Potentially Nonconforming Hemyc and MT Fire Barrier Configurations," dated June 8, 2006. (ADAMS Accession No. ML061660092)

183. Letter from G.R. Horn (Nebraska Public Power District) to U.S. NRC, "Response to Additional Information re GL 92-08, 'Thermo-Lag 330-1 Fire Barriers – 10CFR50.54(f),'" dated February 9, 1994. (NUDOCS Accession No. 9402160296)

184. Letter from J.R. Hall (U.S. NRC) to G.R. Horn (Nebraska Public Power District), "Advises that Util. 930416 Response to GL 92-08, 'Thermo-Lag 330-1 Fire Barriers' Acceptable and Actions Complete," dated May 30, 1995. (NUDOCS Accession No. 9506140408)

185. Letter from G.R. Horn (Nebraska Public Power District) to U.S. NRC, "Forwards Response to GL 92-08, 'Thermo-Lag 330-1 Fire Barriers,' Including Qualification of Thermo-Lag Fire Barriers and Basis for Ampacity Derating Factors Protected by Thermo-Lag Barriers," dated April 16, 1993. (NUDOCS Accession No. 9304210293)

186. Letter from P.M. Beard, Jr. (Florida Power Corporation) to U.S. NRC, "Response to Additional Information Request on Generic Letter 92-08, 'Thermo-Lag Fire Barriers,'

Pursuant to 10 CFR 50.54(f)," dated February 9, 1994. (NUDOCS Accession No. 9402160141)

187. Letter from J.P. Cowan (Florida Power Corporation) to S.J. Collins (U.S. NRC), "Response to NRC Letter 'Confirmatory Order Modifying License' Regarding the Final Implementation Date for the Crystal River Unit 3 (CR-3) Thermo-Lag Resolution Program," dated May 25, 2000. (ADAMS Accession No. ML003722384)

188. Memorandum from J.A. Calvo (U.S. NRC) to F.J. Hedbon (Florida Power Corporation), "Safety Evaluation Report Addressing Thermo-Lag Related Ampacity Derating Issues for Crystal River (TAC No. M91772)," dated November 7, 1997. (NUDOCS Accession No. 9711200235)

189. Letter from J.K. Wood (Centerior Energy) to L.J. Callan (U.S. NRC), "Response to 10CFR50.54(f) Request for Additional Information Regarding Generic Letter 92-08, 'Thermo-Lag 330-1 Fire Barriers' (TAC No. M85542)," dated February 11, 1994. (NUDOCS Accession No. 9402240321)

190. Letter from D. Jacobs (Pacific Gas and Electric Company) to U.S. NRC, "60-Day Response to NRC Generic Letter 2006-03, 'Potentially Nonconforming Hemyc and MT Fire Barrier Configurations,'" dated June 6, 2006. (ADAMS Accession No. ML061720079)

191. Letter from E.E. Fitzpatrick (Indiana Michigan Power Company) to U.S. NRC, "Donald C. Cook Nuclear Plant Units 1 and 2 Response to Generic Letter 92-08, 'Thermo-Lag 330-1 Fire Barriers,'" dated December 27, 1996. (NUDOCS Accession No. 9701060108)

192. Letter from M.A. Pfeifer (Indiana Michigan Power) to U.S. NRC, "60-Day Response to NRC Generic Letter 2006-03: Potentially Nonconforming Hemyc and MT Fire Barrier Configurations," dated June 1, 2006. (ADAMS Accession No. ML061600213)

193. Letter from J.F. Stang (U.S. NRC) to R.P. Powers (Indiana Michigan Power Company), "Closeout of Generic Letter 92-08 Related Issues Involving Ampacity Derating of Thermo-Lag (TAC Nos. MA3387 and MA3388)," dated July 14, 1999. (NUDOCS Accession No. 9907220068)

194. Dresden Nuclear Power Station, "Dresden Station Units 2 and 3, 3M Fire Wrap Qualification Evaluation, NTSC Project 99-40540, Dresden Report No. 12-N208-05," dated December 31, 1999. (ADAMS Accession No. ML011910445)

195. Letter from T.S. O'Neill (AmerGen Energy Company, LLC and Exelon Nuclear) to U.S. NRC, "60-Day Response to Generic Letter 2006-03, 'Potentially Nonconforming Hemyc and MT Fire Barrier Configurations,'" dated June 6, 2006. (ADAMS Accession No. ML061640343)

196. Letter from J.F. Franz, Jr. (IES Utilities Inc.) to S.J. Collins (U.S. NRC), "Forwards DAEC Thermo-Lag Final Resolution Report Which Summarizes Thermo-Lag Installations," dated October 31, 1997. (NUDOCS Accession No. 9711130377)

197. Letter from J.A. Stall (Florida Power and Light Company) to U.S. NRC, "60-Day Response to NRC Generic Letter 2006-03, Potentially Nonconforming Hemyc and MT Fire Barrier Configurations," dated June 9, 2006. (ADAMS Accession No. ML061640269)

198. Letter from J.F. Franz (IES Industries Inc.) to L.J. Callan (U.S. NRC) "Response To NRC Request For Additional Information Regarding Generic Letter 93-08, 'Thermo-Lag 330-1 Fire Barriers', Pursuant To 10CFR50.54(F) – Duane Arnold Energy Center," Dated February 14, 1994. (NUDOCS Accession No. 9403010160)

199. Letter from K.Young (IES Utilities Inc.) to W.T.Russel (U.S. NRC) "Fire Endurance Testing of Thermo-Lag 330-1 Fire Barrier Material," dated November 22, 1994) (9711130377 – letter from J.F.Franz (IES Utilities Inc.) to S.J. Collins (U.S. NRC) "Thermo-Lag Final Resolution Report," dated October 31, 1997. (NUDOCS Accession No. 9412050112)

200. Letter from R.J. Laufer (U.S. NRC) to E.Protsch (IES Utilities) "Safety Evaluation Addressing Thermo-Lag Related Ampacity Derating Issues – Duane Arnold Energy Center (TAC NO. 82809)," Dated January 26, 1999. (NUDOCS Accession No. 9901290062)

201. Letter from D. Morey (Southern Nuclear Operating Company, Inc) to U.S. NRC, "Joseph M. Farley Nuclear Plant Plans to Address Kaowool Issues Kaowoll Fire Barrier Meeting Request," dated May 1, 2000. (ADAMS Accession No. ML003712374)

202. Letter from D.K. Cobb (Detroit Edison Energy) to U.S. NRC, "Fermi 2 Response to Generic Letter 2006-03, Potentially Nonconforming HEMYC and MT Fire Barrier Configurations," dated June 9, 2006. (ADAMS Accession No. ML061660087)

203. Letter from P. Dietrich (Entergy Nuclear Northwest) to U.S. NRC, "Response to Generic Letter 2006-03, Potentially Nonconforming Hemyc and MT Fire Barrier Configurations," dated June 7, 2006. (ADAMS Accession No. ML061650025)

204. Letter from G.S. Vissing (U.S. NRC) to M. Kansler (Entergy Nuclear Operations, Inc), "Exemption from the Requirements of Section III.G.2.c of Appendix R to 10 CFR Part 50 (TAC No. MB0395," dated May 29, 2001. (ADAMS Accession No. ML010790125)

205. Letter from H.J. Faulhaber (Omaha Public Power District) to U.S. NRC, "Fort Calhoun Station Unit No. 1, Supplemental Response to General Letter 2006-03, 'Potentially Nonconforming Hemyc and MT Fire Barrier Configurations,'" dated August 2, 2006. (ADAMS Accession No. ML070850193)

206. Letter from J.M. Heffley (Constellation Energy Group) to U.S. NRC, "Calvert Cliffs and Nine Mile Point, Units 1 and 2 and R.E. Ginna – Response to Generic Letter 2006-03, Potentially Nonconforming Hemyc and MT Fire Barrier Configurations," dated June 9, 2006. (ADAMS Accession No. ML061650026)

207. Letter from B. Vaidya (U.S. NRC) to W.R. Bryan (Entergy Operations, Inc), "Grand Gulf Nuclear Station – Issuance of Amendment Re: Proposed Resolution of Kaowool Issues (TAC No. MC8180)," dated September 29, 2006. (ADAMS Accession No. ML062140354)

208. Letter from C.J. Gannon, Jr. (Progress Energy) to U.S. NRC, "Shearon Harris Unit 1 30-Day Response to NRC Generic Letter 2006-03, 'Potentially Nonconforming Hemyc and MT Fire Barrier Configurations,'" dated April 28, 2006. (ADAMS Accession No. ML061240052)

209. Letter from C.J. Gannon, Jr. (Progress Energy) to U.S. NRC, "Shearon Harris Nuclear Power Plant Unit 1 60-Day Response to NRC Generic Letter 2006-03, 'Potentially Nonconforming Hemyc and MT Fire Barrier Configurations,'" dated June 9, 2006. (ADAMS Accession No. ML061710062)

210. Letter from L.M. Stinston (Southern Nuclear Operating Company, Inc.) to U.S. NRC, "Joseph M. Farley, Edwin I. Hatch, and Vogtle, NRC Generic Letter 2006-03 Response," dated June 9, 2006. (ADAMS Accession No. ML061600376)

211. U.S. NRC, "Edwin I. Hatch Nuclear Power Plant – NRC Integrated Inspection Report 50-321/01-03, 50-366/01-03," dated July 26, 2001. (ADAMS Accession No. ML012080121)

212. Letter from G.P. Barnes (PSEG Nuclear, LLC) to U.S. NRC, "Hope Creek Response to Generic Letter 2006-03, 'Potentially Nonconforming Hemyc and MT Fire Barrier Configurations,'" dated June 7, 2006. (ADAMS Accession No. ML061660080)

213. Letter from M.A. Cunningham (U.S. NRC) to J.A. Grobe (U.S. NRC), "Hemyc and MT Electrical Raceway Fire Barrier System Closeout Actions," dated December 17, 2008. (ADAMS Accession No. ML083090083)

214. Letter from F.R. Dacimo (Entergy) to U.S. NRC, "Indian Point, Units 2 and 3, Response to Generic Letter 2006-03, Potentially Nonconforming Hemyc and MT Fire Barrier Configurations," dated June 8, 2006. (ADAMS Accession No. ML061720091)

215. Letter from G.T. Bischof (Virginia Electric and Power Company) to U.S. NRC, "Kewaunee Unit 1, Millstone Units 2 and 3, North Anna Units 1 and 2, Surry Units 1 and 2, Generic Letter 2006-03, 'Potentially Nonconforming HEMYC and MT Fire Barrier Configurations' Response to Request for Additional Information," dated May 31, 2007. (ADAMS Accession No. ML071520515)

216. Letter from D.J. Chrzanowski (Commonwealth Edison) to T.E. Murley (U.S. NRC), "Forwards Response to GL 92-08, 'Thermo-Lag 330-1 Fire Barriers,'" dated April 16, 1993. (NUDOCS Accession No. 9304260035)

217. Letter from T.S. O'Neill (AmerGen Energy Company, LLC and Exelon Nuclear) to U.S. NRC, "60-Day Response to Generic Letter 2006-03, 'Potentially Nonconforming Hemyc and MT Fire Barrier Configurations,'" dated June 6, 2006. (ADAMS Accession No. ML061640343)

218. Letter from D.M. Skay (U.S. NRC) to I. Johnson (Commonwealth Edison Company), "Completion of Licensing for Generic Letter 92-08, 'Thermo-Lag 330-1 Fire Barriers – LaSall County Station (TAC Nos. M85563 and M85564),'" dated April 23, 1997. (NUDOCS Accession No. 9704290218)

219. Letter from G.A. Hunger, Jr. (PECO Energy Co.) to U.S. NRC, "Peach Bottom, Units 2 and 3 and Limerick, Units 1 and 2, Request for Additional Information Regarding Generic Letter 92-08, 'Thermo-Lag 330-1 Fire Barriers,'" dated February 4, 1994. (ADAMS Accession No. ML041040462)

220. Letter from G.A. Hunger, Jr. (PECO Energy Co.) to U.S. NRC, "Peach Bottom Atomic Power Station, Units 2 and 3, Limerick Generating Station, Units 1 and 2, Response to

NRC Generic Letter 92-08, 'Thermo-Lag 330-1 Fire Barriers,'" dated April 16, 1993. (NUDOCS Accession No. 9304220215)

221. U.S. NRC, "IR 05000352/2001-014, IR 05000353/2001-014, on 12/10-12/21/2001, Exelon Nuclear, Limerick Generating Station, Units 1 and 2. Fire Protection. No Violations Noted," dated January 7, 2002. (ADAMS Accession No. ML020080162)

222. Letter from B.C. Buckley (U.S. NRC) to G.D. Edwards (PECO Energy Co.), "Informs of Completion of Licensing Action for GL 92-08, 'Thermo-Lag 330-1 Fire Barriers,' dated 921217 for Plant Units 1 and 2," dated September 21, 1998. (NUDOCS Accession No. 9809240274)

223. Letter from T.S. O'Neill (AmerGen Energy Company, LLC and Exelon Nuclear) to U.S. NRC, "60-Day Response to Generic Letter 2006-03, 'Potentially Nonconforming Hemyc and MT Fire Barrier Configurations,'" dated June 6, 2006. (ADAMS Accession No. ML061640343)

224. Letter from B.C. Buckley (U.S. NRC) to J.A. Hutton (PECO Energy Co.), "Safety Evaluation Addressing Thermo-Lag Ampacity Derating Issues, Peach Bottom Atomic Power Station, Units 2 and 3, and Limerick Generating Station, Units 1 and 2 (TAC Nos. MA3404, MA3405, MA3872, and MA3873)," dated January 12, 2000. (ADAMS Accession No. ML003677253)

225. Letter from H.B. Barron (Duke Energy) to U.S. NRC, "Response to NRC Generic Letter 2006-03, Potentially Nonconforming Hemyc and MT Fire Barrier Configurations," dated June 7, 2006. (ADAMS Accession No. ML061640310)

226. Letter from G.T. Bischof (Virginia Electric and Power Company) to U.S. NRC, "Generic Letter 2006-03, 'Potentially Nonconforming HEMYC and MT Fire Barrier Configurations,' Response to Request for Additional Information," dated May 31, 2007. (ADAMS Accession No. ML071520515)

227. Letter from R.B. Ennis (U.S. NRC) to J.A. Price (Dominion Nuclear Connecticut, Inc.), "Completion of Staff Review Related to Ampacity Derating Issues Associated with Generic Letter 92-08 (TAC No. MA3392)," dated May 9, 2002. (ADAMS Accession No. ML020700197)

228. Letter from R.B. Samworth (U.S. NRC) to R.O. Anderson (Northern States Power Company), "Generic Letter 92-08 – Thermo-Lag 330-1 Fire Barriers (TAC M85573)," dated May 27, 1993. (NUDOCS Accession No. 9305070350)

229. Letter from C.D. Terry (Niagara Mohawk Power Corporation) to U.S. NRC, "Response to 931222 Request for Additional Information re GL 92-08, 'Thermo-Lag 330-1 Fire Barriers,' per 10CFR50.54(f)," dated February 10, 1994. (NUDOCS Accession No. 9402170065)

230. Letter from C.D. Terry (Niagara Mohawk Power Corporation) to U.S. NRC, "Response to Request for Additional Information Dated September 15, 1994, Regarding Generic Letter 92-08, 'Thermo-Lag 330-1 Fire Barriers,' (TAC No. M85575)" dated December 14, 1994. (NUDOCS Accession No. 9412190124)

231. Letter from G.T. Bischof (Virginia Electric and Power Company) to U.S. NRC, "Generic Letter 2006-03, 'Potentially Nonconforming HEMYC and MT Fire Barrier Configurations,' Response to Request for Additional Information," dated May 31, 2007. (ADAMS Accession No. ML071520515)

232. Meeting Summary, "Summary of May 2, 1997, Meeting Regarding Completion of Installation of Thermal Lag Fire Barriers," dated May, 28, 1997. (NUDOCS Accession No. 9706020134)

233. Letter from T.S. O'Neill (AmerGen Energy Company, LLC and Exelon Nuclear) to U.S. NRC, "60-Day Response to Generic Letter 2006-03, 'Potentially Nonconforming Hemyc and MT Fire Barrier Configurations,'" dated June 6, 2006. (ADAMS Accession No. ML061640343)

234. Letter from E.J. Weinkam (Nuclear Management Company, LLC) to U.S. NRC, "Response to Generic Letter 2006-03: Potentially Nonconforming Hemyc and MT Fire Barrier Configurations," dated June 8, 2006. (ADAMS Accession No. ML061600209)

235. Letter from W.L. Stewart (Arizona Public Service Company) to U.S. NRC, "Response to the Follow-Up to the Request for Additional Information Regarding Generic Letter 92-08, Issued Pursuant to 10CFR50.54(f)," dated December 22, 1994. (NUDOCS Accession No. 9412290093)

236. Letter from J.L. Levine (Arizona Public Service Company) to U.S. NRC, "Summary of Major Work Activities that have Recently Been Completed to Resolve Remaining Open Actions re GL 92-08, 'Thermo-Lag 330-1 Fire Barriers,'" dated December 24, 1997. (NUDOCS Accession No. 9801050050)

237. Letter from M.A. Balduzzi (Entergy Nuclear Operations, Inc.) to U.S. NRC, "Response to NRC Generic Letter 2006-03, 'Potentially Nonconforming Hemyc and MT Fire Barrier Configurations," dated June 6, 2006. (ADAMS Accession No. ML061640132)

238. Letter from E.J. Weinkam (Nuclear Management Company, LLC) to U.S. NRC, "Response to Generic Letter 2006-03: Potentially Nonconforming Hemyc and MT Fire Barrier Configurations," dated June 8, 2006. (ADAMS Accession No. ML061600209)

239. Letter from C.W. Fay (Wisconsin Electric Power Company) to H.R. Denton (U.S. NRC), "Docket Nos. 50-266 and 50-301: Status of Fire Protection Modifications, Point Beach Nuclear Plant, Units 2 and 3," dated March 18, 1981. (NUDOCS Accession No. 8103250393)

240. Letter from E.J. Weinkam (Nuclear Management Company, LLC) to U.S. NRC, "Response to Generic Letter 2006-03: Potentially Nonconforming Hemyc and MT Fire Barrier Configurations," dated June 8, 2006. (ADAMS Accession No. ML061600209)

241. Letter from J.M. Solymossy (Nuclear Management Company, LLC) to U.S. NRC, "Prairie Island Nuclear Generating Plant Safety Evaluation Summary Report," dated December 10, 2003. (ADAMS Accession No. ML033530476)

242. Letter from T.S. O'Neill (Exelon Generation Company, LLC) to U.S. NRC, "60-Day Response to Generic Letter 2006-03, 'Potentially Nonconforming Hemyc and MT Fire Barrier Configurations,'" dated June 6, 2006. (ADAMS Accession No. ML061640343)

243. Letter from J.J. Fargo (Entergy Operations, Inc.) to S.A. Varga (U.S. NRC), "Response to NRC Letter Requesting Additional Information Regarding Generic Letter 92-08, 'Thermo-Lag 330-1 Fire Barriers,' Pursuant to CFR 50.54(f)," dated February 9, 1994. (NUDOCS Accession No. 9402240188)

244. Letter from R.J. King (Entergy Operations, Inc) to U.S. NRC, "NRC Generic Letter 2006-03, dated April 10, 2006, 'Potentially Nonconforming Hemyc and MT Fire Barrier Configurations,'" dated June 1, 2006. (ADAMS Accession No. ML061570394)

245. Letter from J.F. Lucas (Progress Energy) to U.S. NRC, "Response to NRC Generic Letter 2006-03, 'Potentially Nonconforming Hemyc and MT Fire Barrier Configurations,'" dated June 8, 2006. (ADAMS Accession No. ML061640136)

246. Letter from J.A. Stall (Florida Power and Light Company) to U.S. NRC, "Generic Letter 92-08 Corrective Actions," dated June 23, 1998. (NUDOCS Accession No. 9806290310)

247. Letter from C.O. Thomas (U.S. NRC) to T.F. Plunkett (Florida Power and Light Company), "Closeout Report for the Ampacity Derating Issues Related to Generic Letter 92-08, 'Thermo-Lag 330-1 Fire Barriers' – St. Lucie Plants, Units 1 and 2, And Turkey Point, Units 3 and 4," dated March 26, 1999. (NUDOCS Accession No. 9904020172)

248. Meeting Summary (U.S. NRC), "NRC Status Meeting Regarding Raceway Fire Barrier Project," dated April 5, 2001. (ADAMS Accession No. ML011780094)

249. Meeting Summary (U.S. NRC), "Summary of Meeting Between the Nuclear Regulatory Commission (NRC) Staff and PSEG Nuclear LLC on April 5, 2001 to Discuss Status of PSEG's Electrical Cable Raceway Fire Barrier Project," dated June 26, 2001. (ADAMS Accession No. ML011430238)

250. Letter from A.E. Scherer (Southern California Edison) to U.S. NRC, "Response to Request for Additional Information on Response to Generic Letter 2006-03," dated June 18, 2007. (ADAMS Accession No. ML071710548)

251. Letter from T.C. Feigenbaum (North Atlantic Energy Service Corporation) to the U.S. NRC, "Response to Generic Letter 92-08 (TAC No. 85603)," dated March 31, 1993. (NUDOCS Accession No. 9304140290)

252. Letter from S.E. Thomas (South Texas Project Electric Generating Station) to U.S. NRC, "South Texas Project Units 1 and 2 Thermo-Lag Confirmatory Order Completion Notice," dated February 8, 1992. (ADAMS Accession No. ML040990180)

253. Letter from T.W. Alexion (U.S. NRC) to W.T. Cottle (South Texas Project Electric Generating Station), "Completion of Licensing Action for Generic Letter 92-08, Thermo-Lag 330-1 Fire Barriers," dated April 4, 1997. (NUDOCS Accession No. 9704080234)

254. Letter from U.S. NRC to Houston Lighting and Power Company, "Safety Evaluation by the Office of NRR Related to Ampacity Derating Issues, STP Nuclear Operating Company,

Docket Nos. 50-498 and 50-499, South Texas Project Units 1 and 2 (STP)," dated January 19, 1999. (NUDOCS Accession No. 9901250034)

255. Letter from L.M. Padovan (U.S. NRC) to G.J. Taylor (South Carolina Electric and Gas Company), "Thermo-Lag Related Ampacity Derating Issue and Completion of Licensing Action for Generic Letter 92-08, Thermo-Lag 330-1 Fire Barriers, for Virgil C. Summer Nuclear Station," dated May 5, 1998. (NUDOCS Accession No. 9805080208)

256. Letter from G.T. Bischof (Virginia Electric and Power Company) to U.S. NRC, "Kewaunee Unit 1, Millstone Units 2 and 3, North Anna Units 1 and 2, Surry Units 1 and 2, Generic Letter 2006-03, Potentially Nonconforming HEMYC and MT Fire Barrier Configurations, Response to Request for Additional Information," dated May 31, 2007. (ADAMS Accession No. ML071520515)

257. Letter from R.G. Bryam (Pennsylvania Power and Light, Inc.) to U.S. NRC, "Susquehanna Steam Electric Station, Resolution of Thermo-Lag Issues," dated May 4, 1998. (ADAMS Accession No. ML041040513)

258. Letter from R.G. Bryam (Pennsylvania Power and Light, Inc.) to U.S. NRC, "Susquehanna 1 and 2 Status of Fire Protection Corrective Actions, in Response to NRC Confirmatory Order Regarding Fire Barriers," dated April 28, 2000. (ADAMS Accession No. ML003711917)

259. Letter from B.T. McKinney (Pennsylvania Power and Light, Inc.) to U.S. NRC, "Susquehanna Response to Generic letter 2006-03, Potentially Nonconforming Hemyc and MT Barrier Configurations," dated June 9, 2006. (ADAMS Accession No. ML061660076)

260. Letter from T.G. Colburn (U.S. NRC) to J.W. Langenbach (GPU Nuclear Corporation), "Generic Letter 92-08 Closeout Report – Thermo-Lag Related Ampacity Derating Issues for Three Mile Island, Unit 1 (TAC No. MA3340)," dated January 22, 1999. (NUDOCS Accession No. 9901260477)

261. Letter from T.F. Plunkett (Florida Power and Light Company) to L.F. Callan (U.S. NRC), "Response to Request for Additional Information – Generic Letter 92-08 Thermo-Lag 330-1 Fire Barrier," dated February 7, 1994. (NUDOCS Accession No. 9402150406)

262. Letter from R.J. Hovey (Florida Power and Light Company) to U.S. NRC, "Informs that Util. Proceeding With Design and Will Commence Implementation of Fire Barrier Upgrades for Fire Zones. Util. Will Aggressively Pursue Implementation of Thermo-Lag Fire Barrier Upgrades to Be Complete No Later Than 991231," dated June 15, 1998. (NUDOCS Accession No. 9806220249)

263. Letter from C.O. Thomas (U.S. NRC) to T.F. Plunkett (Florida Power and Light Company), "Closeout Report for the Ampacity Derating Issues Related to Generic Letter 92-08, Thermo-Lag 330-1 Fire Barriers – St. Lucie Plants, Units 1 and 2, And Turkey Point, Units 3 and 4," dated March 26, 1999. (NUDOCS Accession No. 9904020172)

264. Letter from D.A. Reid (Vermont Yankee Nuclear Power Corporation) to U.S. NRC, "Addresses Concerns Delineated in GL 92-08 re Thermo-Lag 330 Fire Barriers, Per NRC Bulletin 92-001," dated April 16, 1993. (NUDOCS Accession No. 9304190144)

265. Letter from J/P. Pelletier (Vermont Yankee Nuclear Power Corporation) to U.S. NRC, "Completion of Actions to Address NRC Generic Letter 92-08, Thermo-Lag Fire Barriers," dated June 28, 1993. (NUDOCS Accession No. 9307020175)

266. NRC Inspection Report, "NRC Integrated Inspection Report 05000271-06-005," dated January 19, 2007. (ADAMS Accession No. ML070190286)

267. Letter from J. M. DeVincentis (Vermont Yankee Nuclear Power Station) to U.S. NRC, "Vermont Yankee Nuclear Power Station – Status of Hemyc Fire Barrier Wrap at Vermont Yankee," dated August 17, 2005. (ADAMS Accession No. ML0523405090)

268. Letter from C.K. McCoy (Georgia Power Company) to U.S. NRC, "Response to Request for Additional Information Regarding Generic Letter 92-08, Thermo-Lag Fire Barriers," dated May 10, 1995. (NUDOCS Accession No. 9505180590)

269. Letter from L.M. Stinson (Southern Nuclear Operating Company) to U.S. NRC, "Joseph M. Farley, Edwin I. Hatch and Vogtle, NRC Generic Letter 2006-03 Response," dated June 9, 2006. (ADAMS Accession No. ML061600376)

270. Letter from J.A. Ridgel (Entergy Operations, Inc.) to U.S. NRC, "Response to Generic Letter 2006-03, Potentially Nonconforming Hemyc and MT Fire Barrier Configurations," dated June 7, 2006. (ADAMS Accession No. ML061600210)

271. Letter from R.A. Muench (Wolf Creek Nuclear Operating Corporation) to U.S. NRC, "Additional Information Regarding Thermo-Lag Fire Barriers," dated June 20, 1996. (NUDOCS Accession No. 9606250350)

272. Letter from J. Donohew (U.S. NRC) to R.A. Muench (Wolf Creek Nuclear Operating Corporation), "Closeout of Response to Generic Letter 2006-03, Potentially Nonconforming Hemyc and MT Fire Barrier Configurations," dated June 23, 2006. (ADAMS Accession No. ML061650179)

273. Letter from K.M. Thomas (U.S. NRC) to O.L. Maynard (Wolf Creek Operating Corporation), "Safety Evaluation Addressing Thermo-Lag Related Ampacity Derating Issues for the Wolf Creek Generating Station," dated April 6, 1998. (NUDOCS Accession No. 9804140367)

274. Letter from R.C. Hagan (Wolf Creek Nuclear Operating Corporation) to U.S. NRC, "Supplemental Response to the Follow Up to the Request for Additional Information Regarding Generic Letter 92-08," dated July 27, 1995. (NUDOCS Accession No. 9507030329)

275. Letter from R.W. Brown (Peak Seals, Inc) to P.M. Madden (U.S. NRC), "Listed Documents Re Three Hour Fire Endurance Test on 3M Interam Fire Barrier Wrap Sponsored by Peak Seals," dated August 7, 1995. (NUDOCS Accession No. 9509050173)

276. Fire Endurance Test, "Omega Point Fire Endurance Test of 3M Interam Mat Fire Protective Envelopes (24 in. and 6 in. Cable Trays, 5 in., 3 in., and 1 in. Conduits, 2 in. Air Drop and a 12 in. x 12 in. by 8 in. Junction Box," dated August 2, 1995. (NUDOCS Accession No. 9509070113)

277. Letter from T. Dogan (Vectra) to R. Brown (Peak Seals, Inc.), "E-50 Series Fire Endurance Test Evaluation," dated August 1, 1995. (NUDOCS Accession No. 9709090341)

278. Memorandum from A. Singh (U.S. NRC) to C.E. McCracken (U.S. NRC), "Trip to Omega Point Laboratories, 3M Company Interam 1-Hour Raceway Fire Barrier Fire Endurance Test (TAC No. M82809)," dated May 25, 1995. (NUDOCS Accession No. 9506090198)

279. Test Report provided by Peak Seals Inc., "Test Plan Number CTP-1199, One (1) Hour Fire Endurance Test, 3M Interam Fire Wrap," report date May 4, 1995. (NUDOCS Accession No. 9507140090)

280. Test Report Provided by Omega Point Laboratories, "ASTM E136-94, Behavior of Materials in a Vertical Tube Furnace at 750°C, 3M E-50 Interam Series Mat," report date January 17, 1995. (NUDOCS Accession No. 9705050067)

281. Letter from R.W. Brown (Peak Seals, Inc.) to L.B. Marsh (U.S. NRC), "Inform that Peak Seals has become Master Distributor of 3M Interam Fire Wrap Sys for Commercial Nuclear Power Plants, Response to Specific Questions," dated October 3, 1997. (NUDOCS Accession No. 9802040354)

282. Letter from J.K. Wood (Centerior Energy) to U.S. NRC, "Combustibility Testing of 3M Interam Material," dated February 7, 1997. (NUDOCS Accession No. 9702200311)

283. Letter from Vermont Yankee Nuclear Power Corporation to U.S. NRC, "Completion of Actions to Address NRC Generic Letter 92-08: Thermo-Lag Fire Barriers," dated June 28, 1993. (NUDOCS Accession No. 9307020175)

284. Letter from R. Licht (3M Fire Protection Products) to C.E. McCracken (U.S. NRC), "Acknowledges Receipt of 930504 Letter Requesting Info on 3M Fire Barrier Systems for Protection of Electrical Raceways. Informs that 3M Would Like to Supply Requested Info in Three Parts, Covering Flexible Wrap Systems, Rigid Panel Systems, and FS195 Systems," dated May 18, 1993. (NUDOCS Accession No. 9308310099)

285. Letter from D.R. Coy (3M Ceramic Materials Department), "Discusses NUMARC Meeting 931201-02 re Performance of Certain Fire Barrier materials. 3M Additional Test SVC Program Will Provide Technical Support, Supply of 3M Fire Protection Products, and Fire Testing at Cottage Grove Facility," dated January 14, 1994. (NUDOCS Accession No. 9401310171)

286. Letter from K.W. Howell (Underwriters Laboratories, Inc.) to B.J. Youngblood (U.S. NRC), "Qualification of 3M Fire Wrap," dated October 22, 1984. (NUDOCS Accession No. 8410240232)

287. Test Report provided by Detroit Edison, "TSI Technical Note 42584, Analysis of the Thermal Response of the Junction between a Partially Protected Member Which Protrudes the Thermo-Lag 330 Fire Barrier and a Cable Tray," report date April 1984. (NUDOCS Accession No. 8408080299)

288. Letter from R.L. Tedesco (U.S. NRC) to R.L. Mittl (Public Service Electric and Gas Company), "Salem Cable Tray Fire Barrier Evaluation; Docket Nos. 50-311/272," dated March 18, 1981. (NUDOCS Accession No. 8103240604)

289. Letter from D.R. Coy (3M Ceramic Materials Department), "Informs that 3M Will be Providing Complete Document Package for 3M Interam E-50 Series Material, 1-and-3 Hour Systems, Including Fire Test Reports, per 10CFR50, App R," dated August 3, 1993. (NUDOCS Accession No. 9308120140)

290. Letter from D.R. Coy (3M Ceramic Materials Department), "Advises that 3M Will Continue to Supply Interam E-50 Series Materials and all Peripheral 3M Fire Protection Products Used in Installation Process Under Original Nuclear Product Designations," dated March 1, 1994. (NUDOCS Accession No. 9403170125)

291. Test Report Provided by 3M Fire Protection Products, "3M Fire Test Reports 94-27 and 94-42, re Upgrading TSI Material for 3-Hour Conduit Systems using 3M's Interam E-50 Series Mats," dated March 17, 1994. (NUDOCS Accession No. 9403280162)

292. Letter from Dwight E. Nunn (Tennessee Valley Authority) to U.S. NRC, "Watts Bar Nuclear Plant (WBN) – 3M Fire Barrier Material, Cable Compressive Load Testing," dated June 17, 1994. (NUDOCS Accession No. 9406240089)

293. Letter from D.R. Coy (3M Fire Protection Products) to Nuclear Power Utility Customers, "Letter Advising that Effective 950628 3M Fire Protection Products Certified Nuclear Installer Program Will End and that Peak Seals, Inc Will Become Exclusive Supplier of 3M Fire Protection Products, Effective 950629," dated April 28, 1995. (NUDOCS Accession No. 9505230123)

294. Test Report Provided by 3M Fire Protection Products, "Results of Fire Test Conducted on Latest NRC Criteria With 3M Interam E-50 Series Material," dated May 25, 1995. (NUDOCS Accession No. 9506220297)

295. Test Report Provided by U.S. NRC, "Trip to Omega Point Laboratories – Peak Seals 3M Interam Raceway Fire Barrier Fire Endurance Test Program (April 20, 1995) (TAC No. M82809)," dated May 1, 1995. (NUDOCS Accession No. 9507140050)

296. 3M Fire Protection Products, "3M Advanced Training Program: 3M Interam E-50 Series Fire Protection Systems for the Nuclear Industry," dated June 15, 1993. (NUDOCS Accession No. 9308040286)

297. Meeting Summary, "Meeting on Fire Endurance Test Acceptance Criteria with the Nuclear Management and Resources Council," dated November, 20 1992. (NUDOCS Accession No. 9212080298)

298. Letter from C.E. McCracken (U.S. NRC) to R.R. Licht (3M Fire Protection Products), "NRC Intent to Review Fire Barrier Systems Used by Licensees Re Compliance with NRC Fire Protection Requirements," dated May 4, 1993. (NUDOCS Accession No. 9403170119)

299. SwRI Test Report, "Ampacity Derating of Fire-Protected Cables in Conduit / Cable Trays Using 3M Incorporated's Passive Fire Protection Systems Identified as 3M Interam E-50A,

E-50D, E-53A, and E-50D/E-53A," dated September, 30 1986 (NUDOCS Accession No. 9308040280)

300. 3M Fire Protection Products, "3M Interam E-50 Series Fire Protection Mat, 1-Hour Flexible Wrap System for Electrical Raceways, Installation Booklet Including Quality Assurance Guidelines and Typical Drawings," dated June 19, 1987. (NUDOCS Accession No. 9308040230)

301. Test Report Provided by Twin City Testing Corporation, "Qualification Fire Tests of the 3M Interam E-50D Fire Protection Mat for 3-Hour Rated Electrical Raceways," dated March 1986. (NUDOCS Accession No. 9308040267)

302. SwRI Test Report No. 01-7912, "Qualification Fire Test of a Protective Envelope System," dated June 1984. (NUDOCS Accession No. 9308040243)

303. 3M Fire Protection Products, "3M Interam E-54A Fire Protection Mat, 3-Hour Flexible Wrap System for Electrical Raceways, Installation Booklet Including Quality Assurance Guidelines and Typical Drawings," dated October 27, 1987. (NUDOCS Accession No. 9308040228)

304. 3M Fire Protection Products, "3M Interam E-50 Series 1-hour and 3-hour Flexible Wrap Fire Protection Systems." (NUDOCS Accession No. 9308040224)

305. SwRI Project Report No. 01-7912a(1), "Qualification Fire Test of a Protective Envelope System," dated June 1985. (NUDOCS Accession No. 9308040263)

306. Test Report Provided by Twin City Testing Corporation, "Qualification Fire Tests of the 3M Interam E-50 Series Fire Protection Mat for 1-Hour Rated Electrical Raceways," dated September 1986. (NUDOCS Accession No. 9308040275)

307. SwRI Project Report No. 01-7912(2), "Qualification Fire Test of a Protective Envelope System," dated June 1985. (NUDOCS Accession No. 9308040254)

308. Letter from R. Licht (3M Fire Protection Products) to C.E. McCracken (U.S. NRC), "Response to Questions Noted in 930504 Letter and Provides Info Intended to Validate Use of 3M Interam E-50 Series 1-Hour and 3 Hour Fire Protection Systems," dated June 30, 1993. (NUDOCS Accession No. 9308040054)

309. Letter from R.W. Brown (Peak Seals) to P.M. Madden (U.S. NRC), "Response: 3M Interam Fire Wrap Systems," dated August 7, 1995. (NUDOCS Accession No. 9509050173)

310. Test Report Provided by Central Laboratories Services, "Testing to Determine Ampacity Derating Factors for 3M Fire Barrier Wrapped Conduits and Air Drops, Job Number 94-0357, Revision 0," dated February 3, 1994. (NUDOCS Accession No. 9403030227)

311. SwRI Test Report No. 01-8818-208/-209d, "Ampacity Derating of Fire-Protected Cables in Conduit/Cable Trays Using 3M Incorporated's Passive Fire Protection Systems Identified as 3M Interam E-50A (Verification Tests)," dated October 6, 1986. (NUDOCS Accession No. 9308040249)

312. SwRI Test Report No. 1208-001, "Nuclear Component Qualification Test Report for the Generic Seismic Qualification of 3M Interam E-50D 3-Hour Fire Protection System," dated July 1986. (NUDOCS Accession No. 9308040186)

313. Test Report Provided by 3M Fire Protection Products, "3M Fire Test #92-115: 3-hour Fire Protection on Conduits with the 3M Interam E-50 Series Mats," dated August, 6, 1992. (NUDOCS Accession No. 9308040119)

314. Test Report Provided by 3M Fire Protection Products, "3M Fire Test #87-79: 3-hour Fire Protection on a Cable Tray With the 3M Interam E-50 Series Mats," dated July 24, 1992. (NUDOCS Accession No. 9308040106)

315. Test Report Provided by 3M Fire Protection Products, "3M Fire Test #92-167: 1-hour Fire Protection on 1" Sch. 40 Steel Conduits Using the 3M Interam E-53A Mats," dated June 25, 1993. (NUDOCS Accession No. 9308040101)

316. Test Report Provided by 3M Fire Protection Products, "3M Fire Test #92-141: 1-hour and 3-Hour Fire Protection on 1" Sch. 40 Steel Conduits Using the 3M Interam E-54A Mats," dated August 27, 1992. (NUDOCS Accession No. 9308040099)

317. Test Report Provided by 3M Fire Protection Products, "3M Fire Test #87-40: 1-hour Fire Protection on a Cable Tray With the 3M Interam E-50 Series Materials," dated April 3, 1992. (NUDOCS Accession No. 9308040097)

318. Test Report Provided by 3M Fire Protection Products, "3M Fire Test Report #87-57: 3M Chemolite Building 66, Large Scale Furnace," dated May 27, 1987. (NUDOCS Accession No. 9308040094)

319. Test Report Provided by 3M Fire Protection Products, "3M Fire Test #87-76: 60 Minute Fire Protection on Conduits with Interam E-53A Mat and FireDam 150 Caulk," dated June 25, 1987. (NUDOCS Accession No. 9308040090)

320. Letter from R. Licht (3M Ceramic Materials) to C.E. McCracken (U.S. NRC), "Response to Questions Noted in 930504 Letter and Provides Info Intended to Validate Use of 3M Interam E-50 Series 1-Hour and 3-Hour Fire Protection Systems," dated June 30, 1993. (NUDOCS Accession No. 9308040054)

321. Letter from J.S. Marshall (Texas Utilities Services) to S. Burwell (U.S. NRC), "Comanche Peak Steam Electric Station Fire Qualification Test Report Transmittal," dated October 29, 1981. (NUDOCS Accession No. 8112160281)

322. Letter from R.L. Tedesco (U.S. NRC) to R.J. Gary (Texas Utilities), "Comanche Peak Tray Fire Barrier Evaluation," dated December 1, 1981.(NUDOCS Accession No. 8112300294)

323. Letter from S.C. Black (U.S. NRC) to W.J. Cahill (Texas Utilities), "Thermo-Lag Acceptance Methodology for Comanche Peak Steam Electric Station – Unit 2," dated October 29, 1992. (NUDOCS Accession No. 9211040234)

324. Letter from J.S. Perry (Clinton Power Station) to S.A. Varga (U.S. NRC), "Illinois Power's Response to the Nuclear Regulatory Commissions Request for Additional Information

Regarding Generic Letter 92-09, 'Thermo-Lag 330-1 Fire Barriers,'" dated February 9, 1994. (NUDOCS Accession No. 9403040225)

325. Letter from J.J. Kelly Jr. (TU Electric) to U.S. NRC, "Comanche Peak Steam Electric Station (CPSES) – Unit 1 Fire Endurance Test Report," Dated March 2, 1999. (NUDOCS Accession No. 9903110179)

326. TU Electric sponsored test report performed by OPL, "One Hour Fire Endurance Test Of Articles Protected With Thermo-Lag® Fire Barrier System, Project No. 12340-102571, Test Scheme 13-3," dated February 25, 1999. (NUDOCS Accession No. 9903110227)

327. 45 FR 76602 – Federal Register Vol. 45, No. 225 dated Wednesday, November 19, 1980, "10 CFR Part 50 Fire Protection Program for Operating Nuclear Power Plants."

328. "Thermo-Lag 330-1 Combustibility Evaluation Methodology Plant Screening Guide," NUMARC, dated October 12, 1993 (NUDOCS Accession No. 9310210224).

329. E-mail from P.L. Campbell on behalf of General Electric to NRCREP Resource, "Response from "Comment on NRC Documents," dated November 20, 2009. (ADAMS Accession No. ML093430234)

330. Letter from F.W. Madden (Luminant Generation Company LLC) to M.T. Lesar (U.S. NRC), "Comanche Peak Steam Electric Station; Docket Nos. 50-445 and 50-446; NUREG-1924, "Electric Raceway Fire Barrier Systems (ERFBS) in Nuclear Power Plants, Draft Report for Comment," dated December 2, 2009. (ADAMS Accession No. ML093441088)

331. Letter from J.C. Butler (NEI), to M.T. Lesar (U.S. NRC), "Industry Comments on Draft NUREG 1924, 'Electric Raceway Fire Barrier Systems in U.S. Nuclear Power Plants'," dated December 3, 2009. (ADAMS Accession No. ML093500089)

332. Letter from D.P. Helker (Exelon Nuclear), to M.T. Lesar (U.S. NRC), "Comments Concerning Draft NUREG-1924, 'Electric Raceway Fire Barrier Systems (ERFBS) in Nuclear Power Plants'," dated December 7, 2009. (ADAMS Accession No. ML100050209)

333. E-mail from W. Harper (Energy Northwest) to G.J. Taylor (U.S. NRC), "RE: NUREG 1924 Edits for Columbia Generating Station," dated December 7, 2009. (ADAMS Accession No. ML101030790)

334. E-mail from A. Holder (Nuclear Generation Group) to A. Klein (U.S. NRC), "Draft NUREG 1924," dated December 8, 2009. (ADAMS Accession No. ML101030781)

335. E-mail from F.J. Wyant (Sandia National Laboratories), to G.J. Taylor (U.S. NRC), "Review of Draft NUREG-1924," dated January 7, 2010. (ADAMS Accession No. ML101030775)

336. E-mail from S.P. Nowlen (Sandia National Laboratories), to G.J. Taylor (U.S. NRC), "ERFBS report NUREG-1924," dated January 11, 2010. (ADAMS Accession No. ML101030765)

337. NRC Inspection Report, "Clinton Inspection Report 50-461/98026(DRS) And Notice Of Enforcement Discretion," dated December 13, 1998. (NUDOCS Accession No. 9812180152)

338. OPL Ampacity Derating Test Report, "Ampacity Derating of Fire Protected Cables, Project No. 8610-102166, Electrical Test to Determine the Ampacity Derating of a Protective Envelope for Class 1E Electrical Circuits" dated December 23, 1997. (ADAMS Accession No. ML101300320)

339. OPL Ampacity Derating Test Report, "Ampacity Derating of Fire Protected Cables, Project Nos. 8610-102164 and 102165, Electrical Test to Determine the Ampacity Derating of a Protective Envelope for Class 1E Electrical Circuits" dated December 26, 1997. (ADAMS Accession No. ML101300321)

340. OPL Ampacity Derating Test Report, "Ampacity Derating of Fire Protected Cables, Project No.14540-100770, Electrical Test to Determine the Ampacity Derating of a Protective Envelope for Class 1E Electrical Circuits" dated December 5, 1996. (ADAMS Accession No. ML101300301)

341. OPL Ampacity Derating Test Report, "Ampacity Derating of Fire Protected Cables, Project No. 14540-99074 and 99075, Electrical Test to Determine the Ampacity Derating of a Protective Envelope for Class 1E Electrical Circuits" dated March 25, 1996. (ADAMS Accession No. ML101300295)

342. OPL Combustibility Test Report, "ASTM E136-94 Behavior Of Materials in a Vertical Tube Furnace at 750ºC, 3M E-50 INTERAM™ SERIES MAT, Project No. 14540-99234," dated January 17, 1995 (ADAMS Accession No. ML101300314)

343. Wyle Laboratories Test Report, "Final Report CTP-2010 Design Basis Accident Testing on 3M Ineram™ Flexible Fire Wrap, Report No. 46979-1," dated July 27, 1998. (ADAMS Accession No. ML101300325)

344. OPL Fire Endurance Test Report, "Fire Endurance Test of 3M Interam™ Mat Fire Protective Envelopes (24 in. and 6 in., 2 in. Air Drop and a 12 in. X 12 in. x8 in. Junction Box), Project No. 14540-99123," dated December 5, 1995. (ADAMS Accession No. ML101300310)

345. OPL Fire Endurance Test Report, "Fire Endurance Test of 3M Interam™ Fire Wrap Fire Protective Envelope, Project No. 8610-102570," dated May 19, 1998. (ADAMS Accession No. ML101300322)

346. OPL Fire Endurance Test Report, "Fire Endurance Test of 3M Interam™ Mat Fire Protective Envelopes (6 in. wide and 24 in. by 4 in. Deep Steel Ladder-Back Cable Trays, Project No. 14540-99417," dated September 22, 1997. (ADAMS Accession No. ML101300319)

347. OPL Fire Endurance Test Report, "Fire Endurance Test of 3M Interam™ Mat Fire Protective Envelopes (5 in., 3in., and 1 in. Conduits and a 12 in. x 12 in. x 8 in. Junction Box, Project No. 14540-99416," dated February 26, 1996. (ADAMS Accession No. ML101300318)

348. Letter from R.W. Brown (PCI Promatec) to M.H. Salley (U.S. NRC), "Comments on Draft Report NUREG-1924 Electric Raceway Fire Barrier Systems in U.S. Nuclear Power Plants," dated April 30, 2010. (ADAMS Accession No. ML101300259)

# Appendix A    The Browns Ferry Fire

On March 22, 1975, a major fire occurred at the Browns Ferry Nuclear Plant (BFN), located near Decatur, Alabama. At the time of the fire, it was the largest nuclear power plant (NPP) in the world, having three units with a maximum design power output of about 3195 MWe. Unit 1 began operation on August 1, 1974, while Unit 2 came online on March 1, 1975. Unit 3 was under construction at the time of the fire and had its own control room, while Units 1 and 2 shared a common space for the control of the respective units. It is important to realize that BFN was the first nuclear plant operated by Tennessee Valley Authority (TVA) and, at the time of the fire, Unit 1 had only been in operating for 8 months while Unit 2 had only accumulated 21 days of operation.

The fire lasted for over 7 hours, damaged over 1,600 electrical cables, and resulted in a loss of all Unit 1 and many Unit 2 emergency core cooling systems used to maintain reactor safety. Figure A-1 shows some of the damage caused by the fire. In this particular case, the fire was intense enough to melt the aluminum conduit and insulation around the conductors. All that remained was the bare copper conductors seen at the center of the photo.

**Figure A-1: Photograph of Conduit Damaged from Fire (NUREG/BR-0361)**

The following provides a brief summary of what occurred on March 22, 1975, and the lessons learned from the most severe U.S. NPP fire to date. Please refer to NUREG/BR-0361, "The Browns Ferry Nuclear Plant Fire of 1975 and the History of NRC Fire Regulations," for a detailed discussion of the fire and supporting documentation.

## A.1 The Fire Event

The fire was initiated by a small lit candle being used to check for air leakage between the cable spreading room (CSR) and the reactor building. The fossil fuel plant technicians commonly used this method to check for condenser leakage, and it was passed on to the early nuclear industry as a means to check the effectiveness of the seal operation. At about 12:20 p.m., the candle flame ignited some polyurethane foam used as part of a temporary cable penetration seal. Because of the pressure differential between the reactor building and the CSR, the fire propagated into the Unit 1 reactor building. This resulted in two fires across the fire wall—one in the CSR and another in the Unit 1 reactor building.

In the CRS, the technician attempted to beat out the fire with his flashlight and, when that failed, they discharged several carbon dioxide ($CO_2$) fire extinguishers that also were unsuccessful in extinguishing the fire. The operators in the main control roomf were made aware of the fire about 15 minutes after it initially started. About 40 minutes after fire initiation, the assistant shift engineer initiated the CSR evacuation alarm and then proceeded to actuate the in situ $CO_2$ Cardox fire-extinguishing system. However, safety measures had been intentionally taken to disable the Cardox system while the cable penetration inspections were being conducted. Power had been removed to eliminate the potential for the automatic actuation, and a metal bar had been installed under the break-out glass used for manual actuation. The delay in making the system available resulted in actuating the Cardox system about 50 minutes after the fire began. The Cardox system flushed the CSR with enough $CO_2$ to displace most of the oxygen. However, the Cardox system was actuated three times in over 2 hours and was unable to completely extinguish the fire. The fire in the CSR was finally put out 4 hours after it started by the use of manual fire suppression efforts (extinguishers).

When the plant firefighting personnel found the fire in the reactor building, it was burning in cable trays that were about 6.1 m (20 feet) above the second floor of the reactor building. After setting up a ladder to reach the fire, $CO_2$ extinguishers were unsuccessfully discharged into the fire. The application of the $CO_2$ extinguishers would exhaust the surface flames of the fire but was not able to extinguish the deep-seated burning in the cables. As a result, when extinguishing efforts ceased, the cables would re-ignite and continue burning. Firefighting efforts were further complicated by the loss of ventilation and lighting systems. In addition, a shortage of self contained breathing apparatus existed because plant operators were using a number of these devices while manually aligning valves in an attempt to get the reactor in a shutdown cooling mode. About 7 hours after the fire started, the plant superintendent agreed to use water on the fire (the local fire chief had suggested using water 5 hours prior—at 2:00 p.m.). Shortly after application of the water, the fire was declared out 7.5 hours after it started.

## A.2 Plant Response to the Fire

Twenty minutes after the fire started, Unit 1 operators noted anomalous behavior of controls and instrumentation for systems designed to provide emergency cooling of the reactor core. Over the next several minutes, the fire caused equipment to spuriously actuate, which resulted in a growing number of incidents. These included the automatic actuation of pumps and equipment that, when the operators determined they should be shutdown, would automatically start up again.

The Unit 1 reactor was manually scrammed about 30 minutes after the fire initiated, shutting the reactor down and stopping the nuclear fission chain reaction. However, decay heat continued to be generated by the radioactive decay of the fuel, requiring continuous long-term cooling of the core to ensure core damage would not occur. Shortly after the reactor scram, a number of instrumentation lights indicating the status of the plant went out. Soon after, the main-steam-isolation valves (MSIVs) closed automatically eliminating the current method of decay heat removal. Closure of the MSIVs hampered the core-cooling efforts in two ways—(1) by closing off the methods of removing heat and (2) by stopping the flow of steam to the turbine driven feedwater pump.

The fire had disabled the High Pressure Coolant Injection and Reactor Core Isolation Cooling systems. Consequently, the only method of keeping water on the core was through the use of the control-rod-drive (CRD) system pump at a flow rate of about 400 liters per minute (105 gallons per minute), resulting in the water level in the reactor vessel to begin decreasing. To avoid core damage, the operators had to reduce the pressure within the reactor coolant system and then manually realign the condensate booster system to supply water to the core. The mode of cooling continued to provide adequate cooling until around 6:00 p.m. a loss of control air system prevented manual control of the pressure relief valves. This resulted in the CRD system to again be the only method to supply cooling water to the core.

After the fire was declared out, smoke began to clear and the reliance on breathing apparatus decreased so that a more orderly approach to obtaining shutdown cooling could be taken. Around 9:50 p.m., control to the relief valves was restored, the reactor was depressurized, and the condensate booster pump again pumped cooling water into the reactor. Additional cooling-water makeup was provided throughout the incident, and additional alternatives could have been used to provide makeup water with the reactor at either high or low pressure. It is believed that at no time during the event did the water level drop below the top of active fuel.

### A.3 Lessons Learned from Browns Ferry Fire

The fire caused an estimated physical damage of $10 million and resulted in two operating units to be incapacitated for over a year. In addition to the above-mentioned costs, additional costs of around $10 million were incurred each month for replacement power.

#### Fire Prevention

Grouped electrical cables are more flammable than most people believed prior to the fire. The use of open flames to detect leaks, the frequency of occurrence of small fires as a part of the leak detection process, the ease with which the cable insulation was ignited, and the spread of flames—all constituted a significant fire hazard.

#### Use of Water

The extent of damage caused by the fire is attributable to the length of time the fire burned. The reluctance to use water on the fire (for fear of conductor shorting) was a position held by many licensees at the time—a position they felt would reduce the likelihood of causing spurious equipment operation. However, the failures caused by the fire as it continued to burn were largely responsible for the difficulties encountered in bringing the plant to a safe-stable state, and the fire was extinguished rather quickly when water was finally applied. Hence, the main lesson learned is that, if initial attempts to extinguish a cable fire with non-water means are unsuccessful, water should be used. Water is the best extinguishing agent available for most potentially deep-seated fires in nuclear power plants and the sooner a fire is extinguished, the less total damage results.

**Redundant System Separation and Independence**

The damage to electrical power and control circuits resulted in the loss of redundant subsystems and equipment. This caught the nuclear industry off guard in light of the independence and separation criteria that had been applied in the initial design of the plants. The principal causes of these failures were found to be the failure to recognize potential sources of failure in safety equipment (i.e., interconnection of safety and non-safety equipment) and the identification that use of conduit to isolate cables form their redundant counterparts did not protect the cables adequately.

## A.4 Additional Information

Numerous documents discuss the 1975 BFN fire. The most recent and comprehensive document is NUREG/BR-0361, "The Browns Ferry Nuclear Plant Fire of 1975 and the History of NRC Fire Regulations," dated February 2009. This brochure provides a great overview of the event and supporting details and documents in an enclosed DVD. The brochure is available on the NRC website.

# Appendix B    Ampacity Derating

When current flows in a conductor, heat is produced because every conductor offers some resistance to the flow of current. The National Electric Code® (NEC) (American National Standards Institute (ANSI)/NFPA 70)) defines ampacity as "the current (in amperes) a conductor can carry continuously under the conditions of use without exceeding its temperature rating." The current-carrying capacity of a particular conductor is dictated by its "ampacity" (that is, how many amperes it can handle). Ampacity is a function of the cross-sectional area or diameter of the wire and its material type (e.g., copper or aluminum) and cable insulation condition for basic installation conditions. For more complex installation conditions, Institute of Electrical and Electronic Engineers (IEEE) 835 provides more extensive and detailed tables. For installations involving cables in open cable trays, National Electrical Manufacturers Association (NEMA) WC51-1991/icae P-54-440 should be consulted. Larger-diameter wires have larger cross-section areas and can safely carry more electrical current without overheating. The ampacity rating of a specific conductor may be obtained from tables in the NEC. These tables are based on the size of the wire, the maximum allowable operating temperature of the insulation material, and the installation conditions.

Cables routed in electrical raceways are derated to ensure that systems have sufficient capacity and capability to perform their intended safety functions. The nominal ampacity values include a safety margin that is sufficient for most installations. However, there are instances where application of the NEC ampacity tables is insufficient. Cables routed in raceways enclosed in fire barriers require additional derating because of the insulating effect of the fire barrier materials. For example, although the addition of fire barrier wrap around cable trays and conduits will affect the ampacity of a conductor, the NEC tables do not address this problem. Several inches of fire barrier material can have a significant effect on the ampacity rating specified in the NEC tables. NRC requires that cable derating due to the use of fire retardant coatings be considered by utilities during plant design or when design changes are made to existing electrical system configurations. Since there are no derating tables in the NEC for this kind of situation, calculations must be performed to determine the current carrying capacity of the enclosed cables.

$$\%Ampacity\_Derating = \frac{I_o - I_f}{I_o} \times 100 \qquad \text{(Equation B-1)}$$

Where:   $I_o$ = Current in amperes required to attain a temperature of 90°C for the baseline case
   $I_f$ = Current in amperes required to attain a temperature of 90°C for the system as protected by the passive fire protection system

Four conditions exist which complicates establishing uniform ampacity values:

1. An infinite number of configurations exist when one considers the vast number of cable sizes, cable types, and cable loading in the vast number of electrical raceways with the vast range of ambient temperatures.

2. The heat transfer mechanisms of radiation, conduction, and convection are very interactive and transient in fire protected electrical raceways. Meticulous attention to

detail, patience to allow conditions to stabilize, and conservatism are required for reliable ampacity testing.

3. The National Electric Code, Insulated Cable Engineers Association publications, and other publications list only very general ampacity values which tend to be conservative and do not consider fire protected electrical raceways.

4. No nationally accepted standard ampacity test exists for fire protected cables, although Underwriters Laboratories has a proposed standard, and a different proposed standard was informally submitted at the April 1986 meeting of the Insulated Conductors Committee of the Institute of IEEE.

**IEEE Standard 848**

IEEE standard 848-1996, "IEEE Standard Procedure for the Determination of the Ampacity Derating of Fire-Protected cables," provides a testing procedure for use in establishing the ampacity and ampacity derating factor for cables protected by ERFBS. This method involves maintaining the room temperature constant at 40°C and adjusting the current flowing through the cable to maintain the cable at 90°C. A baseline case without an ERFBS is conducted followed by the same test specimen protected with an ERFBS. The difference in the current flow (ampacity) is calculated in fractional terms and referred to as the ampacity derating factor. The following provides more detail on this testing approach.

The standard requires a 40% cable fill in the cable tray and conduit tests containing 600V rated copper conductor cables with XLPE insulation and a CSPE jacket. Although the standard suggests a 2.54 cm (1 in) and 10.16 cm (4 –in) conduit sizes for testing, if significant ERFBS design differences exist for different size conduits, then tests shall consist of conduits sizes to bound the specific configurations. The average surface emissivity is also required to be documented as the test results can be affected by the conduits emissive properties.

For a cable tray test, a 60.96 cm (24 in) wide cable tray is to be filled with three layers of cable,. Both cable trays and conduits must be at least 12 ft in length and oriented horizontally to represent worst case conditions. Air drop configurations do not require a cable fill fraction, but must be representative of field applications. For cable temperature measurements, Type T (copper/constantan) 24 AWG maximum thermocouples are to be used to measure the temperature of a cables copper conductors at various locations. To accurately measure the conductor temperatures, a small incision in made axially along the cable jacket and insulation and the thermocouple junction is placed in direct contact with the conductor strand. The incisions are closed by placing a single layer wrap of glass-reinforced electrical tape around the cut.

To evaluate the ampacity derating for the ERFBS, a baseline ampacity measurement of an unprotected assembly is needed for comparison to the ampacity measurement with an applied ERFBS. For either scenario, the circuit is energized with a 60 Hz single-phase source sufficient to cause the conductor to reach 90 °C (194 °F) at the central location of the cable within the enclosure. The standard also specifies temperature range limits for the various thermocouple locations. The single-phase source (typically a current source) is adjusted such that the conductor maintains the 90°C (194°F) temperature and these ampacity values are recorded when the system has been determined to reach steady state condition, which is when:

a) A minimum of 3-hours has elapsed since the last adjustment of current level or perturbation of the system occurred.
b) The rate of change of the average of thermocouple readings at the hot spot does not exceed ±0.2°C (±0.36°F) per hours for conduit, tray, and free-air drop.

To ensure that the average thermocouple temperature rise does not exceed the ±0.2°C change per hour, the standard provides a linear regression analysis method using the least-square method.

During the entire test, room enclosure is maintained at a 40 ± 2 °C (104 ± 3.6 °F) temperature. In addition, induced air currents within the room enclosure and radiant energy from the heat source should not impinge upon the test assembly.

At the conclusion of the tests, the final conductor temperature and ambient temperature may not match 90 °C (194 °F) and 40 °C (104 °F) respectively. The standard provides a normalization method which computes the normalized current as follows:

$$I' = I \sqrt{\frac{(T_{c'} - T_{a'})(\alpha + T_c)}{(T_c - T_a)(\alpha + T_{c'})}}$$  (Equation B-2)

where

- I'    is normalized current, amperes
- I    is test current at equilibrium, amperes
- $T_c$    is hottest conductor temperature at center at equilibrium, °C
- $T_a$    is measured enclosure ambient temperature, °C
- $T_c'$    is normalized conductor temperature = 90°C
- $T_a'$    is normalized ambient temperature = 40°C
- α    is 234.5 for copper and 228.1 for aluminum

Following the normalization, the ampacity derating factor can be found by the following equation,

$$ADF = \frac{(I_o - I_f)}{I_o} \cdot 100$$  (Equation B-3)

where

- $I_o$    is the normalized current for the baseline condition, amperes
- $I_f$    is the normalized current for the passive fire-protected cable system or cable penetration fire stop system, amperes
- ADF    is the ampacity derating factor, %

# Appendix C  Summaries of NRC Generic Communications on ERFBS

NRC has issued a number of generic communications regarding ERFBS. Summaries of the issues that were addressed in these generic communications are presented below.

## Bulletins (BLs)

### Bulletin 92-01
Bulletin No. 92-01, "Failure of Thermo-Lag 330 Fire Barrier System to Maintain Cabling in Wide Cable Trays and Small Conduits Free from Fire Damage," June 24, 1992 informed the licensees that NRC has determined that the 1- and 3-hour pre-formed assemblies installed on small conduit and wide cable trays (wider than 36 cm (14 inches)) do not provide the level of safety as required by NRC requirements. The bulletin requested plants that use Thermo-Lag 330-1 to identify areas in the plant where it is used and where it is used to protect either small diameter conduit or wide trays that provide safe shutdown capability, the licensee should implement the appropriate compensatory measures.

### Bulletin 92-01, Supplement 1
Failure of Thermo-Lag 330 Fire Barrier System to Perform its Specified Fire Endurance Function, August 28, 1992, alerts licensees and permit holder of additional apparent failure in fire endurance testing performed for Texas Utilities Electric Company associated with the Thermo-Lag 330 fire barrier system which many plants have installed to protect safe shutdown capability. The bulletin requested identification of areas where Thermo-Lag 330 is used and to implement appropriate compensatory measures until the licensee can declare the fire barriers operable on the basis of applicable tests demonstrating appropriate 1- or 3-hour performance.

## Generic Letters (GLs)

### Generic Letter 86-10
Generic Letter 86-10, "Implementation of Fire Protection Requirements," April 24, 1986, provides guidance as to acceptable methods of satisfying Commission regulatory fire protection requirements. GL 86-10 also contained two enclosure, 1 "Interpretations of Appendix R," provided NRCs interpretation of the Appendix R regulations, and enclosure 2 "Appendix R Questions and Answers," provides NRC answers to questions raised at and subsequent to an Appendix R workshop. Section 3.2 of enclosure 2, "Appendix R Questions and Answers," to GL 86-10 provides NRC acceptance criteria on the 325°F temperature criterion.

### Generic Letter 86-10, Supplement 1
Generic Letter 86-10, Supplement 1, "Fire Endurance Test Acceptance Criteria for Fire Barrier Systems Used to Separate Redundant Safe Shutdown Trains Within the Same Fire Area" March 25, 1994, provides guidance and acceptance criteria as to how NRC will review and evaluate the adequacy of fire endurance tests and fire barrier systems proposed by licensees or applicants. Supplement 1 was issued to (1) clarify the applicability of the test acceptance criteria in GL 86-10 to raceways fire barrier systems, (2) specify a set of fire endurance test acceptance criteria that are acceptable for demonstrating the fire barrier systems can perform the required fire-resistive function and maintain the protected safe shutdown train free of fire damage, (3) specify acceptable options for hose stream testing, and (4) specify criteria for cable

functionality testing when a deviation is necessary, such as when the fire barrier temperature rise criteria are exceeded or the test specimen cables sustain visible damage.

### Generic Letter 92-08
Generic Letter 92-08 "Thermo-Lag 330-1 Fire Barriers, December 17, 1992," requested additional information to verify that licensees using Thermo-Lag 330-1 ERFBS comply with NRC regulations. The issuance of GL 92-08 was stimulated by NRCs concern that the Thermo-Lag 330-1 ERFBS may not provide the level of fire endurance intended by the licensees, and that may results in licensees not meeting the requirements of 10 CFR 50.48 and GDC 3. The GL requested information related to (1) fire endurance qualification testing, (2) ampacity derating of cables enclosed in Thermo-Lag 330-1 barrier, and (3) the licensees evaluation and application of the results of tests conducted to determine the fire endurance ratings and the ampacity derating factors of Thermo-Lag 330-1 barriers.

### Generic Letter 2006-03

Generic Letter 2006-03, "Potentially Nonconforming HEMYC and MT Fire Barrier Configurations," requested licensees to evaluate their facilities to confirm compliance with the existing applicable regulatory requirements in light of the information provided in the GL and, if appropriate, take additional actions.

# Information Notices (INs)

### Information Notice 91-47
Information Notice 91-47, "Failure of Thermo-Lag Fire Barrier Material to Pass Fire Endurance Test," August 6, 1991, gave licensees information on the fire endurance test performed by Gulf States Utilities Company on a Thermo-Lag 330-1 fire barrier installed on wide aluminum cable tray and the associated fire test failure. IN 91-47 alerted licensees with problems that could result from the use of or improper installation of Thermo-Lag material to satisfy the electrical raceway fire protection requirements for safe shutdown components specified in Section III.G.2 of Appendix R to 10 CFR 50. These Thermo-Lag ERFBS problems included not following manufacture installation procedures, missing stress skin and/or structural ribbing, and lack of documentation on qualification for large cable trays.

### Information Notice 91-79
Information Notice 91-79, "Deficiencies in the Procedures for Installing Thermo-Lag Fire Barrier Materials," December 6, 1991, identified problems that could results from improperly installed Thermo-Lag 330-1 fire barriers that are used to satisfy NRC fire protection requirements for safe shutdown components. After NRC review and discussion of the installation details with various Thermo-Lag 330-1 users, numerous variations were identified. At the time of issuance of IN 91-79, NRC had not been able to verify that all of the specific installation variations observed had been qualified by independent qualification testing or engineering analysis.

### Information Notice 92-46
Information Notice 92-46, "Thermo-Lag Fire Barrier Material Special Review Team Final Report Findings, Current Fire Endurance Tests, and Ampacity Calculation," June 23, 1992, presented information on failed Thermo-Lag 330-1 testing conducted by Texas Utilities for large cable trays and small cable conduits, mathematical errors found in the calculation of cable ampacity derating factors for Thermo-Lag fire resistive barriers, and provided the Special Review Team Final Report as an attachment. The attached report identified that some licensees have not

adequately reviewed and evaluated the fire endurance and ampacity test results for applicability to the Thermo-Lag fire barrier systems installed in their facilities. Additional findings included, incomplete or indeterminate fire test results, barrier installations that were not constructed in accordance with vendor-recommended installation procedures, incomplete installation procedures, and as-built fire barrier configurations that may not have been qualified by a valid fire endurance test or evaluated in accordance with the guidance of GL 86-10.

IN 92-46, "Thermo-Lag Fire Barrier Material Special Review Team Final Report Findings, Current Fire Endurance Tests, and Ampacity Calculation Errors," was issued to provide information of the findings of NRC's Thermo-Lag Special Review Team, current Thermo-Lag 330 fire resistance testing being conducted by Texas Utilities and Thermal Science, Inc., and errors found in the calculation of cable ampacity derating factors for Thermo-Lag fire resistive barriers.

### Information Notice 92-55
IN 92-55, "Current Fire Endurance Test Results for Thermo-Lag Fire Barrier Material, July 27, 1992," provided a brief summary of the Thermo-Lag 330 small scale testing results obtained by the National Institute of Standards and Technology (NIST). NIST conducted both 1-hour and 3-hour small scale fire endurance tests to determine the fire resistive properties of Thermo-Lag 330 pre-formed panels. The 1-hour barrier exceeded the temperature rise criteria in 22 minutes, while the 3-hour barrier lasted 2 hours and 20 minutes before exceeding the acceptance criteria.

### Information Notice 92-82
IN 92-82, "Results of Thermo-Lag 330-1 Combustibility Testing, December 15, 1992," provided information on the results of Thermo-Lag 330-1 combustibility testing that was conducted by NIST. Using ASTM E-136 and ASTM E-1354 testing standards, these tests revealed that Thermo-Lag 330-1 fire barrier material is combustible. Each thermo-Lag specimens tested exhibited a weight loss of greater than 50 percent and exhibited flaming beyond 30 seconds.

### Information Notice 93-40
Fire Endurance Test Results for Thermal Ceramics FP-60 Fire Barrier Material, May 26, 1993, alerted addresses of the results of the fire endurance and ampacity derating test report submitted by Thermal Ceramics on the FireMaster FP-60 fire barrier system and the results of NRC staff reviews.

### Information Notice 93-41
One Hour Fire Endurance Test Results for Thermal Ceramics Kaowool, 3M Company FS-195 and 3M Company Interam E-50 Fire Barrier Systems," May 28, 1993. IN 93-41 identified deficiencies associated with various fire barrier testing reviewed during an NRC reverification inspection conducted by NRC inspectors at the Salem Nuclear Generating Station in 1993.

### Information Notice 94-22
IN 94-22, "Fire Endurance and Ampacity Derating Test Results for 3-hour Fire-Rated Thermo-Lag 330-1 ire Barriers," March 16, 1994, provided information on the preliminary results of the fire endurance and ampacity derating tests of Thermo-Lag 330-1 fire barriers conducted by NRC at Underwriter Laboratories, Inc.

### Information Notice 94-34
IN 94-34,"Thermo-Lag 330-660 Flexi-Blanket Ampacity Derating Concerns," dated May 13, 1994, alerted addresses to a potential problem involving the use of non-conservative ampacity derating data by licensees based on previous representations by the vendor, Thermal Science, Inc.

### Information Notice 94-86
"Legal Actions against Thermal Science, Inc. Manufacturer of Thermo-Lag," December 22, 1994. IN 94-86 provided information to addresses on the grand jury indictment of Thermal Science, Inc. and Rubin Feldman, President, for allegedly making false statement to NRC and others concerning the fire barrier material known as Thermo-Lag. The IN also identified that, in a separate legal action, Industrial Testing Laboratories, Inc., and Alan M. Siegel, President, pleaded guilty to making and aiding and abetting the making of false statements concerning Thermo-Lag.

### Information Notice 94-86, Supplement 1
"Legal Actions Against Thermal Science, Inc., Manufacturer of Thermo-Lag," November 15, 1995. Supplement 1 to IN 94-86 provided the Federal jury finding Thermal Science, Inc. (TSI), and its president, Rubin Feldman, not guilty of making false statements about the role of Industrial Testing Laboratories (ITL) in the qualification testing of Thermo-Lag fire barriers.

### Information Notice 95-27
NRC Review of Nuclear Energy Institute, "Thermo-Lag 330-1 Combustibility Evaluation Methodology Plant Screening Guide," May 31, 1995, provided the results of NRC staff review of NEI Combustibility Evaluation Methodology. NRC staff informed NEI that it will not accept the use of the NEI methodology to justify the use of Thermo-Lag materials, or other material such as fire retardant plywood or cable jackets, as noncombustible where noncombustible materials are specified by NRC fire protection requirements.

### Information Notice 95-32
Thermo-Lag 330-1 Flame Spread Test Results, August 10, 1995. IN 95-32 provided the results of NRC flame spread testing of Thermo-Lag 330-1 fire barriers.

### Information Notice 95-49
Seismic Adequacy of Thermo-Lag Panels, October 27, 1995. IN 95-49 identified two specific NRC concerns related to the possibility of varying physical composition of the Thermo-Lag barrier in use across industry and the actual weights of Thermo-Lag use in plants.

### Information Notice 95-49, Supplement 1
Seismic Adequacy of Thermo-Lag Panels, December 10, 1997, provided the results of NRC sponsored mechanical properties testing of Thermo-Lag 330-1 fire barrier material performed by the National Institute of Standards and Technology (NIST). The results of the testing indicated significantly lower mechanical properties than those used by the vendor to demonstrate the seismic adequacy of Thermo-Lag 330-1 panels.

### Information Notice 95-52
"Fire Endurance Test Results for Electrical Raceway Fire Barrier Systems Constructed from 3M Company Interam Fire Barrier Materials," November 14, 1995. IN 95-52 provided the result of industry testing of 3M Interam fire barrier materials.

**Information Notice 95-52, Supplement 1**
Fire Endurance Test Results for Electrical Raceway Fire Barrier Systems Constructed from 3M Company Interam Fire Barrier Materials," March 17, 1998, identified additional fire endurance testing failures associated with a 3-hour 3M Interam Fire barrier system.

**Information Notice 97-59**
Fire Endurance Test Results of Versawrap Fire Barriers, August 1, 1997. IN 97-59 provides a summary of the testing results and several failures identified by NRC staff during a fire endurance test conducted at UL for Transco Products, Inc. of Chicago, Illinois.

**Information Notice 97-70**
Potential Problems with Fire Barrier Penetration Seals, September 19, 1997. IN 97-70 identified problems with installed fire barrier penetration seals that have gone undetected as a result of inadequate surveillance inspection procedures and inadequate acceptance criteria.

**Information Notice 99-17**
Problems Associated with Post-Fire Safe-Shutdown Circuit Analysis, June 3, 1999. IN 99-17 identified potential problems associated with post-fire safe-shutdown circuit analysis. These potential problems could result in a vulnerability to fire-induced circuit failures that could prevent the operation or lead to malfunction of equipment necessary to achieve and maintain post-fire safe shutdown.

**Information Notice 05-07**
Results of Hemyc Electrical Raceway Fire Barrier System Full Scale Fire Testing, April 1, 2005. IN 2005-07 informed addressees of the results of Hemyc electrical raceway fire barrier system (ERFBS) full-scale fire tests. The Hemyc ERFBS did not perform for one hour as designed because shrinkage of the Hemyc ERFBS occurred during the testing.

# Appendix D    Supplemental Test Result Summaries

## APPENDIX D
## TABLE OF CONTENTS

| **Table** | | **Page** |
|---|---|---|
| D.1 | General Caution on use of Information in Appendix D | D-2 |
| D.2 | 3M ERFBS | D-2 |
| D.2.1 | 3M Interam™ E-50 Series | D-2 |
| D.2.2 | 3M Rigid Panel System Testing | D-31 |
| D.3 | Thermo-Lag 330-1 | D-38 |
| D.3.1 | NUMARC Phase 1 Testing | D-38 |
| D.3.2 | NUMARC Phase 2 Testing | D-43 |
| D.3.3 | Texas Utilities (TU) Electric Co. Tests for Comanche Peak Steam Electric Station | D-50 |
| D.3.4 | Tennessee Valley Authority Tests for Watts Bar Nuclear Power Plant | D-66 |
| D.4 | References | D-83 |

## D.1 General Caution on use of information in Appendix D

This section provides information on supplement testing completed on various ERFBS. It is included in this report for completeness. Note that a majority of this testing was conducted prior to publishing of NRC guidance on testing ERFBS. Therefore, the following test results may not represent qualified ERFBS.

## D.2 3M ERFBS

### D.2.1 3M Interam™ E-50 Series

*D.2.1.1 OPL Ampacity Testing Project No. 14540-99074 and 14540-99075*

ERFBS: E54A  
Test Procedure: IEEE P848  
Post GL 86-10 Design

Report Date: March 25, 1996  
Test Dates: October 4 – 25, 1995

Materials used:
E-54A, Interam™ T-49 Tape, Interam™ FireDam™ 150 Caulk, 3M Scotch™ Brand Premium Grade Filament Tape P-898, ½ in x 0.020 in type 304 stainless steel banding straps and clips

1 in diameter galvanized RSC
12 ft long conduit, 13 ft long 4/C 10AWG, 36.36% fill actual, ERFBS Three (3) layers of Interam E-54A @ 0.4" nominal thickness (1.2" total nominal thickness)

12 Thermocouples 24 gauge Type T, Copper-Constantan (±0.5% Limits of Error), in direct contact with the top surface of the cable conductor, located 4@center, 4@36in to left of center, 4@36in to right of center

4in diameter galvanized RSB
12 ft long conduit, 12 3/C 6AWG 13 ft long cables, 42.19% fill actual, ERFBS Three (3) layers of Interam E-54A @ 0.4" nominal thickness (1.2" total nominal thickness)

36 Thermocouples 24 gauge Type T, Copper-Constantan (±0.5% Limits of Error), in direct contact with the top surface of the cable conductor, located 12@center, 12@36in to left of center, 12@36in to right of center (i.e., 3 TC on each cable)

The results of this testing are shown below in Table D-1.

Table D-1. 3M Interam E54A Ampacity Results 1" & 4" Conduits

| Test Item | Current (Amps) | Conductor Temp (°C) | Room Temp (°C) | Corrected Current* (Amps) | Ampacity Derating Factor % |
|---|---|---|---|---|---|
| 1" Conduit Baseline | 31.67 | 90.7 | 40.2 | 31.55 | - - - |
| 1" Conduit with ERFBS | 29.10 | 90.6 | 40.4 | 29.07 | 7.86 |
| 4" Conduit Baseline | 25.60 | 89.3 | 29.2 | 25.55 | - - - |
| 4" Conduit with ERFBS | 22.93 | 89.4 | 40.9 | 23.26 | 8.96 |

\* The corrected current values are calculated using IEEE P848/D16

*D.2.1.2 OPL Ampacity Testing Project No. 14540-100770*

ERFBS: E54A  
Test Procedure: IEEE P848  
Post GL 86-10 Design  

Report Date: December 5, 1996  
Test Dates: August 1 – October 10, 1996

Materials used:
E-54A, Interam™ T-49 Tape, Interam™ FireDam™ 150 Caulk, 3M Scotch™ Brand Premium Grade Filament Tape P-898, ½ in x 0.020 in type 304 stainless steel banding straps and clips

24 in Wide Cable Tray
24in wide x 4in deep cable tray assembly, 12 ft long, 96 3/C 6AWG, 58.90% actual fill, ERFBS 3 layers of Interam E54A @ 0.4" nominal thickness

15 Thermocouples 24 gauge Type T, Copper-Constantan (±0.5% Limits of Error), in direct contact with the top surface of the cable conductor, located 5@center, 5@36in to left of center, 5@36in to right of center and all were within the second layer of cables (there were a total of 3 layers of cables).

The results of this testing are shown below in Table D-2.

Table D-2. 3M Interam™ E54A Ampacity Results 12" Wide Cable Tray

| Test Item | Current (Amps) | Conductor Temp (°C) | Room Temp (°C) | Corrected Current* (Amps) | Ampacity Derating Factor % |
|---|---|---|---|---|---|
| 24" Tray Baseline | 28.34 | 90.2 | 40.7 | 28.49 | - - - |
| 24" Tray with ERFBS | 14.43 | 90.7 | 39.5 | 14.28 | 49.88 |

\* The corrected current values are calculated using IEEE P848/D16

D.2.1.3 OPL Ampacity Test Project No. 8610-102164 and 8610-102165

ERFBS: E54A  
Test Procedure: IEEE P848  
Post GL 86-10 Design

Report Date: December 26, 1997  
Test Dates: November 17 – 25, 1997

Materials used:
E-54A, Interam™ T-49 Tape, Interam™ FireDam™ 150 Caulk, 3M Scotch™ Brand Premium Grade Filament Tape P-898, ½ in x 0.020 in type 304 stainless steel banding straps and clips

1 in diameter galvanized RSC
12 ft long conduit, 13 ft long 4/C 10AWG, 36.36% fill actual, ERFBS five layers of Interam E-54A @ 0.4" nominal thickness (1.2" total nominal thickness)

12 Thermocouples 24 gauge Type T, Copper-Constantan (±0.5% Limits of Error), in direct contact with the top surface of the cable conductor, located 4@center, 4@36in to left of center, 4@36in to right of center

4in diameter galvanized RSB
12 ft long conduit, 12 3/C 6AWG 13 ft long cables, 42.19% fill actual, ERFBS five layers of Interam E-54A @ 0.4" nominal thickness (1.2" total nominal thickness)

36 Thermocouples 24 gauge Type T, Copper-Constantan (±0.5% Limits of Error), in direct contact with the top surface of the cable conductor, located 12@center, 12@36in to left of center, 12@36in to right of center (i.e., 3 TC on each cable)

The results of this testing are shown below in Table D-3.

Table D-3. 3M Interam E54A Ampacity Results

| Test Item | Current (Amps) | Conductor Temp (°C) | Room Temp (°C) | Corrected Current* (Amps) | Ampacity Derating Factor % |
|---|---|---|---|---|---|
| 1" Conduit Baseline | 31.67 | 90.7 | 40.2 | 31.55 | - - - |
| 1" Conduit with ERFBS | 25.12 | 90.0 | 40.1 | 25.15 | 20.29 |
| 4" Conduit Baseline | 29.59 | 90.6 | 39.2 | 29.21 | - - - |
| 4" Conduit with ERFBS | 19.20 | 89.5 | 38.6 | 19.01 | 34.92 |

* The corrected current values are calculated using IEEE P848/D16

*D.2.1.4 OPL Ampacity Testing Project No. 8610-102166*

ERFBS: E54A  
Test Procedure: IEEE P848  
Post GL 86-10 Design

Report Date: December 23, 1997  
Test Dates: October 27 – November 10, 1997

Materials used:
E-54A, Interam™ T-49 Tape, Interam™ FireDam™ 150 Caulk, 3M Scotch™ Brand Premium Grade Filament Tape P-898, ½ in x 0.020 in type 304 stainless steel banding straps and clips

24 in Wide Cable Tray
24in wide x 4in deep cable tray assembly, 12 ft long, 96 3/C 6AWG, 58.90% actual fill, ERFBS six layers of Interam E54A @ 0.4" nominal thickness

15 Thermocouples 24 gauge Type T, Copper-Constantan (±0.5% Limits of Error), in direct contact with the top surface of the cable conductor, located 5@center, 5@36in to left of center, 5@36in to right of center and all were within the second layer of cables (there were a total of 3 layers of cables).

The results of this testing are shown below in Table D-4.

**Table D-4. 3M Interam™ E54A Ampacity Results 12" Wide Cable Tray**

| Test Item | Current (Amps) | Conductor Temp (°C) | Room Temp (°C) | Corrected Current* (Amps) | Ampacity Derating Factor % |
|---|---|---|---|---|---|
| 24" Tray Baseline | 28.34 | 90.2 | 40.7 | 28.49 | - - - |
| 24" Tray with ERFBS | 12.36 | 90.8 | 40.7 | 12.36 | 56.62 |

* The corrected current values are calculated using IEEE P848/D16

*D.2.1.5 OPL Fire Endurance Test Report Project No. 14540-99416*

ERFBS: E54A  
Acceptance Criteria: GL 86-10 Supplement 1  
Post GL 86-10 Design

Report Date: February 26, 1996  
Test Dates: January 22, 1996  
Fire Test Duration: 3-hr

All conduits were supported by a common trapeze type hanger formed from Unistrut P1000 channel. The blockouts in the concrete floor slab were sealed with Five Star Grout, installed to the full thickness of the slab. In addition to the 3M Interam™, Dow Corning 732 adhesive/sealant and FireDam™ FD-150 Caulk were used to fill gaps and seams and T-49 Aluminum Foil Tap to secure overlap joints.

5in Conduit Assembly
5in diameter galvanized RSC and fittings assembled into an "U-shaped" configuration, having an overall horizontal dimension of 102in. and an overall vertical dimension of 76in. at each leg. One leg had a 90 degree radius elbow, while the other used an iron condulet LB. Five layers of 3M Interam™ E54A mat material.

### 3in Conduit Assembly

3in diameter galvanized RSC and fittings assembled into an "U-shaped" configuration, having an overall horizontal dimension of 102in. and an overall vertical dimension of 76in. at each leg. One leg had a 90 degree radius elbow, while the other used an iron condulet LB. The vertical section containing the condulet also transitioned into a steel junction box (12 in x 12 in x 8 in). *Five* layers of 3M Interam™ E54A mat material to protect the conduit and six layers to protect the junction box.

### 1in Conduit Assembly

1in diameter galvanized RSC and fittings assembled into an "U-shaped" configuration, having an overall horizontal dimension of 102in. and an overall vertical dimension of 76in. at each leg. One leg had a 90 degree radius elbow, while the other used an iron condulet LB. Five layers of 3M Interam™ E54A mat material.

**Table D-5. 3M Interam™ E54A Fire Endurance Testing 1", 3", & 5" RSC (OPL)**

| Thermocouple Location | Max. Individual Temperature °C (°F) | Max Average Temperature °C (°F) |
|---|---|---|
| 5" Steel Conduit | | |
|     Conduit Surface | 153 (307) | 117 (243) |
|     Bare #8 Wire in Conduit | 132 (269) | 108 (227) |
| 3" Steel Conduit | | |
|     Conduit Surface | 153 (307) | 114 (237) |
|     Bare #8 Wire in Conduit | 132 (270) | 153 (220) |
| 1" Steel Conduit | | |
|     Conduit Surface | 192 (377) | 153 (307) |
|     Bare #8 Wire in Conduit | 174 (346) | 148 (299) |
| Steel Junction Box | 107 (224) | 101 (213) |

Acceptance Criteria: Single pt. T≤196°C; Average T≤154°C

### Results

All raceway items met the requirements of the TEST PLAN for a fire resistance rating of three hours. The TEST PLAN followed GL 86-10 Supplement 1 guidance as it related to acceptance criteria of the fire exposure and hose stream test.

*D.2.1.6 OPL Fire Endurance Test Report Project No. 14540-99417*

ERFBS: E54A  
Acceptance Criteria: GL 86-10 Supplement 1  
Post GL 86-10 Design  

Report Date: September 22, 1997  
Test Dates: August 21, 1997  
Fire Test Duration: 3-hours  

Both cable tray assemblies were supported by a common trapeze type hanger formed from 3x4.1 steel channel. The blockouts in the concrete floor slab were sealed with Dow Corning 3-6548 RTV silicone foam material. In addition to the 3M Interam™, Dow Corning 732 adhesive/sealant and FireDam™ FD-150 Caulk were used to fill gaps and seams and T-49

Aluminum Foil Tap to secure overlap joints.

6" Cable Tray
B-Line Systems, Inc. 6 in. wide by 4 in. deep, ladder back cable tray, assembled into an "U-shaped" configuration having a horizontal dimension of 104 in. and a vertical dimension of 76 in. at each leg. Six layers of 3M Interam™ E54A mat.

24" Cable Tray
B-Line Systems, Inc. 24 in. wide by 4 in. deep, ladder back cable tray, assembled into a "U-shaped" configuration having a horizontal dimension of 104 in. and a vertical dimension of 76 in. at each leg. Six layers of 3M Interam™ E54A mat.

**Table D-6. 3M Interam™ E54A Fire Endurance Testing Steel Cable Trays (OPL)**

| Thermocouple Location | Max. Individual Temperature °C (°F) | Max Average Temperature °C (°F) |
|---|---|---|
| 6" Cable Tray | | |
| Front Tray Side Rail | 159 (319) | 124 (256) |
| Rear Tray Side Rail | 167 (333) | 129 (264) |
| Bare #8 Wire on Rungs | 151 (303) | 119 (247) |
| 24" Cable Tray | | |
| Front Tray Side Rail | 163 (325) | 133 (271) |
| Rear Tray Side Rail | 157 (315) | 132 (269) |
| Bare #8 Wire on Rungs | 142 (287) | 111 (231) |

Acceptance Criteria: Single pt. T≤214°C; Average T≤172°C

Results

Following the hose stream test, the outermost layer of 3M material on the supports and raceway assemblies had become dislodged by the water hose stream.

For the 6 in wide cable tray, the 6$^{th}$ (outermost) layer of 3M material was completely dislodged by the hose stream test. The 5$^{th}$ layer was also mostly eroded from the water stream. The 4$^{th}$ layer foil covering was intact and the mat material itself uncharred, however the material was completely discolored (green). The 3$^{rd}$ layer foil covering was intake, as well as the mat material, and this layer was only partially discolored. The 2$^{nd}$ layer, the standoff between the 2$^{nd}$ and 1$^{st}$ layer and the 1$^{st}$ layer were completely intake and uncharred. Much of the material covering the supports underneath the tray was washed away during the water hose stream. The material (in all layers and/or areas which survived the water stream) was completely discolored (green) in the vicinity of the support member.

The 24 in. wide cable tray assembly had the 6$^{th}$ (outermost) layer of 3M material completely dislodged by the hose stream test. The 5$^{th}$ layer was also mostly eroded from the water stream. The 4$^{th}$ layer foil covering was intact and the mat material itself uncharred, however the material was completely discolored (green). The 3$^{rd}$ layer foil covering was intact, as well as the mat material, and this layer was only partially discolored. The 2$^{nd}$ layer, the standoff between the 2$^{nd}$ and 1$^{st}$ layer and the 1st layer were completely intact and uncharred. Much of the material covering ths supports underneath the tray was washed away during the water hose stream. The material (in all layers and/or areas which survived the water stream) was completely discolored (green) in the vicinity of the support member.

The report concluded that, "all of the raceway item evaluated in the fire exposure test, clad with 3M Interam™ Mat material met the requirements of the TEST PLAN," which based its acceptance criteria and followed the guidance of GL 86-10 Supplement 1.

### D.2.1.7 OPL Test Report Project No. 8610-102570

ERFBS: E54A  
Acceptance Criteria: GL 86-10 Supplement 1  
Post GL 86-10 Design  

Report Date: May 19, 1998  
Test Dates: March 19, 1998  
Fire Test Duration: 3-hour  

Article 1: Three 4" diameter conduits and three 2" diameter conduits (conduits 1-6). Five layers of 3M Interam™ E54A.

Article 2: Steel junction box of dimensions 36" x 19" x 10" (Junction Box #3). Six layers of 3M Interam™ E54A.

Article 3: Steel junction box 12" x 12" x 8" with one face against the concrete (Junction Box #1). Six layers of 3M Interam™ E54A.

Article 4: 1" diameter RSC with multiple interfering items (conduit #10 (1" diameter), conduits 11 (2" diameter), and conduit 12 (4" diameter)). Five layers of 3M Interam™ E54A.

Article 5: A trapeze hanger configuration consisting of ½"-13 threaded rod and P-1001 Unistrut™. Five layers of 3M Interam™ E54A within 12 in. of critical item and four layers for the remainder of the length.

Article 6: 4" diameter RSC in close proximity to concrete (conduit #9). Five layers of 3M Interam™ E54A.

Article 7: Steel Junction Box 12" x 12" x 8" with one face against the concrete (Junction Box #2). Six layers of 3M Interam™ E54A.

**Table D-7. 3M Interam E54A Fire Endurance Test Conduits & Junction Boxes (OPL)**

| Thermocouple Location | Max. Individual Temperature °C (°F) | Max Average Temperature °C (°F) |
|---|---|---|
| Conduit #1 | | |
| Bare #8 Conductor | 128 (262) | 108 (226) |
| Conduit Surface | 145 (293) | 110 (230) |
| Conduit #2 | | |
| Bare #8 Conductor | 108 (226) | 96 (204) |
| Conduit Surface | 111 (232) | 95 (203) |
| Conduit #3 | | |
| Bare #8 Conductor | 116 (240) | 98 (208) |
| Conduit Surface | 129 (264) | 102 (215) |
| Conduit #4 | | |
| Bare #8 Conductor | 103 (217) | 92 (197) |
| Conduit Surface | 98 (208) | 92 (197) |

Acceptance Criteria: Single pt. T≤198°C; Average T≤155°C

### Table D-7. 3M Interam E54A Fire Endurance Test Conduits & Junction Boxes (OPL) (Continued)

| Thermocouple Location | Max. Individual Temperature °C (°F) | Max Average Temperature °C (°F) |
|---|---|---|
| Conduit #5 | | |
|     Bare #8 Conductor | 128 (262) | 102 (216) |
|     Conduit Surface | 142 (287) | 109 (228) |
| Conduit #6 | | |
|     Bare #8 Conductor | 106 (222) | 94 (202) |
|     Conduit Surface | 109 (229) | 94 (202) |
| Conduit #9 | | |
|     Conduit Surface | 144 (292) | 118 (244) |
| Conduit #10 | | |
|     Bare #8 Conductor | **206 (403)** | 152 (306) |
|     Conduit Surface | **209 (409)** | **164 (328)** |
| Junction Box #1 | 117 (243) | 152 (306) |
| Junction Box #2 | **989 (1813)** | **918 (1684)** |
| Junction Box #3 | 123 (253) | 112 (233) |
| Trapeze Hangers | **753 (1387)** | **579 (1075)** |

Acceptance Criteria: Single pt. T≤198°C; Average T≤155°C

Results

Articles #1-3, and #6 met the requirements of the TEST PLAN and therefore provide a 3-hour fire endurance rating based on criteria specified in GL 86-10 Supplement 1. Article #7 failed prior to the hose stream test when its protective barrier fell off at approximately 68 minutes into the testing. Article 4 failed to meet the temperature rise criteria at approximately 174 minutes of the 180 minute qualification test.

### Table D-8. 3M Interam E54A Test Results (OPL Project No. 8610-102570)

| Test Item | # layers of ERFBS E54A | Fire Endurance Rating (minutes) |
|---|---|---|
| Article 1 (Conduits 1-6) | 5 | 180 |
| Article 2 (Junction Box #3) | 6 | 180 |
| Article 3 (Junction Box #1) | 6 | 180 |
| Article 4 (Conduit #10) | 5 | 174 |
| Article 6 (Conduit #9) | 5 | 180 |
| Article 7 (Junction Box #2) | 6 | 68 |

*D.2.1.8 OPL Test Report Project No. 14540-99123*

ERFBS: E54A  
Acceptance Criteria: GL 86-10 Supplement 1  
Post GL 86-10 Design  

Report Date: December 5, 1995  
Test Dates: November 28, 1995  
Fire Test Duration: 3-hr  

The two cable trays were supported by a common trapeze type hanger formed from 3x4.1 steel channel. The three conduit assemblies were supported on a common trapeze type hanger formed from Unistrut™ P1000 channel. The blockouts in the concrete slab were sealed with Dow Corning 3-6548 RTV silicone foam material. Electrical cables (power, control, and instrumentation) were installed in both cable tray assemblies. Table D-9 identifies the percent cable fill in each cable tray.

### 6" Cable Tray
B-Line Systems Inc., 6 in. wide x 4 in. deep, ladder back cable tray, assembled into a "U-shaped" configuration having a horizontal dimension of 104 in. and a vertical dimension of 76 in. at each leg. Five layers of 3M Interam™ E54A.

### 24" Cable Tray
B-Line Systems Inc., 24 in. wide x 4 in. deep, ladder back cable tray, assembled into a "U-shaped" configuration having a horizontal dimension of 104 in. and a vertical dimension of 76 in. at each leg. A 2in. conduit stub assembly transitioned through the upper steel deck, extending 8 in. below the deck insulation, forming an air drop which transitioned into the center of the horizontal section of the 24 in. cable tray. Five layers of 3M Interam™ E54A.

### 5" Conduit
5 in. diameter galvanized RSC and fittings assembled into a "U-shaped" configuration having an overall horizontal dimension of 102 in. and an overall vertical dimension of 76 in. at each leg. One leg of the conduit assembly transitioned through the upper steel deck into a standard radius 90 degree elbow while the other end used an iron condulet LB. Five layers of 3M Interam™ E54A.

### 3" Conduit
3 in. diameter galvanized RSC and fittings assembled into a "U-shaped" configuration having an overall horizontal dimension of 102 in. and an overall vertical dimension of 76 in. at each leg. One leg of the conduit assembly transitioned through the upper steel deck into a standard radius 90 degree elbow while the other end used an iron condulet LB. The vertical leg connected to the condulet LB also transitioned through a junction box (12 in. x 12 in. x 8in.). The top of the junction box was located 12 in. below the insulated steel deck. Five layers of 3M Interam™ E54A.

### 1" Conduit
1 in. diameter galvanized RSC and fittings assembled into a "U-shaped" configuration having an overall horizontal dimension of 102 in. and an overall vertical dimension of 76 in. at each leg. One leg of the conduit assembly transitioned through the upper steel deck into a standard radius 90 degree elbow while the other end used an iron condulet LB. Five layers of 3M Interam™ E54A.

## Table D-9. 3M Interam E54A Fire Endurance Test Tray & Conduits (OPL)

| Thermocouple Location | Cable Fill (% Actual) | Max. Individual Temperature °C (°F) | Max Average Temperature °C (°F) |
|---|---|---|---|
| <u>6" Cable Tray</u> | 19.43 | | |
| Front Tray Side Rail | | **223 (433)** | **196 (384)** |
| Rear Tray Side Rail | | **225 (437)** | **197 (387)** |
| Bare #8 Wire under Rungs | | **208 (406)** | **173 (344)** |
| Bare #8 Wire on Rungs | | 191 (376) | **159 (318)** |
| <u>24" Cable Tray</u> | 14.72 | | |
| Front Tray Side Rail | | **211 (412)** | **188 (371)** |
| Rear Tray Side Rail | | **215 (419)** | **187 (369)** |
| Bare #8 Wire under Rungs | | 128 (263) | 113 (235) |
| Bare #8 Wire on Rungs | | 142 (287) | 127 (261) |
| <u>5" Steel Conduit</u> | | | |
| Conduit Surface | | **201 (393)** | **163 (325)** |
| Bare #8 Wire | | 189 (373) | 126 (259) |
| <u>3" Steel Conduit</u> | | | |
| Conduit Surface | | **233 (452)** | **214 (418)** |
| Bare #8 Wire | | **224 (436)** | **184 (363)** |
| <u>1" Steel Conduit</u> | | | |
| Conduit Surface | | **269 (517)** | **244 (472)** |
| Bare #8 Wire | | **264 (507)** | **238 (461)** |
| <u>2" Conduit Air Drop</u> | | | |
| Bare #8 Wire in Conduit | | 170 (338) | 146 (294) |
| Conduit Stub | | 172 (342) | **167 (333)** |
| <u>Steel Junction Box</u> | | **209 (409)** | **198 (389)** |

Acceptance Criteria: Single pt. T≤199°C; Average T≤157°C

*D.2.1.9 SwRI Project No. 01-7912 (June 1984)*

ERFBS: E50A  
Test Procedure: ANI/MAERP  
Ambient Temperature:  
Desired Rating: 1-hr  
Test Slab: 10' x 12' x 1'  
Acceptance Temperature:  

<u>Thermocouples</u>: Spaced every 12" embedded in cable bundles (no bare conductor)  
<u>Hose Stream</u>: 2-½" Playpipe, 1-1/8" tip, nozzle pressure 30psi, applied 20' from system, for 2.5 minutes  
<u>Furnace</u>: 8' x 10'  
<u>Furnace control</u>: 7 thermocouples (three above cable tray, four below)  
<u>Cable Type</u>: N/A  

**Table D-10. E-50A 1-hr (SwRI 01-7912)**

| Raceway Type | Barrier Protection | Cable Fill | Rating |
|---|---|---|---|
| 2 in dia. Conduit | Five layers of E-50A | 40% 1 - control | 1 - hr |
| 5 in dia. Conduit | Four layers of E-50A | 40% 3-(p), 9-(c), 34-(i) | 1 – hr |
| Cable Tray No. 1 (Solid bottom) | Four layers of E-50A | Single Layer 7-(p), 13-(c), 25-(i) | 1 – hr |
| Cable Tray No. 2 (Solid bottom) | Four layers of E-50A | 40% 13-(p), 47-(c), 166-(i) | 1 – hr |
| Cable Tray No. 3 (Ladder back) | Four layers of E-50A | Single Layer 7-(p), 13-(c), 25-(i) | 1 – hr [a] |
| Air Drop 1 | Six layers of E-50A | 1-(p), 1-(c), 1-(i) | 1 – hr |
| Air Drop 2 | Six layers of E-50A | 1 – (c) | 1 – hr |
| Air Drop 3 | Six layers of E-50A | 1-(p), 1-(c), 1-(i) | 1 - hr |

[a] Significant heat damage to outer jackets of cables. The insulation on the interior conductors appears intact. Some heat discoloration of tray.  
(p) = power cable    (c) = control cable  (i) = instrument cable

Post hose stream test observations indicated that some of the 3M fire-proofing material was dislodged by the hose stream.

D.2.1.10  SwRI Project 01-7912a(1) (June 1985)

ERFBS: E-50A
Test Procedure: ANI/MAERP
Ambient Temperature: 64°F
Desired Rating: 1-hr
Test Slab: 10' x 13.5' x 1' (3000 psi concrete)

Thermocouples: Spaced every 12" embedded in cable bundles (no bare conductor)
Hose Stream: 2-½" Playpipe, 1-1/8" tip, nozzle pressure 30psi, applied less than 20' from system, for 2.5 minutes.
Furnace control: 7 thermocouples (three above cable tray, four below)
Cable Type: (XLPE/PVC); Power (300 mcm); Control (7/c 12AWG); Instrument (2/c 16 AWG)

**Table D-11. E-50A 1-hr (SwRI 01-7912a(1))**

| Raceway Type | Barrier Protection | Cable Fill | 1-hr Max. Temp °C (°F) | Rating |
|---|---|---|---|---|
| 2 in Conduit | 4 layer of E-50A | 1 – control | 102 (215) | 1-hr |
| Cable tray No 1 (ladder back) | 4 layer of E-50A | 7 – power<br>13 – control<br>25 – instrument | 164 (328) | 1-hr |
| Cable Tray No 2 (ladder back) | 4 layer of E-50A | 13 – power<br>47 – control<br>166 - instrument | 102 (215) | 1-hr |

Note: all supports were protected with 4 layers of E-50A up to 16" from cable tray

At approximately 51 minutes into test, a 300mcm located in Cable Tray No. 2 experienced a short circuit to ground. Post test inspections revealed that a screw used to affix a thermocouple had penetrated the insulation of the cable causing the short. The hose stream testing dislodged some of the 3M fire barrier material. The 2/c #16 AWG instrument cable showed signs of shrinkage in the two cable tray tests. There was no other cable damaged noted in the final test report.

*D.2.1.11 UL R10125, 86NK2919 (July 1986)*

ERFBS: E-50D  
Test Procedure: UL Subject 1724  
Ambient Temperature: 68°F  
Desired Rating: 3-hr  
Test Slab: 8" thick concrete  
Acceptance Temperature:  

<u>Hose Stream</u>: 4 minutes, 1-1/8" dia. nozzle, 30psi, 20 ft from center of test assembly  
<u>Furnace control</u>: 10 Thermocouples located approximately 12-inches from underside of floor assembly.  
<u>Cable Type</u>: (XLPE/PVC); Control (7/c 12AWG); Instrument (2/c 16AWG)

**Table D-12. UL Test of 3M E-50D 3-hr (R10125, 86NK2919)**

| Raceway Type | Barrier Protection | Cable Fill | 1-hr Max Temp Rise °C (°F) | Rating |
|---|---|---|---|---|
| 5" dia. steel conduit | 5 layers of E-50D | 2 – control<br>2 – instrument | 121 (250)<br>conduit surface | 3-hr |
| Junction Box 10"x10"x6" | 5 layers of E-50D | | 133 (271)<br>box surface | 3-hr |
| Air drop | 5 layers of E-50D | 4 – control<br>4 – instrument | 144 (291)<br>bare #8 | 3-hr |

Each conductor in each electrical cable was energized and monitored for circuit integrity throughout the fire endurance test. No through openings developed in the electrical circuit protective systems through which the conduit system, junction box, or cable air drop could be seen. No electrical faults occurred in any of the electrical cables within the ERFBS.

### D.2.1.12  SwRI Project 01-7912(2) (June 1985)

**ERFBS:** E-50A  
**Test Procedure:** ANI/MEARP  
**Ambient Temperature:**  
**Desired Rating:** 1-hr  
**Test Slab:** 10' x 12' x 1' (3000 psi)  
**Acceptance Temperature:**  

**Thermocouples:** embedded into cable bundles and spaced at 12-inch intervals.  
**Hose Stream:** 2-½" Playpipe, 1-1/8" nozzle, 30 psi, 2.5 minutes, 20ft from test assembly  
**Furnace control:** 7 thermocouples (three 12" above cable trays and four 12" below)  
**Cable Type:** Power (300mcm); Control (7/c 12AWG); Instrument (2/c 16AWG)

**Table D-13. E-50A 1-hr (SwRI 01-7912(2))**

| Raceway Type | Barrier Protection | Cable Fill | Rating |
|---|---|---|---|
| 2-Inch Conduit | 5 layers of E-50A | Min. Fill<br>1 – control | 1-hr |
| 5-Inch Conduit | 4 layers of E-50A | 40% Fill<br>3 – power<br>9 – control<br>34 – instrument | 1-hr |
| Cable Tray No 1 (Solid back) | 4 layers of E-50A | Single Layer<br>7 – power<br>13 – control<br>25 – instrument | 1-hr |
| Cable Tray No 2 (Solid back) | 4 layers of E-50A | 40% Fill<br>13 – power<br>47 – control<br>166 – instrument | 1-hr |
| Cable Tray No 3 (Ladder back) | 4 layers of E-50A | Single Layer<br>7 – power<br>13 – control<br>25 – instrument | 1-hr |
| Air Drop No 1 | 6 layers of E-50A | 1 – power<br>1 – control<br>1 – instrument | 1-hr |
| Air Drop No 2 | 6 layers of E-50A | 1 – control | 1-hr |
| Air Drop No 3 | 6 layers of E-50A | 1 – power<br>1 – control<br>1 – instrument | 1-hr |
| Junction Box | 5 layers of E-50A | | 1-hr |
| Supports | 2 layers of E-50A | | 1-hr |

Post test examination immediately after the hose stream test identified that some of the 3M fireproofing material was dislodged by the hose stream. Examination of the cables indicated that cable tray No 3 experienced significant heat damage to the outer jackets of cables. The insulation on the interior conductors appeared to be intact and there was some heat discoloration of the tray.

### D.2.1.13 UL R10125-3, 84NK23288 (May 1986)

ERFBS: E-50A  
Test Procedure: UL Subject 1724  
Ambient Temperature: 72°F  
Desired Rating: 3-hr  
Test Slab: 8" thick concrete  
Acceptance Temperature:  

<u>Thermocouples</u>: Thermocouples were placed on exterior of raceways, exterior of cables and on bare No. 8 AWG conductor  
<u>Hose Stream</u>: 2.5 minutes, 1-1/8 in nozzle, 30psi, applied ~17feet from center of test assembly  
<u>Furnace control</u>: 16 Thermocouples placed 12 inches beneath furnace floor  
<u>Cable Type</u>: Power (300mcm XLPE/XLP); Control (7/c 12AWG XLPE/PVC); Instrument (2/c 14AWG XLPE/PVC)

**Table D-14. UL Subject 1724 Test Results of E-50A 3-hr (R10125-3, 84NK23288)**

| Test Article # | Raceway Type | Barrier Protection | Cable Fill | Max. Temp Rise ΔT °C (°F) | Rating |
|---|---|---|---|---|---|
| 1 | 24" wide Cable Tray Open Ladder Type | 5 Layers of E-50A | Single Layer 10 – Power 13 – Control 26 - Instrument | 113 (235) | 3-hr |
|  | Bare #8 Upper |  |  | 115 (239) |  |
|  | Bare #8 Lower |  |  | 85 (185) |  |
| 2 | 2" dia. Conduit | 5 Layers of E-50A | 1 – Power 1 – Control 1 – Instrument | 134 (274) | 3-hr |
|  | Bare #8 |  |  | 129 (264) |  |

Approximately 1 minute into test, CS-195 used at the floor interface as part of the penetration seal began to burn, the flaming continued for 30 minutes. Results section of report concluded that, no openings developed in the ERFBS through which the cable tray or conduit systems could be seen and no electrical faults occurred in any of the electrical cables within the electrical circuit protective systems.

### D.2.1.14 UL R10125-3, 84NK2919 (June 1986)

ERFBS: E-50D  
Test Procedure: UL Subject 1724  
Ambient Temperature:  
Desired Rating: 3-hr  
Test Slab: 8" thick concrete  
Acceptance Temperature:  

<u>Furnace control</u>: Ten Thermocouples located ~12inches below assembly floor  
<u>Cable Type</u>: (XLPE/PVC); Control (7/c 12AWG); Instrument (2/c 16AWG)

**Table D-15. UL Subject 1724 Test Results for E-50D 3-hr (R10125-3, 84NK2919)**

| Raceway Type | Barrier Protection | Cable Fill | Max. Temp Rise ΔT °C (°F) | Rating |
|---|---|---|---|---|
| 5" dia. steel conduit | 5 Layers of E-50D | 2 – Control<br>2 – Instrument<br>1 – Bare #8 | 121 (250) | 3-hr |
| Junction Box 10"x10"x6" | 5 Layers of E-50D | | 133 (271) | 3-hr |
| Air Drop | 5 Layers of E-50D | 4 – Control<br>4 – Instrument<br>1 – Bare #8 | 144 (291) | 3-hr |

*D.2.1.15 UL R10125, 82NK21937 (March 1985)*

<u>ERFBS</u>: E-50A  
<u>Test Procedure</u>: UL Subject 1724  
<u>Ambient Temperature</u>: 88°F  
<u>Thermocouples</u>: Bare No. 8  
<u>Hose Stream</u>: 30psi, 2.5 minutes  
<u>Furnace control</u>: Not specified  
<u>Cable Type</u>: Bare No. 8 AWG  

<u>Desired Rating</u>: 1-hr  
<u>Test Slab</u>: 8" thick concrete  

**Table D-16. UL Subject 1724 Test Results of E-50A 1-hr (R10125, 82NK21937)**

| Raceway Type | Barrier Protection | Cable Fill | Rating |
|---|---|---|---|
| 2" conduit | 4 layers of E-50A | 1 – bare #8 | 57min |
| Air Drop | 5 layers of E-50A | 1 – bare #8 | 57min |
| Cable Tray | 2 layers of E-50A | 1 – bare #8 | 57min |
| Junction Box 12"x12"x6" | Single layer of M20A<br>Single layer of CS-195 on a steel framework. | 1 – bare #8 | 57min |

The water hose stream test eroded the outer layer of mate wrap on the conduit, air drop, and cable tray system as well as the intumescent sheet enclosure around the junction box. A separation was present at the interface of the conduit wrap and the drop-out cable wrap through which the conduit and the No. 8 AWG bare copper conductor were visible.

D.2.1.16 ULR10125, 82NK21937 (March 1985)

ERFBS: E-50A  
Test Procedure: Subject 1724  
Ambient Temperature: 88°F  
Thermocouples: Cables, cable tray, and support  
Hose Stream: 30psi, 2.5 minutes  
Furnace control: Not specified  
Cable Type: (XLPE/PVC); Power (300mcm); Control (7/c 12AWG); Instrument (2/c 16AWG)

Desired Rating: 1-hr  
Test Slab: 8" thick concrete

Table D-17. UL Testing Results for E-50A 1-hr (R10125, 82NK21937)

| Raceway Type | Barrier Protection | Cable Fill | Rating |
|---|---|---|---|
| 24" Cable Tray | 3 layers of E-50A | 40%<br>21 – Power<br>32 – Control<br>221 – Instrument | 1-hr |

No electrical faults developed in any of the electrical cables within the fire barrier system during the fire exposure test, during the move to the hose stream test area or during the hose stream test. The water hose stream test did erode the outer layer of mat wrap on the cable tray system, but no through opening developed in the ERFBS through which the cable tray or cables could be seen. Within the ERFBS the cable PVC jacketing of the 7/c control cable did experience bubbling near the top of the cable tray inside riser elbow.

D.2.1.17 UL R10125, 82NK21937 (March 1985)

ERFBS: E-50A  
Test Procedure: Subject 1724  
Ambient Temperature: 88°F  
Thermocouples: Cables, cable tray, and support  
Hose Stream: 30psi, 2.5 minutes  
Furnace control: Not specified  
Cable Type: (XLPE/PVC); Power (300mcm); Control (7/c 12AWG); Instrument (2/c 16AWG)

Desired Rating: 1-hr  
Test Slab: 8" thick concrete  
Acceptance Temperature:

Table D-18. E-50A UL Test Results (R10125, 82NK21937)

| Raceway Type | Barrier Protection | Cable Fill | Rating |
|---|---|---|---|
| 24" Cable Tray | 3 layers of E-50A | Single layer<br>10 – Power<br>13 – Control<br>33 – Instrument | 55 min |

No electrical faults developed in any of the electrical cables within the fire barrier system during the fire exposure test, during the move to the hose stream test area or during the hose stream test. The water hose stream test did erode the outer layer of mat wrap on the cable tray system, but no through opening developed in the ERFBS. Within the ERFBS the cable PVC jacketing of the 7/c control cable did experience melting at the inside riser elbow.

### D.2.1.18 UL R10125, 84NK23299 (May 1985)

<u>ERFBS</u>: E-50A  
<u>Test Procedure</u>: Subject 1724  
<u>Ambient Temperature</u>: 57°F  
<u>Thermocouples</u>: conduit, bare No. 8AWG, cables  
<u>Hose Stream</u>: Not specified  
<u>Furnace control</u>: 9 thermocouples located ~12" below test deck floor  
<u>Cable Type</u>: (XLPE/PVC) Instrument (2/c 16 AWG)  
<u>Desired Rating</u>: 1-hour  
<u>Test Slab</u>: 8" thick concrete

**Table D-19. E-50A UL Test Results (R10125, 84NK23299)**

| System No. | Raceway Type | Barrier Protection | Cable Fill | Max. Temp Rise ΔT °C (°F) | Rating |
|---|---|---|---|---|---|
| 1 | 2" dia. steel conduit | 5 layers of E-50A | 1- Instrument 1-Bare No. 8 | 136 (277) | 1-hr |
| 2 | 3" dia. steel conduit | 5 layers of E-50A | 1- Instrument 1-Bare No. 8 | 120 (248) | 1-hr |
| 3 | 3" dia. steel conduit | 5 layers of E-50A | 1- Instrument 1-Bare No. 8 | 169 (336) | 59 min |
| 4 | 3" dia. steel conduit | 4 layers of E-50A | 1- Instrument 1-Bare No. 8 | 238 (460) | 46 min |
| 5 | 3" dia. steel conduit | 6 layers of E-50A | 1- Instrument 1-Bare No. 8 | 126 (259) | 1-hr |

Systems 3-5 were installed with a nominal ½ gap between all of the butt seams and were covered with one layer of Type T-49 tape. This assembly method is no recommended, but provided a worst case scenario for testing.

After 1 minute of fire exposure, steam began issuing form the ends of the conduit fire barrier systems which protruded from the test furnace. By 10 minutes, steam and water droplets were issuing from the ends of the conduit. The steaming and dripping continued throughout the testing. During the last 10 minutes of the fire endurance test, the light emitting diodes (LED's) associated with the conductors and ground of the two-conductor No. 16AWG cables in System 3 and 4 flickered on for a moment and then remained off for the remainder of the fire test.

PVC jacketing on the cables was melted such that the individual insulated conductors were visible in several locations. Beneath the PVC jacket, the XLPE insulation appeared undamaged.

*D.2.1.19  3M Fire Test Report #86-78 (June 1986)*

<u>ERFBS</u>: E-54A with steel banding on only the exterior wrap
<u>Test Procedure</u>: ANI/MAERP, UL#263
<u>Ambient Temperature</u>: 20°C (68°F)
<u>Thermocouples</u>: attached to cables
<u>Hose Stream</u>: 2.5 minutes, 1-1/8" nozzle, 30psi, 20 feet from test assembly
<u>Furnace</u>: 3M's Top Loading Furnace (61" x 47" x 30")
<u>Furnace control</u>: Not specified
<u>Cable Type</u>: (XLPE/PVC); Control (7/c 12AWG); Instrument (2/c 14AWG)

<u>Desired Rating</u>: 3-hr
<u>Test Slab</u>: 70" x 56" x 8" (3000psi) concrete

**Table D-20. E-54A 3M Test Results #86-78**

| Raceway Type | Barrier Protection | Cable Fill | Avg. Temp. °C (°F) | Max Temp °C (°F) | Rating |
|---|---|---|---|---|---|
| 2" steel conduit close to slab | 4 layers of E-54A | 2 – control<br>2 – instrument<br>1 – Bare #8 | 182 (360) | 216 (420) | <3hr |
| Three 2" steel conduits enclosed in common barrier | 4 layers of E-54A | <u>Per Conduit</u><br>2 – control<br>2 – instrument<br>1 – Bare #8 | 143 (289) | 168 (335) | 3hr |

All cables were energized with 120Vdc during the entire test and no loss of circuit integrity was detected. After the 2.5 minutes hose stream test, two of the original four layers of Interam™ E-54A ma were still securely attached to the conduit bundle.

*D.2.1.20  3M Fire Test Report #92-115 (August 1992)*

<u>ERFBS</u>: E-54A
<u>Test Procedure</u>: ASTM E-119
<u>Ambient Temperature</u>: 23°C (73°F)
<u>Thermocouples</u>: Conduit, cable, bare #8 conductor
<u>Hose Stream</u>: None
<u>Furnace control</u>: 4 thermocouples

<u>Desired Rating</u>: 3-hr
<u>Test Slab</u>: Not Specified
<u>Furnace</u>: 3M large scale furnace (63" x 52" x 52")
<u>Cable Type</u>: Control 7/c 12AWG

**Table D-21. E-54A 3M Test Results #92-115**

| Raceway Type | Barrier Protection | Cable Fill | Rating |
|---|---|---|---|
| 4" dia. steel conduit | 5 layers of E-54A | 1 – 7/c 12AWG<br>1 – bare #8 | 160 min |
| 4" dia. steel conduit | 3+2 layer system of E-54A | 1 – 7/c 12AWG<br>1 – bare #8 | 177 min |
| 1" dia. steel conduit | 6 layers of E-54A | 1 – 7/c 12AWG<br>1 – bare #8 | 3-hr |

One of the 4" dia. steel conduit has a 3+2 layer type system installation of the E-54A ERFBS. This method of assembly consisted of applying the 3 layer section first to the conduit and then applying a 2 layer assembly on top of the 3 layers previously applied. FireDam 150 Caulk was used to fill the space between the two layer assemblies and between the butt joints of adjacent layer assemblies.

*D.2.1.21  3M Fire Test Report #87-82 (July 1987)*

ERFBS: E-54A  
Test Procedure: ASTM E-119  
Ambient Temperature: 29°C (83°F)  
Thermocouples: Conduit, cable, bare #8 conductor  
Hose Stream: None  
Furnace: 3M large scale furnace (63" x 52" x 52")  
Furnace control: 4 thermocouples  
Cable Type: Control 7/c 12AWG  

Desired Rating: 3-hr  
Test Slab: Not Specified  
Acceptance Temperature:  

**Table D-22. E-54A 3M Test Results #87-82**

| Raceway Type | Barrier Protection | Cable Fill | Rating |
|---|---|---|---|
| 5" dia. steel conduit | 3+2 layer system of E-54A | 1 – control<br>1 – bare No. 8 | 166 min |

*D.2.1.22  3M Fire Test Report (July 1992)*

ERFBS: E-54A  
Test Procedure: ASTM E-119  
Ambient Temperature: 29°C (84°F)  
Thermocouples: Cable tray, cable, bare #8  
Hose Stream: None  
Furnace: 3M large scale furnace (63" x 52" x 52")  
Furnace control: 4 thermocouples  
Cable Type: Control 7/c 12AWG  

Desired Rating: 3-hr  
Test Slab: Not Specified  
Acceptance Temperature:  

**Table D-23. E-54A 3M Test Results (July 1992)**

| Raceway Type | Barrier Protection | Cable Fill | Rating |
|---|---|---|---|
| 24" wide cable tray open ladder back | 5 layers of E-54A | 1 – control<br>1 – bare No. 8 | 158 min |

### D.2.1.23 3M Test Report #92-167 (June 25, 1993)

ERFBS: E-53A
Test Procedure: ASTM E-119
Ambient Temperature: 22°C (71°F)
Thermocouples: Conduit, cable, bare #8
Hose Stream: Method not specified
Furnace: 3M large scale furnace (63" x 49" x 39")(1.6m x 1.2m x 1.0m)
Furnace control: 4 thermocouples
Cable Type: Control 7/c 12AWG

Desired Rating: 1-hr
Test Slab: 70" x 56" x 4" (3000 psi concrete)

**Table D-24. E-53A 3M Test Results #92-167**

| Raceway Type | Barrier Protection | Cable Fill | Avg. Temp. @1-hr °C (°F) | Max Temp @1-hr °C (°F) | Rating |
|---|---|---|---|---|---|
| 1" dia. steel conduit | 3 layers of E-53A without collars around radius bend seams | 1 – control 1 – bare #8 | 201 (394) | 221 (430) | <1-hr |
| 1" dia. steel conduit | 3 layers of E-53A with 3" collars around radius bend seams | 1 – control 1 – bare #8 | 174 (346) | 212 (413) | <1-hr |

### D.2.1.24 3M Fire Test Report #92-141 (August 1992)

ERFBS: E-54A
Test Procedure: ASTM E-119
Ambient Temperature: 21°C (70°F)
Thermocouples: Conduit, cable, bare #8
Hose Stream: Method not specified
Furnace: 3M large scale furnace (63" x 49" x 39")(1.6m x 1.2m x 1.0m)
Furnace control: 4 thermocouples
Cable Type: Control 7/c 12AWG

Desired Rating: 1- and 3-hours
Test Slab: 70" x 56" x 4.5" (3000 psi concrete)
Acceptance Temperature:

**Table D-25. E-54A 3M Test Results #92-141**

| Raceway Type | Barrier Protection | Cable Fill | Rating |
|---|---|---|---|
| 1" dia. steel conduit | 3 layers of E-54A | 1 – control 1 – bare #8 | 1-hr Failure at 106 min |
| 1" dia. steel conduit | 5 layers of E-54A | 1 – control 1 – bare #8 | 3-hr Failure >180 min |

After the hose stream test, the two innermost layers of E-54A applied to conduit # 2 remained intake. The surface of conduit #1 was exposed after the water hose stream test, however, this barrier was exposed to the full 3-hour exposure, even though it was only designed for the 1-hr rating. This is likely why the exposure of the conduit following the hose stream test.

*D.2.1.25  3M Fire Test Report #87-40 (April 1992)*

ERFBS: E-54A  
Test Procedure: ASTM E-119  
Ambient Temperature: 14°C (58°F)  
Thermocouples: Cable tray, cables  
Hose Stream: Method not specified  
Furnace: 3M large scale furnace (63" x 49" x 39")(1.6m x 1.2m x 1.0m)  
Furnace control: 4 thermocouples  
Cable Type: Control (7/c 12AWG), Power (1/c 250mcm)  

Desired Rating: 1- and 3-hour  
Test Slab: 70" x 56" x 4.5" (3000 psi concrete)

**Table D-26. E-54A 3M Test Results #87-40**

| Raceway Type | Barrier Protection | Cable Fill | Avg. Temp @ 1-hr °C (°F) | Max. Temp @ 1-hr °C (°F) | Rating |
|---|---|---|---|---|---|
| 12" cable tray aluminum | 2 layers of E-54A | 1 – control<br>1 – power | 190 (374) | 222 (432) | <1-hr |

After the water hose stream, the surface of the cable tray remained fully covered by at least one layer of E-54A mat.

*D.2.1.26  3M Fire Test Report #87-57 (May 1987)*

ERFBS: E-53A and E-54A  
Test Procedure: ANI/MAERP, ASTM E-119  
Ambient Temperature: 20°C (68°F)  
Thermocouples: cables  
Hose Stream: 2.5 minute, 30psi, 1-1/8" nozzle, 2-1/2" playpipe, 20 ft from assembly  
Furnace: 3M Large Scale top loading furnace (61" x 47" x 30")  
Furnace control: Not specified (10 thermocouples were located within furnace)  
Cable Type: (XLPE/PVC), Power (250mcm), Control (7/c 12AWG), Instrument (2/c 14AWG)

Desired Rating: 1-hr  
Test Slab: 70" x 56" x 8" Concrete  
Acceptance Temperature:

**Table D-27. E-53A & E-54A 3M Test Results #87-57**

| Raceway Type | Barrier Protection | Cable Fill | Avg. Temp @ 60 min °C (°F) | Rating |
|---|---|---|---|---|
| 5" dia. steel conduit | 3 layers of E-53A | 1 – power<br>2 – control<br>2 – instrument<br>1 – bare #8 | 142 (288) | 1-hr |
| 5" dia. steel conduit | 2 layers of E-54A | 1 – power<br>2 – control<br>2 – instrument<br>1 – bare #8 | 166 (331) | <1-hr |

FireDam 150 was applied prior to installation of mats and after the mat is banded. Both conduits survived the water hose stream test and all electrical circuit continuity paths were intact for the full fire test and at the completion of the hose stream tests.

### D.2.1.27 3M Fire Test Report (June 1987)

ERFBS: E-53A  
Test Procedure: ASTM E-119  
Ambient Temperature: 24°C (75°F)  
Thermocouples: Conduit, cable, bare #8  
Hose Stream: None  
Furnace: 3M Large Scale Furnace (63" x 52" x 52") (1.6m x 1.3m x 1.3m)  
Furnace control: 4 thermocouples  
Cable Type: control (7/c 12AWG)  

Desired Rating: 1-hr  
Test Slab:  
Acceptance Temperature:  

**Table D-28. 3M E-53A Test Results (June 1987)**

| Conduit # | Raceway Type | Barrier Protection | Cable Fill | Rating |
|---|---|---|---|---|
| 1 | 2" dia. steel conduit | 3 layers of E-53A applied as a single mat | 1 – control<br>1 – bare #8 | 51 min |
| 2 | 2" dia. steel conduit | 3 layers of E-53A 1st two layers applied as single mat and final layer applied separately | 1 – control<br>1 – bare #8 | 58 min |
| 3 | 2" dia. steel conduit | Same as Conduit 1 with addition of collars overlapping 2" of mat | 1 – control<br>1 – bare #8 | 61 min (1-hr) |

*D.2.1.28 Twin Cities Testing Corporation Tests #86-17, #86-18, #86-19 (February 5-7, 1986)*

<u>ERFBS</u>: E-50D (equivalent to E-54A) – 5 layers  
<u>Desired Rating</u>: 3-hour  
<u>Test Procedure</u>: ANI/MAERP  
<u>Test Slab</u>: 70" x 56" x 8" concrete  
<u>Thermocouples</u>: Taped to jackets of cables, spaced every 10.5 inches  
<u>Hose Stream</u>: 2-½" Playpipe, 1-1/8" tip, nozzle pressure 30psi, applied less than 20' from system, for 2.5 minutes.  
<u>Furnace</u>: 63" x 49.5" x 28"  
<u>Cable Type</u>: (XLPE/PVC); Power (250mcm); Control (7/c 12 AWG); Instrument (2/c 12 AWG)

**Table D-29. Twin Cities Testing 3M E-50D Test Results**

| Fire Test # | Raceway Type | Barrier Protection | Cable Fill | Avg. Temp @3-hr °C (°F) | Max. Temp @3-hr °C (°F) | Rating |
|---|---|---|---|---|---|---|
| 86-17 | 12" Cable Tray ½ solid, ½ ladder | 5 wraps E-50D | Single Layer 5 - power 7 - control 14 - instrument 1- bare #8 | 134 (274) | 111 (326) | 3-hr |
|  | Air Drop | 5 wraps E-50D | 1 - control 1 - instrument 1- bare #8 | 133 (272) | 101 (291) | 3-hr |
| 86-18 | 5" dia. Conduit | 5 wraps E-50D | 2 - control 2 - instrument 1- bare #8 | 124 (255) | 141 (286) | 3-hr |
|  | Junction Box 10" x 10" x 6" | 5 wraps E-50D | 4 - control 4 - instrument 1- bare #8 | 101 (214) | 144 (214) | 3-hr |
| 86-19 | Cable Tray | 5 wraps E-50D | 40% 16 - power 26 - control 76 - instrument 1- bare #8 | 101 (214) | 163 (231) | 3-hr |

Testing was conducted at 3M facilities with independent inspection of installation and verification of test results don by Twin Cities Testing Corporation of St. Paul, Minnesota. After completion of the water hose testing, at least two of the original five layers remained completely intact on all test articles. Post test inspection of the cables found no blistering, charring, melting, or noticeable deterioration of any insulation.

D.2.1.29 *Twin Cities Testing Corporation Tests #86-79, #86-80, #86-81 (July 1, 1986)*

<u>ERFBS</u>: E-53A & E-54A  <u>Desired Rating</u>: 1-hour
<u>Test Procedure</u>: ANI/MAERP  <u>Test Slab</u>: 70" x 56" 8" concrete
<u>Hose Stream</u>: 2-½" Playpipe, 1-1/8" tip, nozzle pressure 30psi, applied 20' from system, for 2.5 minutes.
<u>Furnace</u>: 61" x 47" x 30" (3M's top load, large scale, propane fired)
<u>Cable Type</u>: (XLPE/PVC); Power (250mcm); Control (7/c 12 AWG); Instrument (2/c 14 AWG)

**Table D-30. Twin Cities Testing 3M E-50 Series 1-hr Test Results**

| Fire Test # | Raceway Type | Barrier Protection | Cable Fill | Avg. Temp. @ 1-hr °C (°F) | Max. Temp @ 1-hr °C (°F) | Rating |
|---|---|---|---|---|---|---|
| 86-79 | 5" dia. aluminum conduit | 3 layers E-53A | 2 - control<br>2 - instrument<br>1 - bare #8 | 87 (188) | 99 (211) | 1-hr |
| | 5" diameter steel conduit | 3 layers E-53A | 2 - control<br>2 - instrument<br>1 - bare #8 | 82 (179) | 103 (218) | 1-hr |
| | Junction Box | 3 layers E-54A | 2 - control<br>2 - instrument<br>2 - bare #8 | 99 (210) | 99 (211) | 1-hr |
| 86-80 | 24" wide aluminum cable tray | 1 layer E-54A + 1 layer E-53A | Single Layer<br>10 - power<br>14 - control<br>28 - instrument<br>2 - bare #8 | 96 (204) | 173 (343) | 1-hr |
| | Aluminum cable tray | 1 layer E-54A + 1 layer E-53A | 40% fill<br>24 - power<br>29 - control<br>108 - instrument<br>2 - bare #8 | 102 (215) | 106 (222) | 1-hr |
| 86-81 | Air drop | 3 layers E-54A | 4 - control<br>4 - instrument<br>1 - bare #8 | 98 (209) | 114 (237) | 1-hr |

*D.2.1.30 SwRI Ampacity Testing of E-50 Series (November 1985)*

For these ampacity tests, three types of cables were installed in a horizontal cable tray. Current was applied to the cables and measurements were taken of the copper conductors temperatures and current flow through each cable type under various ambient temperature conditions. The results from this testing are shown in Table D-31.

### Table D-31. SwRI Ampacity Testing Results at 20°C

| Item # | Test Specimen Description | Ampacity Derating Factor % * | | |
|---|---|---|---|---|
| | | 250mcm | 3/c - #8 | 3/c - #16 |
| 1 | Tray with Solid Top | 22.3 | 19.5 | 16.1 |
| 2 | Tray with Solid Top & T-49 Tape | 22.9 | 19.8 | 16.2 |
| 3 | Tray with Solid Top & Bottom | 33.8 | 31.8 | 27.5 |
| 4 | Tray with Solid Top & Bottom & Tape | 31.8 | 30.4 | 25.7 |
| 5 | Tray + 4 Layers of E-50A with Solid Top & Bottom | 48.6 | 50.7 | 48.6 |
| 6 | Tray + 4 Layers of E-50A | 42.6 | 46.4 | 41.8 |
| 7 | Tray + 4 Layers of E-50A + Tape | 40.4 | 44.7 | 39.4 |
| 8 | Tray + 10 Layers of E-50A | 47.4 | 52.1 | 49.3 |
| 9 | Tray + 10 Layers of E-50A + Tape | 44.6 | 50.1 | 46.7 |

* Ampacity Derating factors are based on ambient temperatures of 20°C.

These results indicated that as the ambient temperature increased, the ampacity derating factor also increased.

*D.2.1.31 SwRI Ampacity Testing (April 1987)*

This testing was conducted to determine if the ampacity derating of two layers of E-54A mat protecting an open-ladder cable tray meets the requirements of Bechtel Power Corporation for use at the South Texas Project (STP). Only one type of power cable was used for this testing—namely, a 3/c #6AWG Okonite Power cable. This test was conducted for 60 minutes.
The baseline (no barrier) configuration resulted in 32.1A of current and when protected with two layers of E-54A ERFBS, the current dropped to 17.3, an ampacity derating factor of 46.1 percent.

$$\frac{32.1-17.3}{32.1} \times 100\% = 46.1\% \text{ ADF}$$

### D.2.1.32 SWRI Testing (June 1986)

SwRI performed ampacity testing on a 10.1 cm (4.0-in) diameter conduit configuration against various 3M ERFBSs. The conduit was filled with 20 3/c #6 AWG cables or 69 percent conduit fills.

**Table D-32. Ampacity Results SwRI Conduits**

| Item # | Test Specimen Description | Final Current (Amps) | Ampacity Derating Factor % |
|---|---|---|---|
| 1 | Baseline | 24.0 | 0.00 |
| 2 | E-50A 1-hr | 18.9 | 21.3 |
| 3 | E-50 Series 1-hr | 18.5 | 22.9 |
| 4 | E-50 Series 1-hr with 3M ECP-2200 black coating | 20.2 | 15.8 |
| 5 | M-20A 1-hr | 14.9 | 37.9 |
| 6 | M-20A 1-hr with 3M ECP-2200 black coating | 15.5 | 35.4 |
| 7 | E-50D 3-hr | 17.0 | 29.2 |

### D.2.1.33 SwRI Ampacity Testing Project No. 01-8818-208/-209(1) (June 1986)

Table D-33 shows ampacity results from SwRI tray and conduit tests.

**Table D-33. Ampacity Results SwRI Tray/Conduit**

| Item # | Test Specimen Description | Final Current (Amps) | Ampacity Derating Factor % |
|---|---|---|---|
| 1 | Baseline – Open Tray | 24.6 | 0.00 |
| 2 | Baseline – Conduit | 23.7 | 0.00 |
| 3 | Baseline – Tray solid bottom | 20.0 | 18.7 |
| 4 | Baseline – Tray solid bottom and top | 14.7 | 40.2 |
| 5 | Tray – E50A 1-hr | 13.1 | 46.7 |
| 6 | Conduit – E-50A 1-hr | 18.8 | 20.7 |
| 7 | Tray – E-50D 3-hr | 11.2 | 54.5 |
| 8 | Conduit – E-50D 3-hr | 17.2 | 27.4 |
| 9 | Tray – E-50D/E-53A 1-hr | 14.5 | 41.1 |
| 10 | Conduit – E-53A 1-hr | 18.6 | 21.5 |
| 11 | Conduit – E-53A 1-hr with ECP 2200 black coating | 20.1 | 15.2 |

*D.2.1.34 SwRI Ampacity Testing (October 1986)*

This test series involved ampacity derating testing of a 61.0 cm (24.0 in) wide steel cable tray ladder-back type and 10.1 cm (4.0 in) diameter steel conduit filled to 100 percent visual fill with 3/c XLPE/CSPE #6 AWG cables. Unfortunately, the report does not document any baseline ampacity values for these configurations to determine the ampacity derating factors to apply for raceways protected with the E-50A ERFBS.

**Table D-34. Ampacity Results SwRI without Baseline**

| Item # | Test Specimen Description | Final Current (Amps) | Ampacity Derating Factor % |
|---|---|---|---|
| 1 | Tray protected with E-50A 1-hr wrap | 13.3 | No data reported |
| 2 | Conduit protected with E-50A 1-wrap | 18.9 | No data reported |

*D.2.1.35 3M Ampacity Report (April 1985)*

3M conducted ampacity testing using a 10.1 cm (4-in) diameter schedule 40 galvanized steel conduit, using numerous E-50A mat wrapping configurations. The conduit was loaded with three 500MCM, 600V copper cables with a maximum operating temperature of 90°C. shows the results of this testing. As you can see from the data below, the addition of the high emissivity black tape actually improves the ampacity rating.

**Table D-35. 3M Ampacity Results Conduit**

| Item # | Test Specimen Description | Final Current (Amps) | Ampacity Derating Factor % |
|---|---|---|---|
| 1 | Baseline | 476 | 0.00 |
| 2 | High emissivity black tape | 506 | - 6 |
| 3 | Five Layers of E-50A mat | 383 | 19.5 |
| 4 | Five Layers of E-50A mat with Black Tape | 404 | 15.1 |
| 5 | Ten Layers of E-50A mat | 358 | 24.8 |
| 6 | Ten Layers of E-50A mat with Black Tape | 374 | 21.4 |

*D.2.1.36 UL Testing E-50A (May-September 1985)*

UL tested the ampacity of cables located in a 45.7 cm (18-in) wide ladder back cable tray protected with various layers of 3M ERFBS. The cable tray fill included 29 3/c # 16 AWG XLPE insulated cables, 13 3/c #8AWG non-jacketed, and 13 single-conductor non-jacketed 250MCM[19] cables. summarizes the results of this testing.

Table D-36. UL Ampacity Test Results for E-50A Cable Tray Configurations

| Item # | Test Specimen Description | Ampacity Derating Factor % * | | |
|---|---|---|---|---|
| | | 250mcm | 3/c - #8 | 3/c - #16 |
| 1 | Baseline | 0.00 | 0.00 | 0.00 |
| 2 | Metal Top | 22.3 | 19.5 | 16.1 |
| 3 | Metal Top with Black Tape | 22.9 | 19.8 | 16.2 |
| 4 | Metal Top & Bottom | 33.8 | 31.8 | 27.5 |
| 5 | Metal Top & Bottom with Black Tape | 31.8 | 30.4 | 25.7 |
| 6 | Four Layers of E-50A mat | 42.6 | 46.4 | 41.8 |
| 7 | Four Layers of E-50A mat with Black Tape | 40.4 | 44.7 | 39.4 |
| 8 | Four Layer of E-50A mat over Metal Top and Bottom | 48.6 | 50.7 | 48.6 |
| 9 | Ten Layers of E-50A mat | 47.4 | 52.1 | 49.3 |
| 10 | Ten Layers of E-50A mat with Black Tape | 44.6 | 50.1 | 46.7 |

* Ampacity Derating factors are based on tests conducted at an ambient temperature of 20°C.

*D.2.1.37 UL Ampacity Testing E-54A (October 1986)*

UL tested the ampacity of 3M E-54A 2 layer system in a 61.0 cm (24-in) wide open ladder-back cable tray. Cable loading included 71 3/c #6 Okonite insulated and Okolon jacketed cable, for a maximum fill depth of 3-5/8 inches. documents the results.

Table D-37. UL Ampacity Test Results for E-54A 1-hour Cable Tray Configurations

| Item # | Test Specimen Description | Final Current (Amps) | Ampacity Derating Factor % |
|---|---|---|---|
| 1 | Baseline | 32.1 | 0.00 |
| 2 | 2 Layers of E-54A mat 1-hr | 17.3 | 46.1 |

---

[19] MCM is equal to 1000 circular mils, where 1 circular mil is a unit of area equal to the area of a circle 1 mil in diameter. Large cable conductors will typically be denoted by MCM, with smaller conductors using the AWG designation in the USA.

### D.2.2 3M Rigid Panel System Testing

*D.2.2.1 UL Project 82NK2193, R10125*

<u>ERFBS</u>: M20-A / CS-195                             <u>Desired Rating</u>: 1 –hour
<u>Thermocouples</u>: Cable jacket

Per the request of 3M, UL reviewed temperature data and made recommendations as to the qualification of previous tests, to the acceptance criteria of the temperature of the largest cable jacket to not exceed the maximum temperature rise of 250°F above initial starting temperature.

Test Date: March 3, 1983
Test Reports:  1.) UL Report R10125-1, -2 dated October 19, 1983,
       2.) UL Letter Report R10125 dated November 2, 1983,

#### Table D-38. UL Interpretations of Previously Completed Test Results

| System No. | Raceway/Cable Type | Barrier Protection | Limiting Temp. °C (°F) | Max. Temp °C (°F) at Time (minutes) | Rating |
|---|---|---|---|---|---|
| A | Steel cable tray open-ladder / 300MCM | 1 layer of M20-A surrounded by single sheet of CS-195 composite sheet secured to steel channel framing | 159 (319) | 159 (319) @ 50<br>167 (332) @ 60 | <1-hr |
| B | Steel cable tray open-ladder / 250MCM | 4 layers of M20-A | 167 (333) | 142 (287) @ 60 | 1-hr |
| C.1 | Rigid steel conduit / 300MCM | 3 layers of M20-A 40 % cable fill | 159 (318) | 88 (190) @ 60 | 1-hr |
| C.2 | Rigid steel conduit / 300MCM | 3 layers of M20-A minimum cable fill | 159 (318) | 141 (286) @ 60 | 1-hr |
| D | Air Drop / 300MCM | 5 layers of M20-A | 162 (323) | 145 (293) @ 60 | 1-hr |
| E | Junction Box / 2/c 14AWG | 2 layers of M20-A surrounded by single sheet of CS-195 composite sheet secured to steel channel framing | 163 (326) | 117 (242) @ 60 | 1-hr |

*D.2.2.2 UL Letter Report dated November 2, 1983*

ERFBS: M20-A (4 layers)  
Test Procedure: ASTM E-119  
Ambient Temperature: 86°F (limiting max. single point temperature 411°F)  
Thermocouples: Bare conductors, cable tray side rails  
Hose Stream: 30 psi for 2.5 minutes  
Cable Type: Bare Copper Conductors (14AWG, 8AWG, 1/0AWG, 250MCM)  
Desired Rating: 1-hour  
Test Slab: 8" thick concrete

**Table D-39. UL Test Report on M20-A 1-hr Cable Tray**

| System No. | Raceway Type | Barrier Protection | Cable Fill | Max. Temp °C (°F) | Rating |
|---|---|---|---|---|---|
| 1 | 24" wide galvanized steel cable tray | 4 layers of M20-A | 4 bare copper conductors | 201 (394) | 1-hr |

Following the hose stream test, it was noted that the strips of intumescent mat at the horizontal member of the trapeze support were washed away, leaving a through opening into the electrical circuit protection system through which the bare copper conductors were visible. No other through opening was present in the electrical circuit protective system. The test report concluded that based on previous testing, had the horizontal support member ERFBS protection been covered with a nominal 10 in. wide section of steel hardware cloth secured in place with steel banding straps.

*D.2.2.3 UL Letter Report Dated January 19, 1984*

ERFBS: M20-A / CS-195  
Test Procedure: ASTM E-119  
Ambient Temperature: 89 °F  
Thermocouples: copper conductor and steel of junction box  
Hose Stream: 30 psi, 2.5 minutes; Protective enclosure was intact at the end of the testing.  
Cable Type: Bare copper conductor # 8 AWG  
Desired Rating: 1-hour  
Test Slab: 8" thick concrete

**Table D-40. UL Results for 3M M20-A/CS-195 1-hr Junction Box**

| Raceway Type | Cable Fill | Barrier Protection | Max. Temp Rise $\Delta T$ °C (°F) | Rating |
|---|---|---|---|---|
| Steel Junction Box (12"x12"x6") with 2" Conduit | Bare copper conductor (#8AWG) | 1 layer of M20-A surrounded by sheet of CS-195 secured to a steel framing | 186 (366) | < 1-hr |
| | | Conduit was wrapped with 4 layers of Interam™ E-50 Series | 257 (494) | < 1-hr |

*D.2.2.4 UL Letter Report Dated April 18, 1984*

ERFBS: M20-A  
Test Procedure: ASTM E-119  
Thermocouples: cable jacket and conduit surface  
Hose Stream: None  
Furnace control: per ASTM E-119  
Cable Type: XLPE / Neoprene  

Desired Rating: 1-hr  
Ambient Temperature: 46 °F  

**Table D-41. UL Report on 3M Testing of 3/4, 2, 3-inch Steel Conduits**

| System No. | Raceway Type | Barrier Protection | Cable Fill | Max. Temp Rise $\Delta T$ °C (°F) | Rating |
|---|---|---|---|---|---|
| 1 | ¾" RSC | 5 layers of M20-A | 1 – 2/c 16AWG | 303 (578) @ 60 minutes Exceeded criteria between 40 and 50 minutes | < 1-hr |
| 2 | 3" RSC | 3 layers of M20-A | 17 – 2/c 16AWG | 210 (410) @ 60 minutes Exceeded criteria between 50 and 60 minutes | < 1-hr |
| 3 | 2" RSC | 5 layers of M20-A | 1 – 2/c 16AWG | 317 (602) @ 60 minutes Exceeded criteria between 30 and 40 minutes | < 1-hr |

The report indicated that at 58 minutes into the test, the light emitting diode (LED) associated with the conductors and ground of the 2/c 16AWG cables in the nominal 3in. diameter conduit protective system commenced glowing dimly. At the completion of the 60 minute test, the LEDs associated with the 3in. diameter conduit were fully illuminated.

*D.2.2.5 UL Letter Report dated August 7, 1984*

ERFBS: M20-A plus CS-195  
Test Procedure: UL 1724  
Ambient Temperature: 76°F  
Thermocouples: cable, bare conductor, steel surface, air space between M20-A and CS-195  
Hose Stream: 30psi, 1-1/8 in. nozzle, 20 ft away for 30 seconds  
Cable Type: (XLPE/PVC) 2/c 14AWG control/power; Bare copper conductor #8AWG  

Desired Rating: 1-hr  
Test Slab: 36" x 36" x 2" concrete  

**Table D-42. UL M20-A 1-hour Test Results**

| Raceway Type | Barrier Protection | Cable Fill | Max. Temp °C (°F) | Rating |
|---|---|---|---|---|
| Junction Box 10" x 10" x 6"  Steel | 2 layers of M20-A 1 sheet of CS-195 attached to steel framing | 1- 2/c control 1 – bare #8 | 154 (310) | 1-hr |

The report states that, "By 1 min, the intumescent sheet enclosure was flaming and the aluminum foil tape was peeling from the corners. The flaming of the intumescent sheet was profuse until approximately 10 min, at which time the flaming commenced to diminish." No opening developed during the hose stream test, however the intumescent sheet located at the bottom of the junction box had eroded away. Examination of the cables inside the junction box reveled that the cable jacket had melted and adhered to the bottom and sides of the steel junction box. Although continuity testing was not conducted during the test, approximately 30 minutes following the test, high voltage withstand testing was conducted and found the conductors insulation resistance to be infinite at 1000Vdc. However, the time between fire testing and electrical testing could have been sufficient to allow the cable to heal any electrical damages and these results are somewhat indeterminate.

*D.2.2.6 3M Test Number 84-10*

ERFBS: M20-A
Test Procedure: ASTM E-119
Thermocouples: Conduit, Unistrut Support
Hose Stream: None
Furnace: 3M Large Scale (8' long x 5' wide x 6' high)
Furnace control: Nine Type K located throughout furnace
Cable Type: None

Desired Rating: 1-hr
Ambient Temperature: 50°F

**Table D-43. 3M Test Results of M20-A Conduit 1-hr**

| System No. | Raceway Type | Barrier Protection | Avg. Temp Rise $\Delta T$ °C (°F) | Rating |
|---|---|---|---|---|
| 1 | 2" RSC without support | 2 layers of M20-A | 201 (393) | < 1-hr |
| 2 | 2" RCS withsupport | 2 layers of M20-A | 221 (429) | < 1-hr |
| 3 | Unistrut support | 2 layers of M20-A in each direction from the conduit/support interface | 289 (553) | < 1-hr |

*D.2.2.7 UL Surface Burning Characteristics of Type FS-195 Barrier (June 8, 1982)*

These tests were conducted in accordance with UL 723, "UL Standard Test Method for Surface Burning Characteristics of Building Materials." Three 24' long samples were prepared by joining nominal 2' x 3' composite fire barrier sheets mechanically joined. The FS-195 composite sheets were nominally ¼" thick intumescent elastomeric material vulcanized to a No. 28 MSG galvanized steel plate on one surface and to an aluminum foil covering on the opposite surface. In addition, prior to vulcanization, a reinforced hexagonal wire mesh (chicken wire) was placed over the elastomeric material, beneath the aluminum foil covering. Each sample was tested with the foil facing of the composite fire barrier sheets downward (fire side). For two of the three tests, a slit was cut in the aluminum foil facing along the longitudinal centerline of the 24' long samples. The results are shown in .

### Table D-44. UL 723 Test Results for FS-195

| Test No. | Product | Flame Spread | Fuel Contributed | Smoke Developed |
|---|---|---|---|---|
| 1 | Foil Slit | 17.53 | 0 | 197 |
| 2 | Foil Intact | 16.57 | 0 | 190 |
| 3 | Foil Slit | 18.88 | 0 | 231 |

*D.2.2.8 3M Test Number 94-27*

ERFBS: Thermo-Lag plus E-54A  
Test Procedure: ASTM E-119  
Ambient Temperature: 14°C (58°F)  
Desired Rating: 3-hour  
Test Slab:  
Acceptance Temperature:  
Thermocouples: Conduit, between barrier layers  
Hose Stream: Not conducted  
Furnace: 3M Large Scale Furnace (52" wide x 67" high x 78" long)  
Furnace control: 4 thermocouples  
Cable Type: None  

### Table D-45. 3M Results of Thermo-Lag upgraded with E-54A 3-hr 2" conduit

| Raceway Type | Cable Fill | Avg. Temp Rise $\Delta T$ °C (°F) | Max. Temp Rise $\Delta T$ °C (°F) | Rating |
|---|---|---|---|---|
| 2" RSC - Straight Section 10' | N/A | 129 (265) | 141 (286) | <3-hr |

The average conduit temperature exceeded the average temperature rise criterion at 169 minutes into the test. Maximum temperature rise was not exceeded. The report indicates that a possible area of failure would be the opening of seams in the TSI preformed sections, as no trowel grade material was used during installation. In addition, the 3M furnace was tested by UL to compare it's thermal environment the UL test furnace. UL determined that 53 minutes in the 3M furnace is equivalent to 60 minutes in the UL furnace. A 3-hour correlation was not conducted but the report suggests that on a mathematical basis the correlation would be 159 minutes in the 3M furnace would equate to 180 minutes in the UL furnace.

### D.2.2.1 3M Test Number 94-42

<u>ERFBS</u>: Thermo-Lag plus E-54A
    3 pre-formed sections of Thermo-Lag
    3 layers of E-54A
<u>Test Procedure</u>: ASTM E-119
<u>Thermocouples</u>: Conduit, between barrier layers
<u>Hose Stream</u>: Not conducted
<u>Furnace</u>: 4 Thermocouples
<u>Furnace control</u>: 3M Large Scale Furnace (52" wide x 67" high x 78" long)
<u>Cable Type</u>: None

<u>Desired Rating</u>: 3-hour

<u>Ambient Temperature</u>: (54°F)

**Table D-46. 3M Results of Thermo-Lag upgraded with E-54A 3-hr 1.5" conduit**

| Raceway Type | Cable Fill | Avg. Temp Rise ΔT °C (°F) | Max. Temp Rise ΔT °C (°F) | Rating |
|---|---|---|---|---|
| 1.5" RSC - Straight Section 10' | N/A | 131 (268) | 167 (333) | < 3-hr |

TVA Ampacity Testing 3M M20A

TVA performed ampacity derating testing on the 3M M20A conduit and air drop configurations. The testing followed draft Institute of Electrical and Electronic Engineers (IEEE) Standard P848, "Procedure for the Determination of the Ampacity Derating of Fire Protected Cables." The testing was conducted at Central Laboratories Services located in Chattanooga, TN. All cables tested were Rockbestos Type PXMJ.  shows the results of this testing.

**Table D-47. TVA Ampacity Derating of 3M M20A**

| Raceway Type | Barrier Protection | Cable Fill | Ampacity Derating Factor % |
|---|---|---|---|
| 1" dia. conduit | 5 layers of 3M M20A | 1 – 4/C #10AWG | 29 |
| Air Drop (small) | 5 layers of 3M M20A | 1 – 4/C #10AWG | 42 |
| Air Drop (large) | 5 layers of 3M M20A | 8 – 3/C #6 AWG | 49 |

### SwRI Ampacity Testing of M20-A/CS-195 ERFBS (Project No. 01-8818-208/-209b)

SwRI test report dated September 29, 1986 documents the results of ampacity derating testing conducted on the M20-A and CS-195 Rigid ERFBS. The cables selected were of a 3/c No. 6AWG cross-linked Polyethylene (XLPE)/chlorosulphonated polyethylene (CSPE) construction. The cable conductor temperatures were measured with 24 gauge Type K thermocouples placed under the jacket and insulation in contact with the copper conductor. The cable trays and conduits were filled to 100-percent visual fill resulting in 122 and 20 cables, respectively. The raceways were 30.5 cm (12.0 in) long with the cables extending 0.3 to 0.6 m (1.0 to 2.0 ft) and protected with fiberglass blanket insulation. Three 10.2 cm (4.0 in) diameter rigid steel conduit (RSC) and three 61.0 cm (24.0 in) wide ladder back cable trays were used. The calculated fill depth for the cable tray was 7.4 cm (2.9 in).

Table D-48. SwRI Ampacity Test Results for M20-A and CS-195

| Item # | Test Specimen Description | Final Current (Amps) | Ampacity Derating Factor % |
|---|---|---|---|
| 1 | Baseline – Open Tray | 24.5 | - - - |
| 2 | Baseline – Tray with solid bottom | 19.7 | 19.6 |
| 3 | Baseline – Tray with solid bottom and top | 14.5 | 40.8 |
| 4 | M-20A (4 Layers) | 9.9 | 59.6 |
| 5 | CS-195 (1 Layer) / M-20A (1 Layer) | 10.0 | 59.2 |
| 6 | M-20A (1 Layer) / CS-195 (1 Layer) with ECP 2200 Coating | 13.4 | 45.3 |
| 7 | Baseline – Conduit | 23.7 | - - - |
| 8 | M-20A (5 Layers) | 14.7 | 37.97 |
| 9 | M20-A (5 Layers) with ECP 2200 Coating | 15.3 | 35.44 |

## D.3 Thermo-Lag

### D.3.1 NUMARC Phase 1 Testing

These tests were full scale tests, with test tray(s) and/or conduit(s) that vertically penetrated the furnace roof, descended into the furnace and bent to a horizontal section several feet long ranging from ~2 to ~4 feet below the furnace roof, then bent to a vertical ascending section that again penetrated the furnace roof 8 to 9 feet horizontally from the first penetration. These "elongated U" shaped test trays and conduits were used in all of NUMARC's Phase 1 Thermo-Lag tests.

*D.3.1.1 NUMARC Project No. 13890-95671, Test 1-1 (October 1993)*

ERFBS: This 1 hr. test evaluated a 36" x 4" steel electric cable tray, with ERFBS constructed using Thermo-Lag 330-1, nom. ½" thickness, with stress skin monolithically adhered to the panels on one face. A ~3/16" layer of trowel grade THERMO-LAG-330-1 covered the side rail splice plates. Pieces of 330-069 stress skin were applied over that, then a ~1/16" thick layer of trowel grade THERMO-LAG-330-1 was applied over the stress skin. Joints were re-enforced in a similar manner. The full width of the tray was then covered with stress skins which overlapped the joint and side rail stress skins. These stress skins were then covered with a ~1/16 thick layer of trowel grade THERMO-LAG.

Test Procedure: Thermal Science, Inc., TEST PLAN, Rev. 5

Test Slab: 13' x 8' x 10 GA (Steel), with 2 layers of 2" ceramic fiber blanket insulation

Ambient Temperature: 88°F

Desired Rating: 1-hour

Thermocouples: Every 6" on 2 bare copper wires, one on tray rungs, the other on an electric cable. Also, every 6" on both side rails of the cable tray.
Hose Stream: A test was applied, but no details are given in available documentation.
Furnace: ~11' x 6' x at least 5' (i.e., a depth sufficient to contain the 3' vertical extent of the test tray)
Furnace control: : Ten (10) thermocouples on probes located throughout the furnace
Cable Type: Approx. 1/3 mix of power, control, and inst. cables, 51 total, 15.5% of total area

**Table D-49. NUMARC Thermo-Lag Test 1-1**

| 10.2 Raceway Type | 10.3 Barrier Protection | Cable Fill | Max. Temp Rise $\Delta T$ °C (°F) | Rating |
|---|---|---|---|---|
| B-Line 36" x 4" steel tray | ½" nom. TL 330-1 | 15.5% | 311 (592) | 54 min |

The average temperature increase parameter ($\Delta T$ = 250°F) was exceeded at 59 min., and the single point temperature increase parameter ($\Delta T$ = 325°F) was exceeded at 54 minutes. Also, a barrier opening was noted on the assembly following the fire endurance and hose stream test. Thus, the assembly did not meet the applicable criteria for a fire resistance period of 60 minutes.

*D.3.1.2 NUMARC Project No. 13890-95673, Test 1-3 (October 1993)*

ERFBS: This 3 hr. test evaluated a 36" x 4" steel electric cable tray, with ERFBS constructed using Thermo-Lag 330-1, nom. 1-1/8" thickness, with stress skin monolithically adhered to the panels on both faces. Extensive upgrades were incorporated into the design, including use of additional 5/8" thick overlay Thermo-Lag V-ribbed panels, use of trowel grade Thermo-Lag, use of 330-69 stress skin material in many places, and lacing panels together with stainless steel wire.

Test Procedure: Thermal Science, Inc., TEST PLAN, Rev. 5

Test Slab: 13' x 8' x 10 GA (Steel), with 2 layers of 2" ceramic fiber blanket insulation

Ambient Temperature: 89°F

Desired Rating: 3-hour

Thermocouples: Every 6" on 2 bare copper wires, one on tray rungs, the other on an electric cable. Also, every 6" on both side rails of the cable tray.
Hose Stream: A test was applied, but no details are given in available documentation.
Furnace: ~11' x 6' x at least 5' (i.e., a depth sufficient to contain the 3' vertical extent of the test tray)
Furnace control: Ten (10) thermocouples on probes located throughout the furnace
Cable Type: Approx. 1/3 mix of power, control, and inst. cables, 51 total, 15.5% of total area

**Table D-50. NUMARC Thermo-Lag Test 1-3**

| 10.4  Raceway Type | 10.5  Barrier Protection | Cable Fill | Max. Temp Rise ΔT °C (°F) | Rating |
|---|---|---|---|---|
| B-Line 36" x 4" steel tray | 1" nom. TL 330-1 | 15.5% | 224 (436) | 2 hr. 47 min |

The average temperature increase parameter (ΔT = 250°F) was exceeded at 2 hours and 54 minutes, and the single point temperature increase parameter (ΔT = 325°F) was exceeded at 2 hours and 47 minutes. Also, a barrier opening was noted across the bottom center of the assembly following the fire endurance and hose stream test. Thus, the assembly did not meet the applicable criteria for a fire resistance period of 3 hours.

*D.3.1.3 NUMARC Project No. 13890-95674, Test 1-4 (November 1993)*

ERFBS: This 3 hr. test evaluated a 24" x 4" steel electric cable tray with 5" steel conduit air drop, with ERFBS constructed using Thermo-Lag 330-1, nom. 1" thickness, with stress skin monolithically adhered to the panels on both faces. Extensive upgrades were incorporated into the design, including use of additional 5/8" thick overlay Thermo-Lag V-ribbed panels, use of trowel grade Thermo-Lag, use of 330-69 stress skin material in many places, and lacing panels together with stainless steel wire. Thermo-Lag 330-1 Baseline Pre-shaped conduit sections were used for the conduit and air drop assembly, upgraded with use of THERMO-LAG-330-1 trowel grade subliming material and 330-69 stress skin pieces. Many pieces were stitched together with stainless steel wire.

Test Procedure: Thermal Science, Inc., TEST PLAN, Rev. 5

Test Slab: 13' x 8' x 10 GA (Steel), with 2 layers of 2" ceramic fiber blanket insulation

Ambient Temperature: 78°F

Desired Rating: 3-hr

<u>Thermocouples</u>: Every 6" on 2 bare copper wires, one on tray rungs, the other on an electric cable, and every 6" on both side rails of the cable tray. In addition, two lengths of bare copper wire, with thermocouples every 6", were extended down thru the 5 inch conduit and air drop assembly into the center of the cable tray to measure temperatures in the air drop and conduit.
<u>Hose Stream</u>: A test was applied, but no details are given in available documentation.
<u>Furnace</u>: ~11' x 6' x at least 5' (i.e., a depth sufficient to contain the 3' vertical extent of the test tray)
<u>Furnace control</u>: Ten (10) thermocouples on probes located throughout the furnace
<u>Cable Type</u>: Approx. 1/3 mix of power, control, and inst. cables, 34 total, 15.4% of total area

**Table D-51. NUMARC Thermo-Lag Test 1-4**

| Raceway Type | Barrier Protection | Cable Fill | Max. Temp Rise ΔT °C (°F) | Rating |
|---|---|---|---|---|
| B-Line 24" x 4" steel tray with 5" steel conduit air drop | 1" nom. TL 330-1 | 15.4% | 521 (969) | Tray – 3-hr<br>Air Drop – 1 hr 44 min |

The 24 in. cable tray assembly, clad in nominal 1 in. thick THERMO-LAG 330-1 material with upgrades briefly noted above, met requirements of the TEST PLAN for a fire resistance rating of three hours. However, the air drop assembly and associated 5 in. conduit stub failed to meet the requirements.

*D.3.1.4 NUMARC Project No. 13890-95675, Test 1-5 (November 1993)*

<u>ERFBS</u>: This 3 hr. test was evaluated a 24" x 4" steel electric cable tray with tee section, with ERFBS constructed using Thermo-Lag 330-1, nom. 1" thickness, with stress skin monolithically adhered to the panels on both faces. Extensive upgrades were incorporated into the design, including use of additional 5/8" thick overlay Thermo-Lag V-ribbed panels, use of trowel grade Thermo-Lag, use of 330-69 stress skin material in many places, and lacing panels together with stainless steel wire
<u>Test Procedure</u>: Thermal Science, Inc., TEST PLAN, Rev. 5
<u>Ambient Temperature</u>: 77°F
<u>Test Slab</u>: 13' x 8' x 10 GA (Steel), with 2 layers of 2" ceramic fiber blanket insulation
<u>Desired Rating</u>: 3-hr
<u>Thermocouples</u>: Every 6" on 2 bare copper wires, one on tray rungs, the other on an electric cable, and every 6" on both side rails of the cable tray. Two additional short sections of TC-instrumented bare wire were looped into the tee section.
<u>Hose Stream</u>: A test was applied, but no details are given in available documentation.
<u>Furnace</u>: ~11' x 6' x at least 5' (i.e., a depth sufficient to contain the 3' vertical extent of the test tray)
<u>Furnace control</u>: Ten (10) thermocouples on probes located throughout the furnace
<u>Cable Type</u>: Approx. 1/3 mix of power, control, and inst. cables, 34 total, 15.4% of total area

### Table D-52. NUMARC Thermo-Lag Test 1-5

| Raceway Type | Barrier Protection | Cable Fill | Max. Temp Rise ΔT °C (°F) | Rating |
|---|---|---|---|---|
| B-Line 24" x 4" steel tray, with tee section | 1" nom. TL 330-1 | 15.4% | 186 (366) | 2 hr. 52 min. |

The individual temperature increase parameters were exceeded on the tray rail at 172 minutes, and a large section of panel was dislodged during the hose stream exposure, creating an opening through which the internal cable tray and its cables were visible.

*D.3.1.5 NUMARC Project No. 13890-95676, Test 1-6 (November 1993)*

ERFBS: This 1 hr. test evaluated a 5" Aluminum Conduit, a 3" Aluminum Conduit, a ¾" Aluminum Conduit, an Aluminum Junction Box, and a 3" Steel Conduit, with ERFBS constructed using Thermo-Lag 330-1, pre-shaped conduit sections, nom. ½" thickness on the 4 conduits. Thermo-Lag 330-1 nom. ½" thickness baseline panels were used on the LB box designs, the junction box, and the support members. Extensive upgrades were incorporated into the design, including use of additional 1/4" thick THERMO-LAG-330-1 pre-shaped conduit overlay sections, use of trowel grade Thermo-Lag, use of 330-69 stress skin material in many places, and use of stainless steel bands throughout as appropriate (there are nine (9) pages describing these upgrades).

Test Procedure: Thermal Science, Inc., TEST PLAN, Rev. 5
Test Slab: 13' x 8' x 10 GA (Steel), with 2 layers of 2" ceramic fiber blanket insulation
Ambient Temperature: 87°F
Desired Rating: 1-hr
Thermocouples: One TC every 6" on one bare copper wire in each of the four conduits, one TC every 6" affixed with glass cloth electrical tape to the bottom surface of each conduit, and TCs clamped with stainless steel round-head screws to the interior surface of the junction box.
Hose Stream: A test was applied, but no details are given in available documentation.
Furnace: ~11' x 6' x at least 5' (i.e., a depth sufficient to contain the 3' vertical extent of the test conduit assemblies)
Furnace control: Ten (10) thermocouples on probes located throughout the furnace
Cable Type: There were no electric cables installed in the conduits or in the junction boxes.

### Table D-53. NUMARC Thermo-Lag Test 1-6

| Raceway Type | Barrier Protection | Cable Fill | Max. Temp Rise ΔT °C (°F) | Rating |
|---|---|---|---|---|
| 5" Aluminum Conduit | ½" nom. TL-330-1 | None | 153 (308) | 1 hr. |
| 3" Aluminum Conduit | ½" nom. TL-330-1 | None | 158 (317) | 1 hr. |
| ¾" Aluminum Conduit | ½" nom. TL-330-1 | None | 83 (181) | 1 hr. |
| Aluminum Junction Box | ½" nom. TL-330-1 | None | 139 (283) | 1 hr. |
| 3" Steel Conduit | ½" nom. TL-330-1 | None | 152 (306) | 1 hr. |

As shown in the above table, all four of the test conduits, and the junction box, met the allowable single point maximum temperature increase criterion of 325°F. Although not shown in the table, the maximum average temperature increase for any of the 5 was 246°F, which met the allowable maximum average temperature increase criterion of 250°F.

### D.3.1.6 NUMARC Project No. 13890-95677, Test 1-7 (November 1993)

ERFBS: This 3 hr. test evaluated a 5" Steel Conduit, a 3" Steel Conduit, a ¾" Steel Conduit, and a Steel Junction Box, with ERFBS constructed using Thermo-Lag 330-1, pre-shaped conduit sections, nom. 1" thickness on the 3 conduits. Thermo-Lag 330-1 nom. 1" thickness baseline panels were used on the LB box designs, the junction box, and the support members. Extensive upgrades were incorporated into the design, including use of additional 5/8" thick THERMO-LAG-330-1 pre-shaped conduit overlay sections, use of trowel grade Thermo-Lag, use of 330-69 stress skin material in many places, and use of stainless steel bands throughout as appropriate (there are nine (9) pages describing these upgrades).

Test Procedure: Thermal Science, Inc., TEST PLAN, Rev. 5
Test Slab: 13' x 8' x 10 GA (Steel), with 2 layers of 2" ceramic fiber blanket insulation
Ambient Temperature: 84°F
Desired Rating: 3-hr
Thermocouples: One TC every 6" on one bare copper wire in each of the four conduits, one TC every 6" affixed with glass cloth electrical tape to the bottom surface of each conduit, and TCs clamped with stainless steel round-head screws to the interior surface of the junction box.
Hose Stream: A test was applied, but no details are given in available documentation.
Furnace: ~11' x 6' x at least 5' (i.e., a depth sufficient to contain the 3' vertical extent of the test conduit assemblies)
Furnace control: Ten (10) thermocouples on probes located throughout the furnace
Cable Type: No electric cables were installed in the conduits or the junction box

**Table D-54. NUMARC Thermo-Lag Test 1-7**

| Raceway Type | Barrier Protection | Cable Fill | Max. Temp Rise ΔT °C (°F) | Rating |
|---|---|---|---|---|
| 5" Steel Conduit | 1" nom. TL-330-1 | None | 566 (1050) | 1 hr. 56 min.[1] |
| 3" Steel Conduit | 1" nom. TL-330-1 | None | 599 (1110) | 1 hr. 52 min. |
| ¾" Steel Conduit | 1" nom. TL-330-1 | None | 136 (277) | 3 hr. |
| Junction Box | 1" nom. TL-330-1 | None | 123 (253) | 3 hr. |

[1]Maximum average temperature increase criterion of 250°F exceeded in 1 hr. 53 min.

As shown in the above table, the two largest test conduits did not meet the allowable single point maximum temperature increase criterion of 325°F. The smallest test conduit and the junction box did meet the allowable single point maximum temperature increase criterion of 325°F, and although not shown on the table, they also met the allowable maximum average temperature increase criterion of 250°F.

### D.3.2 NUMARC Phase 2 Testing

These tests were full scale tests, with test aluminum tray(s) and/or aluminum conduit(s) that vertically penetrated the furnace roof, descended into the furnace and bent to a horizontal section several feet long ranging from approximately 2 to 4 feet below the roof, then bent to a vertical ascending section that again penetrated the furnace roof 8 to 9 feet horizontally from the first penetration. This "elongated U" shape was used in NUMARC Phase 2 Tests 2-1, 2-2, 2-3, and 2-9.

#### D.3.2.1 NUMARC Project No. 13890-96141, Test 2-1 (April 1994)

ERFBS: This 1 hr. test evaluated a 6" Aluminum Conduit, a 4" Aluminum Conduit, a 2" Aluminum Conduit, and a 3/4" Aluminum Conduit, with ERFBS constructed using Thermo-Lag 330-1, pre-shaped conduit sections, nom. ½" thickness on the 4 conduits. Thermo-Lag 330-1 V-ribbed baseline panels, ½" nom. thickness, were used for the LB box designs and the support members. Dow Corning 3-6548 RTV silicone foam material was used to seal the blockout (where the conduits enter/exit thru the furnace roof). 3Therma-Lag 330-1 subliming trowel grade material was used to pre-caulk all joints and seams between the panels. Internal silicone elastomer (Promatec 45B) seal material was installed inside each conduit leg at the level of the furnace top deck, and ½" x 0.020" type 304 stainless steel rolled-edge banding straps with wing seals were used.

Test Procedure: NUMARC Phase 2 Test Program TEST PLAN, Rev. 0
Ambient Temperature: 55°F
Test Slab: 13' x 8' x 10 GA (Steel), with 2 layers of 2" ceramic fiber blanket insulation
Desired Rating: 1-hr
Thermocouples: One TC every 6" on one bare copper wire in each of the four conduits, one TC every 6" affixed with glass cloth electrical tape to the bottom surface of each conduit
Hose Stream: A test was applied, but no details are given in available documentation.
Furnace: ~11' x 6' x at least 6' (i.e., a depth sufficient to contain the ~4' vertical extent of the test conduits)
Furnace control: Ten (10) thermocouples on probes located throughout the furnace
Cable Type: No electric cables were installed in the conduits

#### Table D-55. NUMARC Thermo-Lag Test 2-1

| Raceway Type | Barrier Protection | Cable Fill | Max. Temp Rise ΔT °C (°F)[1] | Rating |
| --- | --- | --- | --- | --- |
| 6" Aluminum Conduit | ½" nom. TL-330-1 | None | 152 (305) | 50 min.[2] |
| 4" Aluminum Conduit | ½" nom. TL-330-1 | None | 160 (320) | 48 min.[2] |
| 2" Aluminum Conduit | ½" nom. TL-330-1 | None | 214 (417) | 39 min.[3] |
| 3/4" Aluminum Conduit | ½" nom. TL-330-1 | None | 524 (976) | 27 min.[4] |

[1] the test was terminated at 50 minutes when the last conduit exceeded its first temp. limit criterion (average temperature increase of 250°F, on the 6" conduit)
[2] failed the maximum average temperature increase criterion of 250°F
[3] failed the maximum average temperature increase criterion of 250°F first
[4] failed the maximum single point temperature increase criterion of 325°F, and the maximum average temperature increase criterion of 250°F, at the same time

Additionally, burn-through was noted on the ¾ inch conduit following the fire endurance and hose stream tests.

### D.3.2.2 NUMARC Project No. 13890-96142, Test 2-2 (April 1994)

<u>ERFBS</u>: This 1 hr. test evaluated the following ERFBS: two aluminum conduit assemblies (2 inch and ¾ inch), each separately clad with a nominal thickness of ½ inch Thermo-Lag 330-1, with 3M Fire Dam 150 Caulking and Top Coat for outdoor applications; and two Box Enclosures of ½ inch Thermo-Lag 330-1, one baseline and one with various upgrades (e.g., use of stainless steel bands and stitching with stainless steel wire described in detail in the original test report), each box enclosure containing three aluminum conduit assemblies (3 inch, 2 inch, and ¾ inch). The upgraded box enclosure met the applicable requirements for a fire exposure period of one hour, but the two conduit assemblies and the non-upgraded box enclosure failed to satisfy these requirements.

<u>Test Procedure</u>: NUMARC Phase 2 Test Program TEST PLAN, Rev. 0
<u>Test Slab</u>: 13' x 8' x 6" reinforced concrete slab
<u>Ambient Temperature</u>: 71°F
<u>Desired Rating</u>: 1-hr
<u>Thermocouples</u>: One TC every 6" on one bare copper wire in each of the six conduits, and one TC every 6" affixed with glass cloth electrical tape to the bottom surface of each conduit.
<u>Hose Stream</u>: A test was applied, but no details are given in available documentation.
<u>Furnace</u>: ~11' x 6' x at least 5' (i.e., a depth sufficient to contain the ~3' vertical extent of some of the test conduits)
<u>Furnace control</u>: Ten (10) thermocouples on probes located throughout the furnace
<u>Cable Type</u>: No electric cables were installed in the conduits

**Table D-56. NUMARC Thermo-Lag Test 2-2**

| Raceway: | Barrier Protection | Cable Fill | Max. Temp Rise $\Delta$T °C (°F) | Rating |
|---|---|---|---|---|
| Individual Conduits Center | | | | |
| 2" Aluminum Conduit | ½" nom. TL-330-1 | None | 348 (658) | 35 min. |
| ¾" Aluminum Conduit | ½" nom. TL-330-1 | None | 731 (1348) | 26 min. |
| Baseline (left) Box Enclosure | | | | |
| 3" Aluminum Conduit | ½" nom. TL-330-1 | None | 151 (303)[1] | Failed[2] |
| 2" Aluminum Conduit | ½" nom. TL-330-1 | None | 146 (295)[1] | Failed[2] |
| ¾" Aluminum Conduit | ½" nom. TL-330-1 | None | 141 (285)[1] | Failed[2] |

[1] These conduits all passed the maximum 325°F single point acceptance criterion; they also passed the 250°F maximum average increase criterion (all 3 increased 249°F)
[2] Although the single point and average temperature increase criteria were met, these conduits did not meet the barrier integrity and hose stream requirements.

Table D-56. NUMARC Thermo-Lag Test 2-2 (Continued)

| Raceway: | Barrier Protection | Cable Fill | Max. Temp Rise ΔT °C (°F) | Rating |
|---|---|---|---|---|
| Upgraded(right) Box Enclosure | | | | |
| 3" Aluminum Conduit | ½" nom. TL-330-1 | None | 113 (235)[3] | 1 hr. |
| 2" Aluminum Conduit | ½" nom. TL-330-1 | None | 117 (243)[3] | 1 hr. |
| ¾" Aluminum Conduit | ½" nom. TL-330-1 | None | 109 (229)[3] | 1 hr. |

[3] These conduits also met the 250°F maximum average temperature increase criterion

### D.3.2.3 NUMARC Project No. 13890-96143, Test 2-3 (April 1994)

ERFBS: This 3 hr. test evaluated three aluminum conduit assemblies (6 in., 3 in., and ¾ in.), covered with Thermo-Lag 330-1, 1 in. nominal thickness Pre-Shaped sections. The joints between sections were pre-caulked with Thermo-Lag 330-1 Trowel Grade material. Stainless steel bands were installed on the miter-cut wedge shaped pieces fitted to the radial bend portions of the conduits. These conduits were removed from the test furnace after all had exceeded the applicable acceptance criteria for fire resistance at 102 minutes.

Test Procedure: NUMARC Phase 2 Test Program TEST PLAN, Rev. 0

Test Slab: 13' x 8' x 10 GA (Steel), with 2 layers of 2" ceramic fiber blanket insulation

Ambient Temperature: 59°F

Desired Rating: 3-hr

Thermocouples: One TC every 6" on one bare copper wire in each of the three conduits, one TC every 6" affixed with glass cloth electrical tape to the bottom surface of each conduit

Hose Stream: A test was applied, but no details are given in available documentation.

Furnace: ~11' x 6' x at least 6' (i.e., a depth sufficient to contain the ~4' vertical extent of the test conduits)

Furnace control: Ten (10) thermocouples on probes located throughout the furnace

Cable Type: No electric cables were installed in the conduits

Table D-57. NUMARC Thermo-Lag Test 2-3

| Raceway Type | Barrier Protection | Cable Fill | Max. Temp Rise ΔT °C (°F)[1] | Rating |
|---|---|---|---|---|
| 6" Aluminum Conduit | 1" nom. TL-330-1 | None | 165 (329) | 102 min. |
| 3" Aluminum Conduit | 1" nom. TL-330-1 | None | 211 (411) | 91 min. |
| 3/4" Aluminum Conduit | 1" nom. TL-330-1 | None | 619 (1146) | 63 min. |

[1] the test was terminated at 102 minutes when the last conduit exceeded its temp. limit criteria (at 102 min., the 6" conduit reached an average temp. increase of 251°F (over its 250°F limit), and a max. temp. increase of 329°F (over its 325°F limit) both at the same time)

### D.3.2.4 NUMARC Project No. 13890-96147, Test 2-7 (April 1994)

This test, the following *Test 2-8*, and *Test 2-10* were full scale tests with test trays that vertically penetrated the furnace roof, descended into the furnace and bent to horizontal three feet below the furnace roof, then preceded four feet horizontally at which point they exited the furnace horizontally thru its front wall.

ERFBS: This 1 hr. test evaluated four aluminum tray assemblies (two 6 in. wide and two 24 in. wide) covered with Thermo-Lag 330-1, ½ inch V-ribbed panels. These trays were intended to be basically similar to as-installed ERFBS, with minimal upgrades, for comparison to the following *Test 2-8* which incorporated upgrades.

Test Procedure: NUMARC Phase 2 Test Program TEST PLAN, Rev. 0
Test Slab: 13' x 8' x 10 GA (Steel), with 2 layers of 2" ceramic fiber blanket insulation
Ambient Temperature: 68°F
Desired Rating: 1-hr
Thermocouples: One TC every 6" on each of two bare copper wires in each of the four cable trays, one wire on the tray's bottom rungs under the electric cables on the tray's centerline, and the other directly above on top of the electric cables (both were secured by standard electrical plastic tie wraps). A third such wire was secured to the outside surface of the cable tray rungs, 1" offset from the tray centerline. In addition, one TC every 6" was affixed to both side rails of each of the trays clamped on by screws.
Hose Stream: A test was applied, but no details are given in available documentation.
Furnace: ~11' x 6' x 80"
Furnace control: Ten (10) thermocouples on probes located throughout the furnace
Cable Type: The trays contained an approximate 1/3 each mix of power, control, and instrument cables with total areas as shown in the table below.

#### Table D-58. NUMARC Thermo-Lag Test 2-7

| Raceway Type | Barrier Protection | Cable Fill | Max. Temp Rise $\Delta T$ °C (°F)[1] | Rating |
|---|---|---|---|---|
| Tray A (24" Aluminum) | ½" nom. TL-330-1 | 15.4% | 397 (746) | 21 min. |
| Tray B (6" Aluminum) | ½" nom. TL-330-1 | 16.1% | 167 (333) | 48 min. |
| Tray C (6" Aluminum) | ½" nom. TL-330-1 | 16.1% | 167 (333) | 48 min. |
| Tray D (24" Aluminum) | ½" nom. TL-330-1 | 15.4% | 372 (701) | 23 min. |

[1] the test was terminated at 48 minutes when the last tray exceeded its temp. limit

### D.3.2.5 NUMARC Project No. 13890-96148, Test 2-8 (April 1994)

This test, the previous *Test 2-7*, and *Test 2-10*, were full scale tests with test trays that vertically penetrated the furnace roof, descended into the furnace and bent to horizontal three feet below the furnace roof, then preceded four feet horizontally at which point they exited the furnace horizontally thru its front wall.

ERFBS: This 1 hr. test evaluated four aluminum tray assemblies (two 6 in. wide and two 24 in. wide) covered with Thermo-Lag 330-1, ½ inch V-ribbed panels. These trays were similar to as-installed ERFBS but with significant upgrades, for comparison to the previous *Test 2-7* which did not incorporate upgrades. Upgrades included application of an approximately 3/16 inch thick layer of trowel grade Thermo-Lag 330-1 over the TL panel pieces covering the side rails and splice plates. Pieces of 330-069 stress skin were folded and stapled over the splice plates, and an approximately 1/16 inch thick skim coat of trowel grade TL was placed over the stress skin. Stainless steel tie wires and circumferential stress skin wraps were also used in numerous parts of the improved ERFBS designs.

Test Procedure: NUMARC Phase 2 Test Program TEST PLAN, Rev. 0

Test Slab: 13' x 8' x 10 GA (Steel), with 2 layers of 2" ceramic fiber blanket insulation

Ambient Temperature: 54°F

Desired Rating: 1-hr

Thermocouples: One TC every 6" on each of two bare copper wires in each of the four cable trays, one wire on the tray's bottom rungs under the electric cables on the tray's centerline, and the other directly above on top of the electric cables (both were secured by standard electrical plastic tie wraps). A third such wire was secured to the outside surface of the cable tray rungs, 1" offset from the tray centerline. In addition, one TC every 6" was affixed to both side rails of each of the trays clamped on by screws.
Hose Stream: A test was applied, but no details are given in available documentation.
Furnace: ~11' x 6' x 80"
Furnace control: Ten (10) thermocouples on probes located throughout the furnace
Cable Type: The trays contained an approximate 1/3 each mix of power, control, and instrument cables with total areas as shown in the table below.

#### Table D-59. NUMARC Thermo-Lag Test 2-8

| Raceway Type | Barrier Protection | Cable Fill | Max. Temp Rise ΔT °C (°F) | Rating |
|---|---|---|---|---|
| Tray A (24" Aluminum) | ½" nom. TL-330-1 | 15.4% | 172 (341) | 57 min.[1] |
| Tray B (6" Aluminum) | ½" nom. TL-330-1 | 16.1% | 99 (211) | 60 min. |
| Tray C (6" Aluminum) | ½" nom. TL-330-1 | 16.1% | 99 (210) | 60 min. |
| Tray D (24" Aluminum) | ½" nom. TL-330-1 | 15.4% | 119 (246) | 60 min. |

[1] individual temperature increase limit (325°F) exceeded only on the right cable tray side rail adjacent to the fire stop (i.e., where the ERFBS passed thru the furnace wall), which the licensee attributed to the fire stop and not to failure of the ERFBS.

*D.3.2.6 NUMARC Project No. 13890-96149, Test 2-9 (April 1994)*

ERFBS: This 1 hr. test evaluated a single 36" wide aluminum tray with upgrades similar to the upgraded 24" tray in test 2-8 that included application of an approximately 3/16 inch thick layer of trowel grade Thermo-Lag 330-1 over the nominal ½" thick TL panel pieces covering the side rails and splice plates. Pieces of 330-069 stress skin were folded and stapled over the splice plates, and an approximately 1/16 inch thick skim coat of trowel grade TL was placed over the stress skin. Stainless steel tie wires and circumferential stress skin wraps were also used in numerous parts of the improved ERFBS designs.

Test Procedure: NUMARC Phase 2 Test Program TEST PLAN, Rev. 0

Test Slab: 13' x 8' x 10 GA (Steel), with 2 layers of 2" ceramic fiber blanket insulation

Ambient Temperature: 68°F

Desired Rating: 1-hr

Thermocouples: One TC every 6" on each of two bare copper wires in the tray, one wire on the tray's bottom rungs under the electric cables on the tray's centerline, and the other directly above on top of the electric cables (both were secured by standard electrical plastic tie wraps). A third such wire was secured to the outside surface of the cable tray rungs, 1" offset from the tray centerline. In addition, one TC every 6" was affixed to both side rails of the tray clamped on by screws.

Hose Stream: A test was applied, but no details are given in available documentation.

Furnace: ~11' x 6' x 80"

Furnace control: Ten (10) thermocouples on probes located throughout the furnace

Cable Type: The tray contained an approximate 1/3 each mix of power, control, and instrument cables with total area 15.5% of the tray's cross-sectional area.

**Table D-60. NUMARC Thermo-Lag Test 2-9**

| Raceway Type | Barrier Protection | Cable Fill | Max. Temp Rise $\Delta T$ °C (°F) | Rating |
|---|---|---|---|---|
| 36" Aluminum Tray | ½" nom. TL-330-1 | 15.5% | 159 (319)[1] | 60 min. |

[1] This is below the 325°F single point temperature increase criterion. The maximum average temperature increase was 212°F, below the 250°F criterion. Additionally, no barrier openings were noted on the assembly following the hose stream test.

*D.3.2.7 NUMARC Project No. 13890-96150, Test 2-10 (April 1994)*

This test, and *Tests 2-7 and 2-8*, were full scale tests with test trays that vertically penetrated the furnace roof, descended into the furnace and bent to horizontal three feet below the furnace roof, then preceded four feet horizontally at which point they exited the furnace horizontally thru its front wall.

ERFBS: This 3 hr. test evaluated four aluminum tray assemblies (two 6 in. wide and two 24 in. wide) covered with Thermo-Lag 330-1, 1 inch thick V-ribbed panels. These trays were intended to be basically similar to as-installed ERFBS with minimal upgrades.

Test Procedure: NUMARC Phase 2 Test Program TEST PLAN, Rev. 0

Test Slab: 13' x 8' x 10 GA (Steel), with 2 layers of 2" ceramic fiber blanket insulation

Ambient Temperature: 57°F

Desired Rating: 3 hr.

Thermocouples: One TC every 6" on each of two bare copper wires in each of the four cable trays, one wire on the tray's bottom rungs under the electric cables on the tray's centerline, and the other directly above on top of the electric cables (both were secured by standard electrical plastic tie wraps). A third such wire was secured to the outside surface of the cable tray rungs, 1" offset from the tray centerline. In addition, one TC every 6" was affixed to both side rails of each of the trays clamped on by screws.
Hose Stream: A test was applied, but no details are given in available documentation.
Furnace: ~11' x 6' x 80"
Furnace control: Ten (10) thermocouples on probes located throughout the furnace
Cable Type: The trays contained an approximate 1/3 each mix of power, control, and instrument cables with total areas as shown in the table below.

### Table D-61. NUMARC Thermo-Lag Test 2-10

| Raceway Type | Barrier Protection | Cable Fill | Max. Temp Rise $\Delta T$ °C (°F)[1] | Rating |
|---|---|---|---|---|
| Tray A (24" Aluminum) | 1" nom. TL-330-1 | 15.4% | 164 (328) | 86 min. |
| Tray B (6" Aluminum) | 1" nom. TL-330-1 | 16.1% | 117 (242) | >86 min.[2] |
| Tray C (6" Aluminum) | 1" nom. TL-330-1 | 16.1% | 112 (233) | >86 min.[2] |
| Tray D (24" Aluminum) | 1" nom. TL-330-1 | 15.4% | >284 (>543) | <86 min |

[1] The test was terminated at 86 minutes when the last 24" tray (Tray A) exceeded its 325°F maximum temperature increase limit.
[2] Not determined; neither temperature criteria was exceeded at end of test.

### D.3.3 Texas Utilities (TU) Electric Co. Tests for Comanche Peak Steam Electric Station

These tests were full scale tests, starting with two tests in which conduits vertically penetrated the furnace roof, descended into the furnace and bent to a horizontal section ~6 to 8 feet long 3 feet below the furnace roof, then bent to a vertical ascending section that again penetrated the furnace roof 8-½ feet horizontally from the first penetration (i.e., they were "elongated U-shaped" test trays and conduits). Such "elongated U-shaped" test trays and conduits were used in the Scheme 9-1 and 9-3 tests below.

*D.3.3.1 TU Electric Report No. 12340-94367a, Scheme 9-1 (November 1992)*

<u>ERFBS</u>: This 1 hr. test used ½" nominal thickness Thermo-Lag 330-1 Flat panels and V-ribbed panels to construct assemblies (e.g., hangers, LBD boxes, radial bends) with 5", 3", and ¾" diameter conduits clad with ½" nominal thickness Thermo-Lag 330-1 pre-shaped conduit sections. Upgrades were used similar to those described previously for the Numarc tests.

<u>Test Procedure</u>: Texas Utilities Electric TEST PLAN, Rev. 8
<u>Test Slab</u>: 13' x 8' x 10 GA (Steel), with 2 layers of 2" ceramic fiber blanket insulation
<u>Ambient Temperature</u>: 72°F
<u>Desired Rating</u>: 1 hr.

<u>Thermocouples</u>: One TC every 6" on one of the power cables, one of the control cables, and one of the instrument cables in each conduit (except for the ¾" conduit which had only one instrumented cable) taped to the top surface of the cable with a double wrap of glass fiber reinforced electrical tape, plus one TC every 12" taped to the top outside surface of each conduit with a short piece of glass cloth electrical tape.
<u>Hose Stream</u>: Passed
<u>Furnace</u>: ~11' x 6' x 80"
<u>Furnace control</u>: Ten (10) thermocouples on probes located throughout the furnace
<u>Cable Type</u>: The conduits contained a ~1/3 mix of power, control, and instrument cables with total fill areas as shown in the table below.

**Table D-62. TU Electric Thermo-Lag Test 9-1**

| Raceway Type | Barrier Protection | Cable Fill | Max. Temp Rise $\Delta T$ °C (°F)[1] | Rating[2] |
|---|---|---|---|---|
| ¾" Steel Conduit | ½" nom. TL-330-1 | 39.8% | 283 (542) | 60 min. |
| 3" Steel Conduit | ½" nom. TL-330-1 | 44.2% | 229 (444) | 60 min. |
| 5" Steel Conduit | ½" nom. TL-330-1 | 34.9% | 284 (543) | 60 min. |

[1] The max. $\Delta T$ values shown in the table are all from thermocouples on the outside of the steel conduits, between the steel surface and the Thermo-Lag fire barrier material. The laboratory personal attributed the "excessively high temperatures" measured outside the steel conduits were due to "electro-chemical reactions caused by saturation of the fiberglass thermocouple insulation grading by condensate accumulated on the conduit steel."

[2] All thermocouples on the electric cables inside the conduits showed $\Delta T$s below the acceptance criteria of 250°F maximum average temperature increase and 325°F maximum single point increase. The licensee based the stated rating on those values (i.e., the applicable 60 min. criteria were met based on cable temperature rises within the conduit). This rating is based on site-specific acceptance criteria accepted by the NRC in its October 29, 1992 letter.

### D.3.3.2 TU Electric Report No. 12340-94367j, Scheme 9-3 (December 1992)

ERFBS: This 1 hr. test used ½" nominal thickness Thermo-Lag 330-1 Flat panels and V-ribbed panels to construct assemblies (e.g., hangers, LBD boxes) with 2", 1½", and ¾" diameter conduits clad with ½" nominal thickness Thermo-Lag 330-1 pre-shaped conduit sections (with the exception of the ¾" conduit, which used ¾" nominal thickness Thermo-Lag 330-1 pre-shaped conduit sections). Other upgrades were used similar to those described previously for the Numarc tests.

Test Procedure: Texas Utilities Electric TEST PLAN, Rev. 8

Test Slab: 13' x 8' x 10 GA (Steel), with 2 layers of 2" ceramic fiber blanket insulation

Ambient Temperature: 65°F

Desired Rating: 1 hr.

Thermocouples: One TC every 6" on one of the power cables, on one of the control cables, and on one of the instrument cables in each conduit (except for the ¾" conduit which had only one instrumented cable) taped to the top surface of each cable with a double wrap of glass fiber reinforced electrical tape, plus one TC every 12" taped to the top outside surface of each conduit with a short piece of glass cloth electrical tape.

Hose Stream: Failed.

Furnace: ~11' x 6' x 80"

Furnace control: Ten (10) thermocouples on probes located throughout the furnace

Cable Type: The conduits contained a ~1/3 mix of power, control, and instrument cables with total fill areas as shown in the table below.

**Table D-63. TU Electric Thermo-Lag Test 9-3**

| Raceway Type | Barrier Protection | Cable Fill | Max. Temp Rise ΔT °C (°F)[1] | Rating[2] |
|---|---|---|---|---|
| ¾" Steel Conduit | ¾" nom. TL-330-1 | 39.8% | 444 (831) | Failed |
| 1-½" Steel Conduit | ½" nom. TL-330-1 | 49.5% | 538 (1000) | Failed |
| 2" Steel Conduit | ½" nom. TL-330-1 | 35.8% | 424 (796) | Failed |

[1] The maximum ΔT values shown in the table are all from thermocouples on the outside of the steel conduits, between the steel surface and the Thermo-Lag fire barrier material. The laboratory personal attributed the "excessively high temperatures" measured outside the steel conduits were due to "electro-chemical reactions caused by saturation of the fiberglass thermocouple insulation grading by condensate accumulated on the conduit steel." The corresponding maximum ΔT values for cables inside the conduits were 457°F for the ¾", 413°F for the 1-½", and 358°F for the 2" conduits. Thus the three ΔTs for the steel conduits given in the table, and these three ΔTs for cables within the conduits, all exceed the maximum allowable single point ΔT of 325°F.

[2] No loss of circuit integrity occurred during the test and post-fire cable insulation resistance tests were within limits. However, the ΔT values for cables inside the conduits exceeded allowable limits; and visible cable damage and fire barrier burn through occurred for the 1-1/2" and 2" conduits. In addition, the max. single point temperature increases all exceeded by a significant margin the maximum allowable 325°F at one hour, so all three conduits failed to qualify for one hour. The exact times (less than one hour) at which they exceeded the 325°F criterion were not given in the available documentation. The licensee did not credit the results of this fire test as part of the Thermo-Lag ERFBS qualification basis at CPSES.

### D.3.3.3 TU Electric Report No. 12340-94367c, Scheme 10-1 (December 2, 1992)

<u>ERFBS</u>: This 1 hr. test evaluated an assembly of two parallel 3" conduits with centerlines 8" apart that descended vertically thru the furnace roof to two condulets 3' below the roof, then horizontally 3-½ feet thru a single horizontal 1'-6" x 1' x 6" junction box, then horizontally another 3-½ feet to two condulets, then rose vertically into a vertical 1'-6" x 1' x 6" junction box, then passed thru the furnace roof. The condulet and junction box covers were constructed using Thermo-Lag 330-1 Flat Panels of ½" nominal thickness; Thermo-Lag 330-1 Subliming Trowel Grade Material was used to pre-caulk all joints, seams, and upgraded areas; and Thermo-Lag 330-1 Pre-Shaped Conduit Sections of ½" nominal thickness were used on the conduits. Other upgrades were used similar to those described previously for the Numarc tests.

<u>Test Procedure</u>: Texas Utilities Electric TEST PLAN, Rev. 8

<u>Test Slab</u>: 13' x 8' x 10 GA (Steel), with 2 layers of 2" ceramic fiber blanket insulation

<u>Ambient Temperature</u>: 63°F

<u>Desired Rating</u>: 1 hr.

<u>Thermocouples</u>: One TC every 6" on one of the power cables, on one of the control cables, and on one of the instrument cables in each conduit taped to the top surface of each cable with a double wrap of glass fiber reinforced electrical tape, and one TC every 12" taped to the top outside surface of each conduit with a short piece of glass cloth electrical tape, plus several TCs inside the junction boxes.

<u>Hose Stream</u>: Passed.

<u>Furnace</u>: ~11' x 6' x 80"

<u>Furnace control</u>: Ten (10) thermocouples on probes located throughout the furnace

<u>Cable Type</u>: The conduits contained a ~1/3 mix of power, control, and instrument cables with total fill area 43.4% of the available area inside the conduits.

**Table D-64. TU Electric Thermo-Lag Test 10-1**

| Raceway Type | Barrier Protection | Cable Fill | Max. Temp Rise ΔT °C (°F) | Rating |
|---|---|---|---|---|
| Front 3" Steel Conduit | ½" nom. TL-330-1 | 43.4% | 581 (1078)[1] | 60 min.[2] |
| Rear 3" Steel Conduit | ½" nom. TL-330-1 | 43.4% | 348 (659)[1] | 60 min.[2] |
| Horiz. J-Box | ½" nom. TL-330-1 | N/A | 51 (123) | 60 min. |
| Vert. J-Box | ½" nom. TL-330-1 | N/A | 57 (135) | 60 min. |

[1] The max. ΔT values shown in the table for the conduits are from thermocouples on the outside of the steel conduits, between the steel surface and the Thermo-Lag fire barrier material. The laboratory personal attributed the "excessively high temperatures" measured outside the steel conduits were due to "electro-chemical reactions caused by saturation of the fiberglass thermocouple insulation grading by condensate accumulated on the conduit steel."

[2] All thermocouples on the electric cables inside the conduits showed ΔTs below the acceptance criteria of 250°F maximum average temperature increase and 325°F maximum single point increase. The licensee based the stated rating for the conduits on those values (i.e., the applicable 60 min. criteria "were met" based on cable temperature rises within the conduit). This rating is based on site-specific acceptance criteria accepted by the NRC in its October 29, 1992 letter.

*D.3.3.4 TU Electric Report No. 12340-94367d, Scheme 10-2 (December 16, 1992)*

Based on the available documentation, this test was apparently a near-repeat of "Scheme 10-1" (above) except for differences in their upgrades (e.g., materials and methods used to apply a second layer or TL to the junction boxes). The results were not substantially different for the two tests.

ERFBS: This 1 hr. test evaluated an assembly of two parallel 3" conduits with centerlines 8" apart that descended vertically thru the furnace roof to two condulets 3' below the roof, then horizontally 3-½ feet into a single horizontal 1'-6" x 1' x 6" junction box, then horizontally another 3-½ feet to two condulets, then rose vertically into a vertical 1'-6" x 1' x 6" junction box, then passed thru the furnace roof. The condulet and junction box covers were constructed using Thermo-Lag 330-1 Flat Panels of ½" nominal thickness; Thermo-Lag 330-1 Subliming Trowel Grade Material was used to pre-caulk all joints, seams, and upgraded areas; and Thermo-Lag 330-1 Pre-Shaped Conduit Sections of ½" nominal thickness were used on the conduits. Other upgrades were used similar to those described previously for the Numarc tests.

Test Procedure: Texas Utilities Electric TEST PLAN, Rev. 8

Test Slab: 13' x 8' x 10 GA (Steel), with 2 layers of 2" ceramic fiber blanket insulation

Ambient Temperature: 69°F

Desired Rating: 1 hr.

Thermocouples: One TC every 6" on one of the power cables, on one of the control cables, and on one of the instrument cables in each conduit taped to the top surface of each cable with a double wrap of glass fiber reinforced electrical tape, and one TC every 12" taped to the top outside surface of each conduit with a short piece of glass cloth electrical tape, plus several TCs inside the junction boxes.

Hose Stream: Passed.

Furnace: ~11' x 6' x 80"

Furnace control: Ten (10) thermocouples on probes located throughout the furnace

Cable Type: The conduits contained a ~1/3 mix of power, control, and instrument cables with total fill area 43.4% of the available area inside the conduits.

**Table D-65. TU Electric Thermo-Lag Test 10-2**

| Raceway Type | Barrier Protection | Cable Fill | Max. Temp Rise ΔT °C (°F)[1] | Rating |
|---|---|---|---|---|
| Front 3" Steel Conduit | ½" nom. TL-330-1 | 43.4% | 511 (951)[1] | 60 min.[2] |
| Rear 3" Steel Conduit | ½" nom. TL-330-1 | 43.4% | 818 (1504)[1] | 60 min.[2] |
| Horiz. J-Box | ½" nom. TL-330-1 | N/A | 147 (297) | 60 min. |
| Vert. J-Box | ½" nom. TL-330-1 | N/A | 129 (265) | 60 min. |

[1] The max. ΔT values shown in the table for the conduits are from thermocouples on the outside of the steel conduits, between the steel surface and the Thermo-Lag fire barrier material. The laboratory personal attributed the "excessively high temperatures" measured outside the steel conduits were due to "electro-chemical reactions caused by saturation of the fiberglass thermocouple insulation grading by condensate accumulated on the conduit steel."

[2] All thermocouples on the electric cables inside the conduits showed ΔTs below the acceptance criteria of 250°F maximum average temperature increase and 325°F maximum single point

increase. The testing laboratory based the stated rating on those values (i.e., the applicable 60 min. criteria were met based on cable temperature rises within the conduit). This rating is based on site-specific acceptance criteria accepted by the NRC in its October 29, 1992 letter.

### D.3.3.5 TU Electric Report No. 12340-94367e, Scheme 11-1 (January 1993)

ERFBS: This 1 hr. test evaluated an assembly consisting of a 24" cable tray with air drops into the tray from 5", 3", 2", and 1" conduits. The fire barriers protecting these trays and conduits were constructed using Thermo-Lag 330-1 flat and V-Ribbed panels that were ½" nominal thickness with factory-applied 350 Topcoat, Thermo-Lag 330-660 Flexi-Blanket sheets that were 3/8" nominal thickness, 330-69 stress skin sheets, 330-660 Subliming Trowel Grade material, and 330-1 Trowel Grade subliming compound. Conduits were covered with Thermo-Lag 330-1 Pre-Shaped Conduit Sections nominally ½" thick.

Test Procedure: Texas Utilities Electric TEST PLAN, Rev. 8

Test Slab: 13' x 8' x 10 GA (Steel), with 2 layers of 2" ceramic fiber blanket insulation

Ambient Temperature: 72°F

Desired Rating: 1 hr.

Thermocouples: One TC every 6" on one of the power cables, on one of the control cables, and on one of the instrument cables from each of the conduits that led into the tray (on those cables both through the conduits and in the tray) taped to the top surface of each cable with a double wrap of glass fiber reinforced electrical tape, and one TC every 12" screwed to the tray rails. In addition, two TCs were screwed to each of the four conduit protrusions from the tray (i.e., on the junction between each conduit and the tray).

Hose Stream: Passed.
Furnace: 11' x 6' x 80"
Furnace control: Ten (10) thermocouples on probes located throughout the furnace
Cable Type: The tray and conduits contained a ~1/3 mix of power, control, and instrument cables with total fill area as shown in the table below.

### Table D-66. TU Electric Thermo-Lag Test 11-1

| Raceway Type | Barrier Protection | Cable Fill % | Max. Temp Rise ΔT °C (°F) | Rating |
|---|---|---|---|---|
| 1" steel conduit air drop | ½" nom. TL-330-1 | 36.6 | 111 (232) | 60 min.[1] |
| 2" steel conduit air drop | ½" nom. TL-330-1 | 56.1 | 97 (207) | 60 min.[1] |
| 3" steel conduit air drop | ½" nom. TL-330-1 | 52.8 | 98 (209) | 60 min.[1] |
| 5" steel conduit air drop | ½" nom. TL-330-1 | 29.6 | 104 (219) | 60 min.[1] |
| 24" steel cable tray | ½" nom. TL-330-1 | varied | 109 (229) | 60 min.[1] |

[1] This rating is based on site-specific acceptance criteria accepted by the NRC in its October 29, 1992 letter.

*D.3.3.6 TU Electric Report No. 12340-95766, Scheme 11-2 (August 1993)*

ERFBS: This 1 hr. test evaluated an assembly consisting of a 24" cable tray with air drops into the tray from a 2" and a 1-½" conduit. The fire barriers protecting these trays and conduits were constructed using Thermo-Lag 330-1 flat and V-Ribbed panels that were ½" nominal thickness with factory-applied 350 Topcoat, Thermo-Lag 330-660 Flexi-Blanket sheets that were ¼" nominal thickness, 330-69 stress skin sheets, 330-660 Subliming Trowel Grade material, and 330-1 Trowel Grade subliming compound. Conduits were protected with Thermo-Lag 330-1 Pre-Shaped Conduit Sections nominally ½" thick.

Test Procedure: Texas Utilities Electric TEST PLAN, Rev. 11

Test Slab: 13' x 8' x 10 GA (Steel), with 2 layers of 2" ceramic fiber blanket insulation

Ambient Temperature: 92°F

Desired Rating: 1 hr.

Thermocouples: One TC every 6" on one of the power cables, on one of the control cables, and on one of the instrument cables within the tray, taped to the top surface of each cable with a double wrap of glass fiber reinforced electrical tape, and one TC every 12" screwed to the tray rails. Within the conduits, the TCs were fastened every 6" to a #8 bare copper wire that was loosely wrapped around the cables to be monitored that were pulled into each of the conduits that led into the tray. In addition, two TCs were screwed to both of the conduit protrusions from the tray (i.e., on the junctions between each conduit and the tray).

Hose Stream: Passed.

Furnace: ~11' x 6' x 80"

Furnace control: Ten (10) thermocouples on probes located throughout the furnace

Cable Type: The conduits contained a ~1/3 mix of power, control, and instrument cables with total fill area as shown in the table below.

**Table D-67. TU Electric Thermo-Lag Test 11-2**

| Raceway Type | Barrier Protection | Cable Fill % | Max. Temp Rise $\Delta T$ °C (°F) | Rating |
|---|---|---|---|---|
| 1-½" steel conduit air drop | ½" nom. TL-330-1 | 49.5 | 113 (235) | 60 min. |
| 2" steel conduit air drop | ½" nom. TL-330-1 | 43.5 | 175 (347) | 60 min.[1] |
| 24" steel cable tray | ½" nom. TL-330-1 | ~15 | 103 (217) | 60 min. |

[1] The single point $\Delta T$ parameter was exceeded on the power cable in the 2" air drop bundle at the 59-minute mark. However, the test laboratory and the licensee based the 60 minute rating of the tested Thermo-Lag ERFBS configurations on the site-specific acceptance criteria accepted by the NRC in its October 29, 1992 letter.

### D.3.3.7 TU Electric Report No. 12340-95767, Scheme 11-4 (October 1993)

ERFBS: This 1 hr. test evaluated an assembly containing a box design air drop between a bank of cast-in-concrete conduit stubs and two nested 24' wide cable trays clad with nominal ½" Thermo-Lag 330-1 with various upgrades. The nested trays were the typical elongated "U" shape, with the outside/lower tray deeper and wider than the inside/upper tray. The parts of the trays that formed the bottom parts of the nested "Us" were horizontal, with the concrete stubs perpendicular to the trays (also horizontal). A single fire barrier enclosure included both trays, and butted up against the concrete structure in which the stubs were embedded.

Test Procedure: Texas Utilities Electric TEST PLAN, Rev. 11

Test Slab: 13' x 8' x 10 GA (Steel), with 2 layers of 2" ceramic fiber blanket insulation

Ambient Temperature: 91°F

Desired Rating: 1 hr.

Thermocouples: One TC every 6" on one of the power cables, on one of the control cables, and on one of the instrument cables within the tray, taped to the top surface of each cable with a double wrap of glass fiber reinforced electrical tape. Also, to monitor temperatures in the air drop area, bare #8 AWG stranded copper wires instrumented with TCs were wrapped loosely around the cables in the volume where electric cables were looped from the trays out of the furnace and back (see "Cable Type" below).

Hose Stream: Failed.

Furnace: 11' x 6' x 80"

Furnace control: Ten (10) thermocouples on probes located throughout the furnace

Cable Type: The trays contained a ~1/3 mix of power, control, and instrument cables with total fill area as shown in the table below. In each of the two trays, two groups of cables (each group consisting of one power, one control, and one instrumentation cable) were looped out of the tray, out of the furnace thru one of the conduit stubs, back into the furnace thru an adjacent conduit stub, and back into the tray. In the lower tray, the looped cable groups exited and entered the tray over its side rail. In the upper tray, the cable groups exited and entered the tray between the rungs in the bottom of the tray.

**Table D-68. TU Electric Thermo-Lag Test 11-4**

| Raceway Type | Barrier Protection | Cable Fill | Max. Temp Rise ΔT °C (°F) | Rating |
|---|---|---|---|---|
| Outer cable tray | ½" nom. TL-330-1 | 16.5% | 118 (244) | 60 min. |
| Inner cable tray | ½" nom. TL-330-1 | 16.5% | 105 (221) | 60 min. |
| Box design air drop volume | ½" nom. TL-330-1 | variable | 91 (196) | 60 min.[1] |

[1] A through opening in the air drop box design portion of the ERFBS occurred during the hose stream test. However the test laboratory and the licensee based the 60 minute rating of the tested Thermo-Lag ERFBS configuration on the site-specific acceptance criteria accepted by the NRC in its letter dated October 29, 1992.

*D.3.3.8 TU Electric Report No. 12340-95768, Scheme 11-5 (August 1993)*

ERFBS: This 1 hr. test evaluated three full scale 24" wide steel trays that vertically penetrated the furnace roof, descended into the furnace to a radial bend, ran horizontally three feet below the furnace roof, and exited the furnace thru its front wall about six feet (measured horizontally) from their entrance location. The fire barriers protecting these trays were constructed using Thermo-Lag 330-1 flat and V-Ribbed panels that were ½" nominal thickness with factory-applied 350 Topcoat, 330-69 stress skin sheets, and 330-1 Trowel Grade subliming compound. Various upgrade techniques were used in constructing the three trays.

Test Procedure: Texas Utilities Electric TEST PLAN, Rev. 11

Test Slab: 13' x 8' x 10 GA (Steel), with 2 layers of 2" ceramic fiber blanket insulation

Ambient Temperature: 92°F

Desired Rating: 1 hr.

Thermocouples: Each of the three trays had one TC every 6" on one of the power cables, one of the control cables, and one of the instrument cables taped to the top surface of the cable with a double wrap of glass fiber reinforced electrical tape, plus one TC every 12" along both side rails clamped under a screw head.
Hose Stream: Right and center trays Passed, left tray Failed.
Furnace: 11' x 6' x 80"
Furnace control: Ten (10) thermocouples on probes located throughout the furnace
Cable Type: The trays contained a ~1/3 mix of power, control, and instrument cables with total fill area as shown in the table below.

**Table D-69. TU Electric Thermo-Lag Test 11-5**

| Raceway Type | Barrier Protection | Cable Fill | Max. Temp Rise ΔT °C (°F) | Rating |
|---|---|---|---|---|
| Right Cable Tray | ½" nom. TL-330-1 | 16.5% | 132 (270)[1] | 60 min.[2] |
| Center Cable Tray | ½" nom. TL-330-1 | 16.5% | 191 (376) | Failed |
| Left Cable Tray | ½" nom. TL-330-1 | 16.5% | 236 (457) | Failed |

[1] Below the maximum single point temperature rise criterion of 325°F. The maximum average temperature rise in this tray was 210°F, below the 250°F acceptance criterion
[2] The test laboratory and licensee based the 60 minute rating of the tested Thermo-Lag ERFBS configuration on the site-specific acceptance criteria accepted by the NRC in its letter dated October 29, 1992.

*D.3.3.9 TU Electric Report No. 12340-94367i, Scheme 12-1 (December 1992)*

ERFBS: This 1 hr. test evaluated a 30" wide steel tray that vertically penetrated the furnace roof, descended to a radial bend that bent it to a horizontal direction three feet below the roof, proceeded four feet in a horizontal section, then bent to a vertically ascending section that penetrated the furnace roof ~8 feet horizontally from the first penetration. The fire barrier protecting this tray was constructed using Thermo-Lag 330-1 flat and V-Ribbed panels that were ½" nominal thickness with factory-applied 350 Topcoat, 330-69 stress skin sheets, and 330-1 Trowel Grade subliming compound. Various upgrade techniques were used in constructing the tray.

Test Procedure: Texas Utilities Electric TEST PLAN, Rev. 8

Test Slab: 13' x 8' x 10 GA (Steel), with 2 layers of 2" ceramic fiber blanket insulation

Ambient Temperature: 71°F

Desired Rating: 1 hr.

Thermocouples: The tray had one TC every 6" on one of its power cables, one of its control cables, and one of its instrument cables taped to the top surface of the cable with a double wrap of glass fiber reinforced electrical tape, plus one TC every 12" along both side rails clamped under a screw head.
Hose Stream: Passed.
Furnace: 11' x 6' x 80"
Furnace control: Ten (10) thermocouples on probes located throughout the furnace
Cable Type: The tray contained a ~1/3 mix of power, control, and instrument cables with total fill area as shown in the table below.

**Table D-70. TU Electric Thermo-Lag Test 12-1**

| Raceway Type | Barrier Protection | Cable Fill | Max. Temp Rise $\Delta T$ °C (°F) | Rating |
|---|---|---|---|---|
| 30" steel cable tray | ½" nom. TL-330-1 | 17.1% | 144 (292)[1] | 60 min.[2] |

[1] This is below the 325°F single point maximum temperature rise criterion. The maximum average temperature rise was 201°F, which is below the 250°F acceptance criterion.
[2] The test laboratory and licensee based the 60 minute rating of the tested Thermo-Lag ERFBS configuration on the site-specific acceptance criteria accepted by the NRC in its letter dated October 29, 1992.

*D.3.3.10 TU Electric Report No. 12340-94367h, Scheme 12-2 (December 1992)*

ERFBS: This 1 hr. test evaluated a 24" wide steel tray that vertically penetrated the furnace roof, descended to a radial bend that bent it to a horizontal direction three feet below the roof, proceeded in a horizontal section to a squared bend, then vertically ascended through the furnace roof ~8-½ feet (measured horizontally) from the first penetration. In the middle portion of the horizontal section, a Tee assembly was installed that could allow connection of a second horizontal 24" tray perpendicular to the first tray (such a tray was not actually installed – the Tee fitting ended in a firestop). The fire barrier protecting this tray and Tee was constructed using Thermo-Lag 330-1 flat and V-Ribbed panels that were ½" nominal thickness with factory-applied 350 Topcoat, 330-69 stress skin sheets, and 330-1 Trowel Grade subliming compound. Various upgrade techniques were used in constructing the tray.

Test Procedure: Texas Utilities Electric TEST PLAN, Rev. 8

Test Slab: 13' x 8' x 10 GA (Steel), with 2 layers of 2" ceramic fiber blanket insulation

Ambient Temperature: 69°F

Desired Rating: 1 hr.

Thermocouples: The tray had one TC every 6" on one of its power cables, one of its control cables, and one of its instrument cables taped to the top surface of the cable with a double wrap of glass fiber reinforced electrical tape, plus one TC every 12" along both side rails clamped under a screw head.
Hose Stream: Failed.
Furnace: 11' x 6' x 80"

Furnace control: Ten (10) thermocouples on probes located throughout the furnace
Cable Type: The tray contained a ~1/3 mix of power, control, and instrument cables with total fill area as shown in the table below. In the area of the Tee section, the cables were looped toward the mouth of the Tee, producing a higher cable loading.

**Table D-71. TU Electric Thermo-Lag Test 12-2**

| Raceway Type | Barrier Protection | Cable Fill | Max. Temp Rise $\Delta T$ °C (°F) | Rating |
|---|---|---|---|---|
| 24" steel cable tray | ½" nom. TL-330-1 | 16.6% | 140 (284)[1] | 60 min.[2] |

[1] This is below the 325°F single point maximum temperature rise criterion. The maximum average temperature rise was 213°F, which is below the 250°F acceptance criterion.

[2] Although no burn through of the ERFBS occurred, a barrier opening occurred during the hose stream test where the bottom panel on the Tee section interfaced with the fire stop. However, the testing laboratory and licensee based the 60 minute rating of the tested Thermo-Lag ERFBS configuration on the site-specific acceptance criteria accepted by the NRC in its letter dated October 29, 1992.

*D.3.3.11 TU Electric Report No. 12340-94367I, Scheme 13-1 (December 1992)*

ERFBS: This 1 hr. test evaluated a 12" wide steel tray that vertically penetrated the furnace roof, descended to a radial bend that bent it to a horizontal direction three feet below the roof, proceeded in a horizontal section to another radial bend, then vertically ascended through the furnace roof ~8-½ feet (measured horizontally) from the first penetration. The fire barrier protecting this tray was constructed using Thermo-Lag 330-1 flat and V-Ribbed panels that were ½" nominal thickness with factory-applied 350 Topcoat, 330-69 stress skin sheets, and 330-1 Trowel Grade subliming compound. Various upgrade techniques were used in constructing the tray.

Test Procedure: Texas Utilities Electric TEST PLAN, Rev. 8
Test Slab: 13' x 8' x 10 GA (Steel), with 2 layers of 2" ceramic fiber blanket insulation
Ambient Temperature: 69°F
Desired Rating: 1 hr.
Thermocouples: The tray had one TC every 6" on one of its power cables, one of its control cables, and one of its instrument cables taped to the top surface of the cable with a double wrap of glass fiber reinforced electrical tape, plus one TC every 12" along both side rails clamped under a screw head.
Hose Stream: Passed.
Furnace: 11' x 6' x 80"
Furnace control: Ten (10) thermocouples on probes located throughout the furnace
Cable Type: The tray contained a ~1/3 mix of power, control, and instrument cables with total fill area as shown in the table below.

**Table D-72. TU Electric Thermo-Lag Test 13-1**

| Raceway Type | Barrier Protection | Cable Fill | Max. Temp Rise ΔT °C (°F) | Rating |
|---|---|---|---|---|
| 12" steel cable tray | ½" nom. TL-330-1 | 14.7% | 127 (261)[1] | 60 min.[2] |

[1] This is below the 325°F single point maximum temperature rise criterion. The maximum average temperature rise was 209°F, which is below the 250°F acceptance criterion.

[2] The testing laboratory and licensee based the 60 minute rating of the tested Thermo-Lag ERFBS configuration on the site-specific acceptance criteria accepted by the NRC in its letter dated October 29, 1992.

*D.3.3.12 TU Electric Report No. 12340-95769, Scheme 13-2 (August 1993)*

ERFBS: This 1 hr. test evaluated a 12" wide steel tray and a separate 2" conduit that both vertically penetrated the furnace roof, descended to radial bends that bent them to horizontal directions three feet below the roof, proceeded in horizontal sections to another pair of radial bends, then vertically ascended through the furnace roof ~8-½ feet (measured horizontally) from the first penetrations. The separate fire barriers protecting the tray and conduit were constructed using Thermo-Lag 330-1 flat and V-Ribbed panels that were ½" nominal thickness with factory-applied 350 Topcoat, 330-69 stress skin sheets, 330-1 Trowel Grade subliming compound, and Thermo-Lag 330-1 Pre-Shaped conduit sections with ½" nominal thickness. Various upgrade techniques were used in constructing the separate tray and conduit barriers.

Test Procedure: Texas Utilities Electric TEST PLAN, Rev. 11

Test Slab: 13' x 8' x 10 GA (Steel), with 2 layers of 2" ceramic fiber blanket insulation

Ambient Temperature: 92°F

Desired Rating: 1 hr.

Thermocouples: The tray and the conduit each had one TC every 6" on one of their power cables, one of their control cables, and one of their instrument cables taped to the top surface of the cable with a double wrap of glass fiber reinforced electrical tape. In addition, the tray had one TC every 12" along both side rails clamped under a screw head, and the conduit had one TC every 12" along its top outside surface, held in position by short pieces of Glass Cloth Electrical tape.

Hose Stream: Failed.

Furnace: 11' x 6' x 80"

Furnace control: Ten (10) thermocouples on probes located throughout the furnace

Cable Type: The tray and conduit each contained ~1/3 mixes of power, control, and instrument cables with total fill areas shown in the table below.

**Table D-73. TU Electric Thermo-Lag Test 13-2**

| Raceway Type | Barrier Protection | Cable Fill | Max. Temp Rise °C (°F)[1] | Rating |
|---|---|---|---|---|
| 12" Cable Tray | ½" nom. TL-330-1 | 14.7% | 179 (355) | Failed[2] |
| 2" Conduit | ½" nom. TL-330-1 | 43.5% | 234 (454) | Failed[2] |

[1] These temperature rises are above the 325°F single point maximum temperature rise criterion, therefore both the tray and the conduit failed to achieve the desired 60 minute qualification.

[2] Minor areas of burn through were observed on the two assemblies prior to hose

stream testing. The licensee does not credit the results of this fire test as part of the Thermo-Lag ERFBS qualification basis at CPSES.

### D.3.3.13 TU Electric Report No. 12340-102571, Scheme 13-3 (February 1999)

ERFBS: This 1 hr. test evaluated two full-scale 2" steel conduit assemblies (A & B), two full-scale 12" wide steel ladder back cable tray assemblies (C & D), and a 12" wide steel ladder back cable tray segment (E). Each conduit assembly penetrated the furnace roof, descended into the furnace to a 90° condulet fitting, ran horizontally three feet below the furnace roof, and exited the furnace through its front wall about six feet (measured horizontally) from their entrance location. Similarly, each full-scale cable tray assembly penetrated the furnace roof, descended into the furnace to 90° radial bend fittings, ran horizontally three feet below the furnace roof, and exited the furnace through its front wall about six feet (measured horizontally) from their entrance location. The 12" cable tray segment simply extended vertically downward into the furnace for a three foot distance. The fire barriers protecting the conduits and cable trays were constructed using Thermo-Lag 330-1 flat and v-rib panels and pre-shaped conduit half round sections that were ½" nominal thickness with Thermo-Lag 331 topcoat, 330-69 stress skin sheets, and 330-1 trowel grade subliming compound. The joints on the condulet fitting enclosure for each conduit were upgraded with standard stress skin and trowel grade reinforcement. The overall external surface of Conduit A received an additional ¼" thickness of trowel grade material, while Conduit B did not. Similarly, the bottom and side rail surfaces of Cable Tray C received an additional ¼" thick trowel grade build-up, while Cable Tray D did not. Finally, the bottom of the ERFBS installed on Tray E was sealed with a 12" deep silicone foam fire stop.

Test Procedure: Texas Utilities Electric TEST PLAN, Rev. 1 (October 26, 1998)

Test Slab: 13' x 8' 10 GA (Steel), with 2 layers of 2" ceramic fiber blanket insulation

Ambient Temperature: 74°F

Desired Rating: 1 hr.

Thermocouples: The conduits each had one TC every 6" along the outside surface clamped under a screw head, and one TC every 6" on the single 5/C 12 AWG cable routed inside the conduits. The cable trays each had one TC every 6" on both side rails clamped under a screw head. The cable trays also had one TC every 6" on one power, control, and instrument cables.

The TCs on cables were secured to the top of the cables with a double wrap of glass fiber reinforced electrical tape. Additionally, the trays each had one TC installed every 6" on a single #8 AWG bare copper conductor routed along the longitudinal centerline of the trays on top of the enclosed cables. The fire stop assembly had three TCs on its unexposed: one located 1" from a side rail, one located 1" from a penetrating cable, one located in the fire stop center.

Hose Stream: Conduit B – Failed; Conduit A, Trays C, D, and E Passed
Furnace: 11' x 6' x 80"
Furnace control: Ten (10) thermocouples on proves located throughout the furnace
Cable Type: Conduits A & B each contained a single control cable. Trays C & D each contained 19 instrumentation cables, and Tray E contained 40 instrumentation cables that penetrated through the fire stop.

**Table D-74. TU Electric Thermo-Lag Test 13-3**

| Raceway Type | Barrier Protection | Cable Fill | Max. Temp Rise ΔT °C (°F) | Rating[2] |
|---|---|---|---|---|
| 2" Conduit A | 5/8" nom. TL-330-1 | 8.6% | 162 (323) | 60 min. |
| 2" Conduit B | 1/2" nom. TL-330-1 | 8.6% | 729 (1344)[1] | Failed |
| 12" Cable Tray C | 5/8" nom. TL-330-1 | 5.5% | 126 (258)[1] | 59 min. |
| 12" Cable Tray D | 1/2" nom. TL-330-1 | 4.6% | 192 (378)[1] | Failed |
| 12" Cable Tray E | 1/2" nom. TL-330-1 | 11.4% | 124 (256) | 60 min. |

[1] These temperature rises are above the 325°F single point maximum temperature rise criterion, therefore Conduit B and Cable Trays C & D failed to achieve the desired 60 minute qualification.
[2] For Cable Tray-C, the side rail and bare #8 bare conductor ΔTs exceeded allowable limits at 59 minutes. However, the following acceptance parameters were met: 1) no fire barrier burn through after fire and hose stream test; 2) no visible cable damage; 3) the post-fire cable insulation resistance tests were well within allowable limits. The licensee credited the results of this fire test for Conduit A, Cable Tray C, and Cable Tray E (fire stop) as part of the Thermo-Lag ERFBS qualification basis at CPSES.

### D.3.3.14 TU Electric Report No. 12340-94367m, Scheme 14-1 (December 1992)

ERFBS: This 1 hr. test evaluated a 30" wide steel tray that vertically penetrated the furnace roof, descended to a radial bend that bent it to a horizontal direction three feet below the roof, proceeded in a horizontal section to a squared bend, then vertically ascended through the furnace roof ~8 feet (measured horizontally) from the first penetration. In the middle portion of the horizontal section, a Tee assembly was installed that could allow connection of a second horizontal 30" tray perpendicular to the first tray (such a tray was not actually installed – the Tee fitting ended in a firestop). The fire barrier protecting this tray and Tee was constructed using Thermo-Lag 330-1 flat and V-Ribbed panels that were ½" nominal thickness with factory-applied 350 Topcoat, 330-69 stress skin sheets, and 330-1 Trowel Grade subliming compound. Various upgrade techniques were used in constructing the tray.

Test Procedure: Texas Utilities Electric TEST PLAN, Rev. 8

Test Slab: 13' x 8' x 10 GA (Steel), with 2 layers of 2" ceramic fiber blanket insulation

Ambient Temperature: 70°F

Desired Rating: 1 hr.

Thermocouples: The tray had one TC every 6" on one of its power cables, one of its control cables, and one of its instrument cables taped to the top surface of the cable with a double wrap of glass fiber reinforced electrical tape, plus one TC every 12" along both side rails clamped under a screw head.
Hose Stream: Passed.
Furnace: 11' x 6' x 80"
Furnace control: Ten (10) thermocouples on probes located throughout the furnace
Cable Type: The tray contained a ~1/3 mix of power, control, and instrument cables with total fill area as shown in the table below.

**Table D-75. TU Electric Thermo-Lag Test 14-1**

| Raceway Type | Barrier Protection | Cable Fill | Max. Temp Rise ΔT °C (°F) | Rating |
|---|---|---|---|---|
| 30" steel cable tray | ½" nom. TL-330-1 | 17.3% | 166 (331)[1] | 60 min.[2] |

[1] A single thermocouple (TC 91) located on the front cable tray side rail nearest the fire stop assembly exceeded the allowable 325°F single point ΔT criteria by 6°F during the last 1 minute of the fire endurance portion of the test. The test laboratory and licensee based the 60 minute rating of the tested Thermo-Lag ERFBS configuration on the site-specific acceptance criteria accepted by the NRC in its letter dated October 29, 1992.

*D.3.3.15 TU Electric Report No. 12340-95100a, Scheme 15-1 (March 1993)*

ERFBS: This 1 hr. test evaluated a 36" wide steel tray that vertically penetrated the furnace roof, descended to a radial bend that bent it to a horizontal direction three feet below the roof, proceeded four feet in a horizontal section, then bent to a vertically ascending section that penetrated the furnace roof ~8 feet horizontally from the first penetration. The fire barrier protecting this tray was constructed using Thermo-Lag 330-1 flat and V-Ribbed panels that were ½" nominal thickness with factory-applied 350 Topcoat, 330-69 stress skin sheets, and 330-1 Trowel Grade subliming compound. Various upgrade techniques were used in constructing the tray.

Test Procedure: Texas Utilities Electric TEST PLAN, Rev. 9
Test Slab: 13' x 8' x 10 GA (Steel), with 2 layers of 2" ceramic fiber blanket insulation
Ambient Temperature: 68°F
Desired Rating: 1 hr.

Thermocouples: The tray had one TC every 6" on one of its power cables, one of its control cables, and one of its instrument cables taped to the top surface of the cable with a double wrap of glass fiber reinforced electrical tape, plus one TC every 12" along both side rails clamped under a screw head.
Hose Stream: Passed.
Furnace: 11' x 6' x 80"
Furnace control: Ten (10) thermocouples on probes located throughout the furnace
Cable Type: The tray contained a ~1/3 mix of power, control, and instrument cables with total fill area as shown in the table below.

**Table D-76. TU Electric Thermo-Lag Test 15-1**

| Raceway Type | Barrier Protection | Cable Fill | Max. Temp Rise $\Delta T$ °C (°F) | Rating |
|---|---|---|---|---|
| 36" steel cable tray | ½" nom. TL-330-1 | 17.4% | 107 (224)[1] | 60 min.[2] |

[1] This temperature rise is below the 325°F single point maximum temperature rise criterion, therefore the tray achieved the desired 60 minute qualification.
[2] The testing laboratory and licensee based the 60 minute rating of the tested Thermo-Lag ERFBS configuration on the site-specific acceptance criteria accepted by the NRC in its letter dated October 29, 1992.

*D.3.3.16 TU Electric Report No. 12340-95770, Scheme 15-2 (October 1993)*

<u>ERFBS</u>: This 1 hr. test evaluated two air drops with W-008 power cables, each wrapped with two layers of Thermo-Lag 330-660 Flexi-blanket (each layer ~¼" thick), that were laid in the same 36" wide x 3" deep (non-ERFBS wrapped) cable tray. The outer Flexi-blanket layer of each ERFBS was pre-caulked with a layer of Thermo-Lag 330-660 Trowel Grade material, and both layers were stainless-steel banded every 6". Also placed in the tray (for cable fill purposes) were three W-020 cables which were neither Thermo-Lag protected nor temperature monitored.

<u>Test Procedure</u>: Texas Utilities Electric TEST PLAN, Rev. 11

<u>Test Slab</u>: 13' x 8' x 10 GA (Steel), with 2 layers of 2" ceramic fiber blanket insulation

<u>Ambient Temperature</u>: 92°F

<u>Desired Rating</u>: 1 hr.

<u>Thermocouples</u>: One TC every 6" was taped to the top surface of each W-008 cable with a double wrap of glass fiber reinforced electrical tape. Also, a bare #8 AWG stranded copper wire (instrumented with TCs every 6") was placed in both ERFBS-protected (air drop) areas with its W-008 cable.

<u>Hose Stream</u>: Passed.

<u>Furnace</u>: 11' x 6' x 80"

<u>Furnace control</u>: Ten (10) thermocouples on probes located throughout the furnace

<u>Cable Type</u>: The tray contained the two W-008 power cables each separately protected within the previously described ERFBS, and three unprotected W-020 cables wrapped together in Siltemp (a high temperature cloth). The fraction of the unprotected 36" x 3" cable tray's area filled with those cables was 4.5%.

### Table D-77. TU Electric Thermo-Lag Test 15-2

| Raceway Type | Barrier Protection | Cable Fill[2] | Max. Temp Rise $\Delta T$ °C (°F)[1] | Rating |
|---|---|---|---|---|
| Front W-008 ERFBS | ½" nom. Thermo-Lag 330-660 Flexi-blanket | 2.25% | 329 (625) | 60 min.[3] |
| Rear W-008 ERFBS | ½" nom. Thermo-Lag 330-660 Flexi-blanket | 2.25% | 257 (494) | 60 min.[3] |

[1] The $\Delta T$ values listed above were recorded on the bare #8 AWG stranded wire that was positioned inside the Thermo-Lag ERFBS protection each W-008 750 kCM cables.

[2] Fraction of cable tray filled by the W-008 cable; total fraction of cable tray filled = 4.5%

[3] Although the $\Delta T$ criterion was exceeded on the bare #8 AWG conductor that was positioned inside the ERFBS protecting each W-008 cable, the test laboratory and licensee based the 60 minute rating of the tested Thermo-Lag ERFBS configuration on the site-specific acceptance criteria accepted by the NRC in its letter dated October 29, 1992. However, to ensure complete protection of large power cables that are wrapped in exposed cable trays, the licensee upgraded the installed configurations by adding a third layer of Thermo-Lag 330-660

## D.3.4 Tennessee Valley Authority Tests for Watts Bar Nuclear Power Plant

### D.3.4.1 TVA Project No. 11210-94554c, Test 6.1.1 (January 1993)

ERFBS: This 1 hr. test evaluated two 5" diameter conduit configurations, two 1" diameter conduit configurations, and two 2" diameter conduit air drop configurations. The conduit configurations were the elongated "U" type previously described, each penetrating the furnace roof in two locations about 6 feet apart and descending about 3 feet down into the furnace. The air drops were each two conduits penetrating the furnace roof in two locations about 3 feet apart, one terminating about 7" below the roof, and the other terminating in a radial bend just below the furnace roof, with only a bare, TC-instrumented wire between the two terminations (the "air drop"). The ERFBS were constructed using Thermo-Lag (TL) 330-1 V-ribbed panels ~5/8" thick with stress skin monolithically adhered to one face, TL 330-1 Pre-Formed Conduit Sections ~5/8" thick, and TL 330-1 Subliming Trowel Grade material. Construction techniques and upgrades were used similar to those previously described.

Test Procedure: TVA TEST PLAN RD 139599

Test Slab: 13' x 8' x 10 GA (Steel), with 2 layers of 2" ceramic fiber blanket insulation

Ambient Temperature: 65°F

Desired Rating: 1 hr.

Thermocouples: A single bare #8 AWG stranded copper wire was instrumented with TCs and pulled through each conduit configuration. The TCs were spaced 6" apart on the wires in the 1" and 2" conduits, and every 12" on the wires in the 5" conduits. Also, TCs were placed every 12" along the bottom surface of each conduit, held in position by clamping under the head of a #8 x 32 x ¼" long stainless steel screw in a drilled and threaded hole at each location.

Hose Stream: A test was applied, but no details are given in available documentation.

Furnace: 11' x 6' x 80"

Furnace control: Ten (10) thermocouples on probes located throughout the furnace

Cable Type: No electric cables were installed in the conduits and air drops.

Table D-78. TVA Thermo-Lag Test 6.1.1

| Raceway Type | Barrier Protection | Cable Fill | Max. Temp Rise ΔT °C (°F) | Rating |
|---|---|---|---|---|
| 5" Conduit (#1) | ~5/8" TL 330-1 | None | 207 (404)[1] | 60 min.[1] |
| 1" Conduit (#2) | ~5/8" TL 330-1 | None | 866 (1591)[2] | Failed |
| 5" Conduit (#3) | ~5/8" TL 330-1 | None | 137 (278)[2] | 60 min. |
| 1" Conduit (#4) | ~5/8" TL 330-1 | None | 675 (1247)[2] | Failed |
| 2" Air Drop (#5) | ~5/8" TL 330-1 | None | 414 (777)[2] | Failed |
| 2" Air Drop (#6) | ~5/8" TL 330-1 | None | 206 (402)[2] | Failed |

[1] This temperature rise was from a TC located on a bare #8 wire inside a raceway configuration where the licensee argued it was subjected to moisture saturation, which caused its inaccurate, artificially high reading. All measurements on conduit steel were below the single point maximum acceptable 325°F, and all steel conduits maximum average measurements were below the acceptable 250°F.

[2] These temperature rises were recorded by TCs on the conduit steel, and were considered to be accurate.

### D.3.4.2 TVA Project No. 11210-94554a, Test 6.1.2 (January 1993)

ERFBS: This 1 hr. test was a repeat of Test 6.1.1 with certain upgrades, including the addition of a second layer of TL. One each of the 1" and 5" conduits, and one of the 2" air drops, had a second ~3/8" thick layer of TL added using additional V-ribbed panels and preformed conduit sections. A thin coating of TL-330-1 Trowel Grade material was applied over the additional layer, with stainless steel (stainless steel) tie wires every 4 to 6 inches on top of that. The second layer applied to the remaining three configurations consisted of their being wrapped with stainless steel knitted wire mesh held in place with stainless steel tie wires, then covered with an ~3/8" thick layer of TL-330-1 Trowel Grade material, with no tie wire or banding material applied over the trowel material.

Test Procedure: TVA TEST PLAN RD 139599

Test Slab: 13' x 8' x 10 GA (Steel), with 2 layers of 2" ceramic fiber blanket insulation

Ambient Temperature: 61°F

Desired Rating: 1 hr.

Thermocouples: A single bare #8 AWG stranded copper wire was instrumented with TCs and pulled through each conduit configuration. The TCs were spaced 6" apart on the wires in all test specimens.

Hose Stream: A test was applied, but no details are given in available documentation.

Furnace: 11' x 6' x 80"

Furnace control: Ten (10) thermocouples on probes located throughout the furnace

Cable Type: No electric cables were installed in the conduits and air drops.

**Table D-79. TVA Thermo-Lag Test 6.1.2**

| Raceway Type | Barrier Protection | Cable Fill | Max. Temp Rise ΔT °C (°F) | Rating |
|---|---|---|---|---|
| 5" Conduit (#1) | ~1" TL 330-1 | None | 66 (150) | 60 min. |
| 1" Conduit (#2) | ~1" TL 330-1 | None | 96 (204) | 60 min. |
| 5" Conduit (#3) | ~1" TL 330-1 | None | 58 (136) | 60 min. |
| 1" Conduit (#4) | ~1" TL 330-1 | None | 73 (164) | 60 min. |
| 2" Air Drop (#5) | ~1" TL 330-1 | None | 66 (150) | 60 min. |
| 2" Air Drop (#6) | ~1" TL 330-1 | None | 67 (152) | 60 min. |

### D.3.4.3 TVA Project No. 11210-94943a, Test 6.1.3 (April 1993)

ERFBS: This 1 hr. test evaluated four conduit configurations (1 in., 2 in., 3 in., and 4 in. diameter). They were fire-protected with a nominal thickness of 3/8 in. or 5/8 in. Thermo-Lag 330-1 with various upgrades. All were configured as the elongated "Us" previously described, each penetrating the furnace roof in two locations 9 feet apart and descending 3 feet down into the furnace. The

ERFBS were constructed using Thermo-Lag (TL) 330-1 V-ribbed panels,~3/8" or ~5/8" thick, with stress skin monolithically adhered to one face, TL 330-1 Pre-Formed Conduit Section panels ~3/8" thick for the 3" and 2" conduits and ~5/8" thick for the 4" and 1" conduits, and TL 330-1 Subliming Trowel Grade material.

<u>Test Procedure</u>: TVA TEST PLAN RD 328886

<u>Test Slab</u>: 13' x 8' x 10 GA (Steel), with 2 layers of 2" ceramic fiber blanket insulation

<u>Ambient Temperature</u>: 76°F

<u>Desired Rating</u>: 1 hr.

<u>Thermocouples</u>: A single bare #8 AWG stranded copper wire was instrumented with TCs spaced 6" apart and pulled through each conduit configuration. Also, TCs were placed every 6" along the bottom surface of each conduit, held in position by clamping under the head of a #8 x 32 x ¼" long stainless steel screw in a drilled and threaded hole at each location.

<u>Hose Stream</u>: A test was applied, but no details are given in available documentation.

<u>Furnace</u>: 11' x 6' x 80"

<u>Furnace control</u>: Ten (10) thermocouples on probes located throughout the furnace

<u>Cable Type</u>: No electric cables were installed in the conduits and air drops.

### Table D-80. TVA Thermo-Lag Test 6.1.3

| Raceway Type | Barrier Protection | Cable Fill | Max. Temp Rise $\Delta T$ °C (°F) | Rating[1] |
|---|---|---|---|---|
| 3" Conduit (#1) | ~3/8" TL 330-1 | None | 72 (162) | 60 min. |
| 2" Conduit (#2) | ~5/8" TL 330-1 | None | 77 (170) | 60 min. |
| 1" Conduit (#3) | ~5/8" TL 330-1 | None | 80 (176) | 60 min. |
| 4" Conduit (#4) | ~5/8" TL 330-1 | None | 143 (289) | 60 min. |

[1] All maximum average temperature rises were less than the 250°F limit

*D.3.4.4 TVA Project No. 11210-94943b, Test 6.1.4 (April 1993)*

<u>ERFBS</u>: This 1 hr. test evaluated three conduit configurations (3 in. steel, 3 in. aluminum, and 1-½ in. steel) and two generic tube steel support members (2 in. and 4 in.). They were fire-protected with a nominal thickness of 3/8 in. or 5/8 in. Thermo-Lag 330-1 with various upgrades. The conduits were configured as the elongated "Us" previously described, each penetrating the furnace roof in two locations 9 feet apart and descending 3 feet down into the furnace. The support members each penetrated the furnace roof and down into the furnace 3 feet, at which point they made a sharp 90° bend and then extended horizontally 30". The ERFBS were constructed using Thermo-Lag (TL) 330-1 V-ribbed panels,~3/8" or ~5/8" thick, with stress skin monolithically adhered to one face, TL 330-1 Pre-Formed Conduit Section panels ~5/8" thick for the 3" conduits and ~3/8" thick for the 1-½ in. conduit, and TL 330-1 Subliming Trowel Grade material.

<u>Test Procedure</u>: TVA TEST PLAN RD 328886

<u>Test Slab</u>: 13' x 8' x 10 GA (Steel), with 2 layers of 2" ceramic fiber blanket insulation

<u>Ambient Temperature</u>: 75°F

<u>Desired Rating</u>: 1 hr.

Thermocouples: A single bare #8 AWG stranded copper wire was instrumented with TCs spaced 6" apart and pulled through each conduit configuration. Also, TCs were placed every 6" along the bottom surface of each conduit, held in position by clamping under the head of a #8 x 32 x ¼" long stainless steel screw in a drilled and threaded hole at each location. Thermocouples were similarly affixed to the tube steel supports – they were fastened to the top surface of the horizontal sections of the tube steel at 2" intervals starting 12" from the exposed ends. Thermocouples were also fastened at 6" intervals along the vertical sections of the steel tubes.

Hose Stream: A test was applied, but no details are given in available documentation.
Furnace: 11' x 6' x 80"
Furnace control: Ten (10) thermocouples on probes located throughout the furnace
Cable Type: No electric cables were installed in the conduits and air drops.

### Table D-81. TVA Thermo-Lag Test 6.1.4

| Raceway Type | Barrier Protection | Cable Fill | Max. Temp. Rise ΔT °C (°F) | Rating |
|---|---|---|---|---|
| 3" Conduit (#1) | ~5/8" TL 330-1 | None | 167 (332) | Failed |
| 1-½" Conduit (#2) | ~5/8" TL 330-1 | None | 124 (255) | 60 min.[1] |
| 3" Conduit (#3) | ~5/8" TL 330-1 | None | 188 (370) | Failed |

[1] All maximum average temperature rises were less than the 250°F limit

The test results for the tubular steel members support the "18 in. rule," which indicates that tubular steel support members should be ERFBS protected starting no less than 18 inches from the ERFBS-protected raceways they support. At that location in this test, the 4" support was clearly below the 325°F temperature rise limit, and the 2" support was ~2+ °F over that limit, a negligible amount given the repeatability of this test method.

*D.3.4.5 TVA Project No. 11210-94943d, Test 6.1.5 (May 1993)*

ERFBS: This 1 hr. test evaluated four steel conduit configurations (1 in., 2 in., 3 in., and 5 in.), and five steel junction boxes of varying sizes. All junction boxes were NEMA 12 and were fastened directly to the underside of concrete slab furnace roof. All conduits were positioned as close to the concrete as was feasible and practical. The ERFBS were constructed using Thermo-Lag (TL) 330-1 V-ribbed panels, ~3/8" or ~5/8" thick, with stress skin monolithically adhered to one face, TL 330-1 Pre-Formed Conduit Section panels ~5/8" thick for the 5" conduits and the first layer of the 1" conduit, and ~3/8" thick for first and second layer of the 3" and 2" conduits and the overlay for the 1" conduit, and TL 330-1 Subliming Trowel Grade material.

Test Procedure: TVA TEST PLAN RD 328886
Test Slab: 13' x 8' x 10 GA (Steel), with 2 layers of 2" ceramic fiber blanket insulation
Ambient Temperature: 73°F
Desired Rating: 1 hr.

Thermocouples: A single bare #8 AWG stranded copper wire was instrumented with TCs spaced 6" apart and pulled through each conduit configuration. Also, TCs were placed every 6" along the bottom surface of each conduit, held in position by clamping under the head of a #8 x 32 x ¼" long stainless steel screw in a drilled and threaded

hole at each location. TCs were similarly affixed to the interior of the junction boxes, with at least on TC in each square foot of area on each face of the boxes (except for each box's face on the concrete slab).
Hose Stream: A test was applied, but no details are given in available documentation.
Furnace: 11' x 6' x 80"
Furnace control: Ten (10) thermocouples on probes located throughout the furnace.
Cable Type: No electric cables were installed in the conduits and air drops.

### Table D-82. TVA Thermo-Lag Test 6.1.5

| Raceway Type | Barrier Protection[1] | Cable Fill | Max. Temp Rise $\Delta T$ °C (°F) | Rating |
|---|---|---|---|---|
| 1" Conduit (#1) | Thermo-Lag 330-1 | None | 66 (150) | 60 min. |
| 3" Conduit (#2) | Thermo-Lag 330-1 | None | 108 (226) | 60 min. |
| 2" Conduit (#3) | Thermo-Lag 330-1 | None | 77 (170) | 60 min. |
| 5" Conduit (#4) | Thermo-Lag 330-1 | None | 97 (207) | 60 min. |
| Junction B (#1) | Thermo-Lag 330-1 | None | 57 (135) | 60 min. |
| Junction B (#2) | Thermo-Lag 330-1 | None | 79 (175) | 60 min. |
| Junction B (#3) | Thermo-Lag 330-1 | None | 88 (191) | 60 min. |
| Junction B (#4) | Thermo-Lag 330-1 | None | 64 (147) | 60 min. |
| Junction B (#5) | Thermo-Lag 330-1 | None | 105 (221) | 60 min. |

[1] Varied thicknesses and materials – too detailed for inclusion in this summary

#### D.3.4.6 TVA Project No. 11210-94943e, Test 6.1.6 (May 1993)

ERFBS: This 1 hr. test evaluated three 4 in. steel conduit configurations and one large steel junction box (48 x 36 x 12 in.). The ERFBS were constructed using Thermo-Lag (TL) 330-1 V-ribbed panels, ~3/8" or ~5/8" thick, with stress skin monolithically adhered to one face, TL 330-1 Pre-Formed Conduit Section panels ~5/8" thick, and TL 330-1 Subliming Trowel Grade material.

Test Procedure: TVA TEST PLAN RD 328886
Ambient Temperature: 72°F
Test Slab: 13' x 8' x 10 GA (Steel), with 2 layers of 2" ceramic fiber blanket insulation
Desired Rating: 1 hr.

Thermocouples: A single bare #8 AWG stranded copper wire was instrumented with TCs spaced 6" apart and pulled through each conduit configuration. Also, TCs were placed every 6" along the bottom surface of each conduit, held in position by clamping under the head of a #8 x 32 x ¼" long stainless steel screw in a drilled and threaded hole at each location. TCs were similarly affixed to the interior of the junction box, with at least on TC in each square foot of area on each face of the box.
Hose Stream: A test was applied, but no details are given in available documentation.
Furnace: 11' x 6' x 80"
Furnace control: Ten (10) thermocouples on probes located throughout the furnace.
Cable Type: Ten (10) thermocouples on probes located throughout the furnace.

### Table D-83. TVA Thermo-Lag Test 6.1.6

| Raceway Type | Barrier Protection[1] | Cable Fill | Max. Temp Rise ΔT °C (°F) | Rating |
|---|---|---|---|---|
| 4" Conduit (#1) | Thermo-Lag 330-1 | None | 121 (250) | 60 min. |
| 4" Conduit (#2) | Thermo-Lag 330-1 | None | 118 (244) | 60 min. |
| 4" Conduit (#3) | Thermo-Lag 330-1 | None | 130 (266) | 60 min. |
| Junction Box | Thermo-Lag 330-1 | None | 57 (134) | 60 min. |

[1] Varied thicknesses and materials – too detailed for inclusion in this summary

*D.3.4.7 TVA Project No. 11960-97185, Test 6.1.7 (November 1994)*

ERFBS: This 1-hr. test evaluated three cable tray configurations (18 in. x 4 in., steel) and one conduit (3 in. steel). The test qualified installation of Thermo-Lag on 3 cable trays with adjustable risers at 90° and a conduit with a radial bend. The raceways have a horizontal to vertical orientation. They were fire-protected with a nominal thickness of 5/8 in. Thermo-Lag 330-1 fire barrier material. Top panels of the cable trays had V-Ribs running parallel with the side rails. The vertical section of the conduit and vertical support members were covered with 3M M20A fire wrap, overlapping the installed Thermo-Lag by at least 3 inches. Support connectors and voids in channels were filled with Thermo-Lag 330-1 Trowel Grade/Putty.

Test Slab: 13' x 8' x 10 GA (Steel), with 2 layers of 2" ceramic fiber blanket insulation

Desired Rating: 1 hr.

Ambient Temperature: 83°F

Thermocouples: A single bare #8 AWG stranded copper wire was instrumented with TCs spaced 6" apart and fastened to cable tray side rails and conduits by clamping under the head of a #8 x 32 x ¼" long stainless steel screw. Thermocouples were placed in the following locations: one on top of the cable bundle in the left tray, one secured to the bottom of the tray rungs in the left tray, one on top of the cable bundle in the center tray, one secured to the bottom of the tray rungs in the center tray, one on top of the cable tray rungs in the right tray and one pulled through the conduit assembly.

Hose Stream: The test platform was sprayed with water at a 40° hose angle at a pressure of 75 psi. The test setup was sprayed with a 75 gpm water flow rate from a distance of 5 feet for a minimum of 5 minutes.

Furnace: 12' x 7' x 79"

Furnace control: Ten (10) thermocouples on probes located throughout the furnace

Cable Type: Left Tray – 289, 4C #16 AWG 600V (69.36 lbs. cable/linear foot)
Center Cable Tray – 26, 4C #16 AWG 600V (6.24 lbs. cable/linear foot)
Right Tray – 0% Cable Fill

**Table D-84. TVA Thermo-Lag Test 6.1.7**

| Raceway Type | Barrier Protection | Cable Fill | Max. Temp. Rise ΔT °C (°F) | Rating |
|---|---|---|---|---|
| 18" Left Cable Tray | ~5/8" TL 330-1 | 289, 3C #16 AWG | 67 (153) | 60 min. |
| 18" Center Cable Tray | ~5/8" TL 330-1 | 26, 4C #16AWG | 118 (244) | 60 min. |
| 18" Right Cable Tray | ~5/8" TL 330-1 | None | 134 (274) | 60 min. |
| 3" Conduit | ~5/8" TL 330-1 | None | 87 (188) | 60 min. |

The average initial temperature for all thermocouples at the start of the test was 90°F, allowing a maximum allowable individual thermocouple temperature of 415°F in accordance with ASTM E119-88. The cable tray configuration in this test procedure involving Thermo-Lag 330-1 met the 1-hour fire resistance rating. The test provided the effects of cable mass on the performance of an 45.72 cm (18 in) steel, ladder back cable tray, protected with a TA designed nominal 5/8 inch Thermo-Lag 330-1 barrier. The test proved that 1.33 lbs/foot of cable is required in order for the 18" cable tray ERFBS to be acceptable.

*D.3.4.8 TVA Project No. 11960-97186, Test 6.1.8 (September 1994)*

ERFBS: This 1-hr. test evaluated two cable tray configurations (18 in. x 4 in., steel) connected with a tray cross fitting. They were fire-protected with a nominal thickness of 5/8 inch Thermo-Lag 330-1 V-Rib Panel with exterior stress skin overlay. TVA provided a cross fitting with two 18 in. wide by 4 in. deep cable trays connected to a double cross. Two sections of 18 in. by 4 in. ladderback cable tray with 6 in. rung spacing were fitted to one side of the cross fitting. Overall tray and double cross dimensions were 144 in. by 72 in. The cable tray sections were held in place by a "trapeze" type hanger using 3 in. steel channels bolted and welded together 12 in. from the free ends of the tray sections. The cable trays had adjustable risers at 90° from horizontal to vertical and the 3" conduit had a radial bend. Support connectors and voids in channels were filled with Thermo-Lag 330-1 Trowel Grade/Putty.

Test Slab: 13' x 8' x 10 GA (Steel), with 2 layers of 2" ceramic fiber blanket insulation

Desired Rating: 1 hr.

Ambient Temperature: 90°F

Thermocouples: A single bare #8 AWG stranded copper wire was instrumented with TCs spaced 6" apart and fastened to cable tray rails by clamping under the head of a #8 x 32 x ¼" long stainless steel screw. Thermocouples were placed in the following locations: one extended along the left 18 in. cable tray, on the surface of the tray rungs, and right to left, through the special tray fitting. One extended along the right 18 in. cable tray, on the surface of the tray rungs, and right to left, through the special tray fitting. One passed front to rear, through the left side of the special tray fitting, on the surface of the tray rungs, and one passed front to rear, through the right side of the special tray fitting, on the surface of the tray rungs.

Hose Stream: The test platform was elevated to an elevation of 6 ft. and spun at 6-8 revolutions per minute. The test setup was sprayed with a minimum 75 psi pressure at 75 gpm from a distance of 5 feet for a 5 minute duration. The test deck was lowered

after 2 minutes of exposure to the hose stream to allow spray to reach the top of the test platform. The deck was raised after two additional minutes to spray the bottom again.
Furnace: 12' x 7' x 79"
Furnace control: Ten (10) thermocouples on probes located throughout the furnace
Cable Type: No electric cables were installed in the cable trays.

**Table D-85. TVA Thermo-Lag Test 6.1.8**

| Raceway Type | Barrier Protection | Cable Fill | Max. Temp. Rise ΔT °C (°F) | Rating |
|---|---|---|---|---|
| Front Tray 18" x 4" | ~5/8" TL 330-1 | None | 94 (202) | 60 min. |
| Rear Tray 18" x 4" | ~5/8" TL 330-1 | None | 98 (208) | 60 min. |
| Cross Fitting 72" x 72" x 4" | ~5/8" TL 330-1 | None | 82 (180) | 60 min. |

All thermocouples located on the cable tray system met test criteria. The average initial temperature for all thermocouples at the start of the test was 90°F, allowing a maximum allowable individual thermocouple temperature of 415°F in accordance with ASTM E119-88. The "double cross" cable tray configuration in this test procedure involving Thermo-Lag 330-1 met the 1-hour fire resistance rating.

*D.3.4.9 TVA Project No. 11960-97187, Test 6.1.9 (September 1994)*

ERFBS: This 1-hr. test evaluated three stacked cable trays (18 in. x 4 in., steel) clad with nominal 5/8 in. Thermo-Lag 330-1 panel with 1/2 in a common enclosure, one 18 in. by 4 in. steel ladderback cable tray with a solid cover of nominal 5/8 in. Thermo-Lag 330-1 Rib Panels, one 5" air drop covered with a first layer of 5/8 in. and a second layer of 3/8 in. Thermo-Lag 330-1, and one 1 in. air drop covered with a first layer of 5/8 in. and a second layer of 3/8 in. Thermo-Lag 330-1. This test included a "U" shape for the 18 in. cable tray (vertical to horizontal to vertical), three nested 18 in. "U" shaped cable trays, and one multiple to single cable tray transition. 3M M20A fire material overlaps the Thermo-Lag applied on vertical supports by 6 in., and no collars are installed.

Test Slab: 13' x 8' x 10 GA (Steel), with 2 layers of 2" ceramic fiber blanket insulation

Desired Rating: 1 hr.

Ambient Temperature: 78°F

Thermocouples: A single bare #8 AWG stranded copper wire was instrumented with TCs spaced 6" apart and fastened to cable tray side rails and air drop walls by clamping under the head of a #8 x 32 x ¼" long stainless steel screw. Thermocouples were installed on tray rungs, centered between the tray rails, in each of the three stacked cable trays. Similar wires were also installed under the tray rungs, centered between the tray side rails, on the rear cable tray, and in each of the two air drops.

Hose Stream: The test platform was sprayed with water at a 30° hose angle at a pressure of 75 psi while spinning at 6-8 revolutions per minute. The test setup was sprayed with a 75 gpm water flow rate from a distance of 5 feet for 5 minutes.
Furnace: 12' x 7' x 79"
Furnace control: Ten (10) thermocouples on probes located throughout the furnace

Cable Type: 18" Cable tray with raised cover (#2) – 321, 4C #16 AWG 60
5" Air drop (#3) – 44, 4C #16 AWG 60
#1 and 4 did not have any installed electrical cables

### Table D-86. TVA Thermo-Lag Test 6.1.9

| Raceway Type | Barrier Protection | Cable Fill | Max. Temp. Rise ΔT °C (°F) | Rating |
|---|---|---|---|---|
| 3-Cable Trays | 5/8" TL 330-1 | None | 129 (264) | 60 min. |
| 1-Cable Tray | 5/8" TL 330-1 | 321, 4C #16 AWG | 92 (198) | 60 min. |
| 5" Air Drop | 5/8", 3/8" TL 330-1 | 44, 4C # 16 AWG | 48 (118) | 60 min. |
| 1" Air Drop | 5/8", 3/8" TL 330-1 | None | 57 (134) | 60 min. |

The three stacked cable trays, both in the individually wrapped and common enclosure configurations, and the covered rear tray and both air drops evaluated in this test procedure, clad with Thermo-Lag 330-1 material and various upgrades such as different thicknesses and 3M materials as mentioned, met the requirements of the acceptance criterion for a fire resistance rating of one hour.

*D.3.4.10 TVA Project No. 11960-97257, Test 6.1.10 (November 1994)*

ERFBS: This 1-hr. test evaluated three different configurations: One small, 2-sided box with (2) 1 in. steel conduits protected with nominal 5/8 in. Thermo-Lag 330-1 panels supported by a Unistrut frame. One inch conduits exited box and were wrapped individually with preformed Thermo-Lag 330-1 designs. The second configuration was a large 2-sided box with (8) 4 in. aluminum conduits protected with 5/8" nominal Thermo-Lag 330-1 Panels supported by a Unistrut frame. The conduits were also individually wrapped and exited both sides of the box. The third configuration consisted of 4 in. aluminum conduits wrapped individually with 5/8" Thermo-Lag 330-1 outside the two-sided box. Panels were reinforced with 3" wide panels and a 3" wide piece of stress skin and then covered with Thermo-Lag 330-1 Trowel Grade.

Test Slab: 13' x 8' x 10 GA (Steel) with #7 steel rebar and 10" steel I-beam supports    Desired Rating: 1 hr.

Ambient Temperature: 66°F

Thermocouples: A single bare #8 AWG stranded copper wire was instrumented with TCs spaced 6" apart and fastened to conduits by clamping under the head of a #8 x 32 x ¼" long stainless steel screw. Thermocouples were installed in the interior of each of the installed conduits and to the surfaces of the cable tray supports and short cable tray sections. Those located in the support system were used to monitor the heat flow into a protected cable tray from an unprotected cable tray.

Hose Stream: The test platform was sprayed with water at a 30° hose angle at a pressure of 75 psi while spinning at 6-8 revolutions per minute. The test setup was sprayed with a 75 gpm water flow rate from a distance of 5 feet for 5 minutes.

Furnace: 12' x 7' x 79"

Furnace control: Ten (10) thermocouples on probes located throughout the furnace

Cable Type: Electric cables were not utilized in this test.

**Table D-87. TVA Thermo-Lag Test 6.1.10**

| Raceway Type | Barrier Protection | Cable Fill | Max. Temp. Rise $\Delta$T °C (°F) | Rating |
|---|---|---|---|---|
| Small Box with Conduits | 5/8", 3/8" TL 330-1 | None | 67 (153) | 60 min. |
| Large Box with Conduits | 5/8" TL 330-1 | None | 47 (117) | 60 min. |
| 4" Aluminum Conduits | 5/8" TL 330-1 | None | 79 (175) | 60 min. |

Each of the two-sided multiple conduit enclosures clad with Thermo-Lag 330-1 fire barrier and upgrades met the requirements of the acceptance criterion and qualify for a fire resistance rating of one hour. Testing also showed that the maximum temperature of the ambient temperature plus 250°F on the configuration's support system was reached at a distance of 9 inches away from unprotected cable tray. This distance is conservative since it does not consider the additional thermal mass of the protected cable tray assembly.

*D.3.4.11 TVA Project No. 11960-97258, Test 6.1.11 (October 1994)*

ERFBS: This 1-hr. test evaluated four different configurations: The first configuration was a group of (7) aluminum conduits ranging from 2-3 in., covered with a 3-sided enclosure constructed using the Separate Piece Method and a single layer of 5/8 in. nominal Thermo-Lag 330-1 V-Rib Panels. A group of (2) 1 in. steel conduits were also tested and covered with a 3-sided enclosure constructed from a single layer of 5/8 in. Thermo-Lag 330-1 V-Rib Panels using the Score and Fold Method. The third test group was (3) 3 in. aluminum conduits with a 3-sided enclosure constructed using a single layer of Thermo-Lag 330-1 V-Rib Panels using the Score and Fold Method. The last test group consisted of one large junction box (5' x 3' x 2') covered with 5/8 in. nominal Thermo-Lag 330-1 V-Rib panels. The 3-sided ERFBS were installed against concrete constructed using anchor bolts and Thermo-Lag 330-1 Trowel Grade Material that exceeded its 6-month shelf life but demonstrated acceptable testing quality according to the vendor.

Test Slab: 136" x 151" x 10 GA (Steel) with #5 steel rebar supports      Desired Rating: 1 hr.

Ambient Temperature: 62°F

Thermocouples: A single bare #8 AWG stranded copper wire was instrumented with TCs spaced 6" apart and fastened to conduits and junction boxes by clamping under the head of a #8 x 32 x ¼" long stainless steel screw. Thermocouples were affixed to the interior of the junction boxes in each square foot of area on each face of the boxes with exception of the face toward the concrete slab.

Hose Stream: The test platform was sprayed with water at a 30° hose angle at a pressure of 75 psi while spinning at 6-8 revolutions per minute. The test setup was sprayed with a 75 gpm water flow rate from a distance of 5 feet for 5 minutes. The spray technician was lifted with a platform to allow spray on underside and vertical surfaces.

Furnace: 10' x 10' x 48"

Furnace control: Ten (10) thermocouples on probes located throughout the furnace

Cable Type: Electric cables were not utilized in this test.

### Table D-88. TVA Thermo-Lag Test 6.1.11

| Raceway Type | Barrier Protection | Cable Fill | Max. Temp. Rise $\Delta T$ °C (°F) | Rating |
|---|---|---|---|---|
| (7) 2-3" Aluminum Conduits | 5/8" TL 330-1 | None | 45 (113) | 60 min. |
| (2) 1" Steel Conduits | 5/8" TL 330-1 | None | 48 (118) | 60 min. |
| (3) 3" Aluminum Conduits | 5/8" TL 330-1 | None | 51 (124) | 60 min. |
| (1) 5' x 3' x 2' Junction Box | 5/8" TL 330-1 | None | 71 (160) | 60 min. |

Each of the multiple conduit enclosures and the junction box enclosure, clad with Thermo-Lag 330-1 material and upgrades met the acceptance criteria for a fire resistance rating of one hour. The maximum temperatures recorded from the thermocouples did not exceed the ambient temperature of 62°F plus maximum temperature rise of 250°F (total 312°F).

#### D.3.4.12 TVA Project No. 11960-97259, Test 6.1.12 (October 1994)

ERFBS: This 1-hr. test evaluated four different configurations: The first configuration was a group of (4) 1 in. steel conduits arranged in two rows of two and covered with a box made of 5/8 in. nominal Thermo-Lag 330-1 V-Rib Panels, Separate Piece Method with stress skin overlay. The second configuration was a group of (4) 3 in. steel conduits arranged in two rows of two and covered with a box made of 5/8 in. Thermo-Lag 330-1 V-Rib Panels, Score and Fold construction with stress skin overlay. The third configuration was a group of (8) 4 in. aluminum conduits arranged in 4 rows of 2 and covered with a box made of 5/8" nominal Thermo-Lag 330-1 V-Rib Panels. The fourth configuration was a 4 in. steel conduit with junction box (12" x 12" x 60") covered with 5/8 in. nominal Thermo-Lag 330-1 Panels and preformed conduit sections.

Test Slab: 136" x 151" x 10 GA (Steel) with #5 steel rebar supports    Desired Rating: 1 hr.

Ambient Temperature: 78°F

Thermocouples: A single bare #8 AWG stranded copper wire was instrumented with TCs spaced 6" apart and fastened to conduits by clamping under the head of a #8 x 32 x ¼" long stainless steel screw. Wires were installed on the interior of each conduit, and additionally at 12 in. intervals to the sides of the rear conduits in the group of 1 in. steel

conduits and the group of 3 in. steel conduits per request by USNRC representatives. Thermocouples were also placed on every square foot within the junction box.
Hose Stream: The test platform was sprayed with water at a 30° hose angle at a pressure of 75 psi while spinning at 6-8 revolutions per minute. The test setup was sprayed with a 75 gpm water flow rate from a distance of 5 feet for 5 minutes. The spray technician was lifted with a platform to allow spray on underside and vertical surfaces.
Furnace: 10' x 10' x 48"
Furnace control: Ten (10) thermocouples on probes located throughout the furnace
Cable Type: Electric cables were not utilized in this test.

**Table D-89. TVA Thermo-Lag Test 6.1.12**

| Raceway Type | Barrier Protection | Cable Fill | Max. Temp. Rise ΔT °C (°F) | Rating |
|---|---|---|---|---|
| (4) 1" Steel Conduits (#1) | 5/8" TL 330-1 | None | 78 (172) | 60 min. |
| (4) 3" Steel Conduits (#2) | 5/8" TL 330-1 | None | 71 (160) | 60 min. |
| (8) 4" Aluminum Conduits (#3) | 5/8" TL 330-1 | None | 72 (161) | 60 min. |
| 4" Steel Conduit, Junction Box (#4) | 5/8" TL 330-1 | None | 100 (212) | 60 min. |

Each of the multiple conduit enclosures and the 4 in. conduit/junction box enclosure, clad with Thermo-Lag 330-1 material and upgrades met the acceptance criteria for a fire resistance rating of one hour. The maximum temperatures recorded from the thermocouples did not exceed the ambient temperature of 78°F plus maximum temperature rise of 250°F (total 328°F).

*D.3.4.13 TVA Project No. 11960-97260, Test 6.1.13 (October 1994)*

ERFBS: This 1-hr. test evaluated three different configurations: The first configuration was a group of (7) 4 in. steel conduits covered with 5/8 in. nominal Thermo-Lag 330-1 Panels and pre-shaped conduit sections. The second configuration was a 3/4 in. aluminum conduit covered with one layer of 5/8 in. nominal Thermo-Lag 330-1 and an upgrade of 3/8 in. nominal Thermo-Lag 330-1 fire barrier material. The third configuration included on 3/4 in. steel conduit covered with one layer of 5/8 in. nominal Thermo-Lag 330-1 and an upgrade of 3/8 in. nominal Thermo-Lag 330-1. Installations were positioned in a vertical to horizontal to vertical fashion. The score and bend principle was applied at 90° radial bends and voids filled with Thermo-Lag 330-1 Trowel Grade and putty. The test deck was covered with a 10 GA deck steel with 4 in. dia. pipe sockets welded to 3 in. dia. steel pipe legs to hold the assembly at proper level.

Test Slab: 13' x 8' x 10 GA (Steel), with 2 layers of 2" ceramic fiber blanket insulation

Desired Rating: 1 hr.

Ambient Temperature: 80°F

Thermocouples: A single bare #8 AWG stranded copper wire was instrumented with TCs and fastened to conduits by clamping under the head of a #8 x 32 x ¼" long stainless steel screw. Thermocouples were positioned nominally every 6 in. along the interior of the conduits to get a realistic measurement of the temperatures.
Hose Stream: The test platform was sprayed with water at a 30° hose angle at a

pressure of 75 psi while spinning at 6-8 revolutions per minute. The test setup was sprayed with a 75 gpm water flow rate from a distance of 5 feet for 5 minutes.
Furnace: 12' x 7' x 79"
Furnace control: Ten (10) thermocouples on probes located throughout the furnace
Cable Type: Electric cables were not utilized in this test.

### Table D-90. TVA Thermo-Lag Test 6.1.13

| Raceway Type | Barrier Protection | Cable Fill | Max. Temp. Rise ΔT °C (°F) | Rating |
|---|---|---|---|---|
| (7) 4" Steel Conduits | 5/8" TL 330-1 | None | 96 (205) | 60 min. |
| (1) 3/4" Aluminum Conduit | 5/8", 3/8" TL 330-1 | None | 63 (145) | 60 min. |
| (1) 3/4" Steel Conduit | 5/8", 3/8" TL 330-1 | None | 67 (153) | 60 min. |

Each of the ganged conduit enclosures and two individually clad 3/4 in. conduits, clad with Thermo-Lag 330-1 material and upgrades met the acceptance criteria for a fire resistance rating of one hour. The maximum temperatures recorded from the thermocouples did not exceed the ambient temperature of 80°F plus maximum temperature rise of 250°F (total 330°F).

*D.3.4.14 TVA Project No. 11210-98892, Test 6.1.14 (January 1996)*

ERFBS: This 1-hr. test evaluated six different configurations: The first configuration was a 1 in. aluminum conduit protected with a minimum 1/2 in. New Old Stock (NOS) Thermo-Lag 330-1 displaying a variance in Thermogravimetric Analysis plus a nominal 3/8 in. Thermo-Lag 330-1 upgrade preformed section with acceptable TGA. The second configuration was a 3 in. aluminum conduit protected with nominal 5/8 in. Thermo-Lag 330-1 displaying a slight variance in TGA. The third and fourth configurations were 3 in. aluminum conduit protected with a minimum 1/2 in. NOS Thermo-Lag 330-1 displaying a variance in TGA. The fifth configuration was a 4 in. aluminum conduit protected with a minimum 1/2 in. NOS Thermo-Lag 330-1 displaying a variance in TGA. The final configuration was a 4 in. aluminum conduit protected with a minimum 1/2 in. NOS Thermo-Lag 330-1 displaying acceptable TGA. This test not only tested the six separate configurations of Thermo-Lag clad conduits, but also whether using stainless steel bands, re-soaked pre-formed sections, damaged pre-formed sections, and re-certified Trowel Grade Material are acceptable.

Test Slab: 13' x 8' x 10 GA (Steel), with 2 layers of 2" ceramic fiber blanket insulation   Desired Rating: 1 hr.

Ambient Temperature: 65°F

Thermocouples: A single bare #8 AWG stranded copper wire was instrumented with TCs and fastened to conduits by clamping under the head of a #8 x 32 x ¼" long stainless steel screw. Thermocouples were positioned nominally every 6 in. along the interior of the conduits to get a realistic measurement of the temperatures. Additional thermocouples were installed on the outside of the conduit at predetermined banding locations to record thermal effects of stainless steel bands. Two wraps of fiberglass reinforced electrical tape were placed on the conduit, at the thermocouple location to isolate the thermo-junction from the conduit.

Hose Stream: The test platform was sprayed with water at a 30° hose angle at a pressure of 75 psi while spinning at 6-8 revolutions per minute. The test setup was sprayed with a 75 gpm water flow rate from a distance of 5 feet for 5 minutes.
Furnace: 12' x 7' x 79"
Furnace control: Ten (10) thermocouples on probes located throughout the furnace
Cable Type: Conduit "C" – 5, 7C #16 AWG 600V (0.85 lbs. cable/linear foot)
Conduit "D" – 16, 7C #16 AWG 600V (2.70 lbs. cable/linear foot)

**Table D-91. TVA Thermo-Lag Test 6.1.14**

| Raceway Type | Barrier Protection | Cable Fill | Max. Temp. Rise ΔT °C (°F) | Rating |
|---|---|---|---|---|
| 1" Aluminum Conduit (A) | 1/2", 3/8" TL330-1 | None | 72 (161) | 60 min. |
| 3" Aluminum Conduit (B) | 5/8" TL330-1 | None | 149 (300) | 60 min. |
| 3" Aluminum Conduit (C) | 1/2" TL330-1 | 5, 7C # 16 AWG | 135 (275) | 60 min. |
| 3" Aluminum Conduit (D) | 1/2" TL330-1 | 16, 7C #16 AWG | 114 (238) | 60 min. |
| 4" Aluminum Conduit (E) | 1/2" TL330-1 | None | 99 (210) | 60 min. |
| 4" Aluminum Conduit (F) | 1/2" TL330-1 | None | 110 (230) | 60 min. |

The six conduit configurations evaluated in this test procedure, clad with Thermo-Lag 330-1 material and upgrades, met the acceptance criteria for a fire resistance rating of one hour. The average starting temperature was 65°F and when added to the maximum temperature difference of 250°C, none of the test results exceeded the maximum overall temperature of 315°F. This test also determined that 1/2 in. stainless steel bands are not desirable, but still acceptable on existing 3 in. and larger single conduit ERFBS constructed with a single layer of old vintage Thermo-Lag 330-1. Additionally, the test proved that soaking a preformed section of Thermo-Lag and reinstalling it on a conduit performs well, but shall not be used in the future as it is an unacceptable practice. The testing also showed that Thermo-Lag 330-1 Trowel Grade Material which passed its 6-month shelf life could still be mixed and used in a normal manner if TVAN procedures for recertification are followed.

## D.3.4.15 TVA Project No. 11960-97555, Test 6.2.1 (December 1994)

<u>ERFBS</u>: This 3-hr. test evaluated six different configurations: The first configuration consisted of (1) 12 in. wide by 4 in. deep steel ladderback cable tray covered with one layer of 1-1/4 in. nominal Thermo-Lag 330-1 V-Rib Panel, reinforced in select locations with one layer of stress skin, and upgraded with two layers of 3/8 in. nominal Thermo-Lag 770-1 Mat. The second configuration was (1) 24 in. wide by 4 in. deep steel ladderback cable tray, covered with one layer of 1-1/4 in. nominal Thermo-Lag 330-1 V-Rib Panel, reinforced in select locations with one layer of stress skin, and upgraded with two layers of 3/8 in. nominal Thermo-Lag 770-1 Mat. The third configuration was a 12 in. by 12 in. by 60 in. steel junction box covered with one layer of nominal 1-1/4 in. Thermo-Lag 330-1 V-Rib Panels, reinforced with one layer of stress skin, and upgraded with two layers of nominal 3/8 in. Thermo-Lag 770-1 Mat. The two ladderback cable tray configurations were positioned from horizontal to vertical, and the junction box was positioned horizontally.

<u>Test Slab</u>: 13' x 8' x 10 GA (Steel), with 2 layers of 2" ceramic fiber blanket insulation

<u>Desired Rating</u>: 3 hr.

<u>Ambient Temperature</u>: 68°F

<u>Thermocouples</u>: A single bare #8 AWG stranded copper wire was instrumented with TCs and fastened to conduits and junction box by clamping under the head of a #8 x 32 x ¼" long stainless steel screw. Thermocouples were installed every six inches on the top surface of the cable tray rungs in both trays, on the cable tray side rails, and in every square foot of junction box surface area.

<u>Hose Stream</u>: The test platform was sprayed with water at a 30° hose angle at a pressure of 75 psi while spinning at 6-8 revolutions per minute. The test setup was sprayed with a 75 gpm water flow rate from a distance of 5 feet for 5 minutes with minimal exposure to the top surface of the experimental setup.

<u>Furnace</u>: 12' x 7' x 79"

<u>Furnace control</u>: Ten (10) thermocouples on probes located throughout the furnace

<u>Cable Type</u>: Electric cables were not utilized in this test.

### Table D-92. TVA Thermo-Lag Test 6.1.15

| Raceway Type | Barrier Protection | Cable Fill | Max. Temp. Rise $\Delta T$ °C (°F) | Rating |
|---|---|---|---|---|
| 12" x 4" Steel Cable Tray | 1-1/4" TL330-1 (2) 3/8" TL770-1 | None | 69 (156) | 180 min. |
| 24" x 4" Steel Cable Tray | 1-1/4" TL330-1 (2) 3/8" TL770-1 | None | 64 (147) | 180 min. |
| 12" x 12" x 60" Steel Junction Box | 1-1/4" TL330-1 (2) 3/8" TL770-1 | None | 62 (143) | 180 min. |

Both of the cable tray configurations (12 in. and 24 in.) and the junction box configuration evaluated in this test, clad with Thermo-Lag 330-1 material and upgraded with Thermo-Lag 770-1 material met the acceptance criteria for a fire resistance rating of over three hours (213 minutes). All cable trays and the junction box assembly were tested with a worst case design baseline installation of 1-1/4 in. Thermo-Lag 330-1 material. The testing proved that 3-hour EFRBS are achievable as an upgrade to an existing system or as a completely new installation, and based upon the resulting time and temperature readings, this installation could have reduced quantities of material applied or a longer duration rating.

### D.3.4.16 TVA Project No. 11960-97553, Test 6.2.2 (January 1995)

ERFBS: This 3-hr. test evaluated six different configurations: The first configuration was (1) 24 in. by 4 in. steel ladderback cable tray, covered with one layer of 1-1/4 in. nominal Thermo-Lag 330-1 V-Rib Panels, reinforced in select locations with a layer of stress skin and upgraded with two layers of 3/8 in. nominal Thermo-Lag 770-1 Mat. The second configuration is identical to the first, except that the ladderback cable tray's dimensions are 12" wide by 4" deep. The third configuration is (1) 1 in. galvanized steel conduit covered with one layer of 1-1/4 in. nominal Thermo-Lag 330-1 preformed conduit sections and upgraded with three layers of 3/8 in. nominal Thermo-Lag 770-1 Mat. The fourth configuration is identical to the third except the steel conduit is 2 in. instead of 1 in. The fifth configuration is identical to the third and fourth, but is 5 in. thick instead of 1 and 2 inches. The final configuration is (1) 2 in. air drop covered with one layer of 1-1/4 in. nominal Thermo-Lag 330-1 Preformed conduit sections, reinforced with one layer of stress skin, and upgraded with three layers of 3/8 in. nominal Thermo-Lag 770-1 Mat. All configurations in this test were aligned in a horizontal to vertical orientation. External stress skin was applied on top of a layer of Thermo-Lag 770-1 Trowel Grade.

Test Slab: 13' x 8' x 10 GA (Steel), with 2 layers of 2" ceramic fiber blanket insulation

Desired Rating: 3 hr.

Ambient Temperature: 64°F

Thermocouples: A single bare #8 AWG stranded copper wire was instrumented with TCs and fastened to conduits and junction box by clamping under the head of a #8 x 32 x ¼" long stainless steel screw. Thermocouples were installed every six inches on the top surface of the cable tray rungs in both trays, in the interiors of the three conduits, along the cable tray side rails, and the bottom conduit surfaces.

Hose Stream: The test platform was sprayed with water at a 30° hose angle at a pressure of 75 psi while spinning at 6-8 revolutions per minute. The test setup was sprayed with a 75 gpm water flow rate from a distance of 5 feet for 5 minutes with minimal exposure to the top surface of the experimental setup.

Furnace: 12' x 7' x 79"

Furnace control: Ten (10) thermocouples on probes located throughout the furnace

Cable Type: Electric cables were not utilized in this test.

**Table D-93. TVA Thermo-Lag Test 6.2.2**

| Raceway Type | Barrier Protection | Cable Fill | Max. Temp. Rise ΔT °C (°F) | Rating |
|---|---|---|---|---|
| 24" x 4" Steel Cable Tray | 1-1/4" TL330-1 (2) 3/8" TL770-1 | None | 61 (142) | 180 min. |
| 12" x 4" Steel Cable Tray | 1-1/4" TL330-1 (2) 3/8" TL770-1 | None | 68 (155) | 180 min. |
| 1" Galvanized Steel Conduit | 1-1/4" TL330-1 (3) 3/8" TL770-1 | None | 77 (170) | 180 min. |
| 2" Galvanized Steel Conduit | 1-1/4" TL330-1 (2) 3/8" TL770-1 | None | 116 (241) | 180 min. |
| 5" Galvanized Steel Conduit | 1-1/4" TL330-1 (2) 3/8" TL770-1 | None | 74 (166) | 180 min. |
| 2" Air Drop | 1-1/4" TL330-1 (3) 3/8" TL770-1 | None | 55 (131) | 180 min. |

Both of the cable tray configurations (12 in. and 24 in.), all three conduit configurations (5 in., 2 in., and 1 in.) and the 2 in. air drop configuration evaluated in the test procedure, clad with Thermo-Lag 330-1 and Thermo-Lag 770-1 upgrades, met the acceptance criteria for a fire resistance rating of over four hours (250 minutes). The tests proved that 3-hour ERFBS are achievable as an upgrade to an existing system or as a completely new installation and this installation, based upon the resulting time and temperature readings, could have reduced quantities of material applied or a longer duration rating.

## D.4 References

NEI Application Guide for Evaluation of Thermo-Lag 330 Fire Barrier Systems, NEI, 1994. (NUDOCS Accession No. 9603150034)

SwRI Test Report Project No. 01-7912(2), "Qualification Fire Test of a Protective Envelope System," dated June 1985. (NUDOCS Accession No. 9308040254)

Dresden Nuclear Power Station, "Dresden Station Units 2 and 3, 3M Fire Wrap Qualification Evaluation, NTSC Project 99-40540, Dresden Report No. 12-N208-05," dated December 31, 1999. (ADAMS Accession No. ML011910445)

Letter from R.W. Brown (Peak Seals, Inc) to P.M. Madden (U.S. NRC), "Listed Documents Re Three Hour Fire Endurance Test on 3M Interam Fire Barrier Wrap Sponsored by Peak Seals," dated August 7, 1995. (NUDOCS Accession No. 9509050173)

Fire Endurance Test, "Omega Point Fire Endurance Test of 3M Interam Mat Fire Protective Envelopes (24 in. and 6 in. Cable Trays, 5 in., 3 in., and 1 in. Conduits, 2 in. Air Drop and a 12 in. x 12 in. by 8 in. Junction Box," dated August 2, 1995. (NUDOCS Accession No. 9509070113)

Letter from T. Dogan (Vectra) to R. Brown (Peak Seals, Inc.), "E-50 Series Fire Endurance Test Evaluation," dated August 1, 1995. (NUDOCS Accession No. 9709090341)

Memorandum from A. Singh (U.S. NRC) to C.E. McCracken (U.S. NRC), "Trip to Omega Point Laboratories, 3M Company Interam 1-Hour Raceway Fire Barrier Fire Endurance Test (TAC No. M82809)," dated May 25, 1995. (NUDOCS Accession No. 9506090198)

Test Report provided by Peak Seals Inc., "Test Plan Number CTP-1199, One (1) Hour Fire Endurance Test, 3M Interam Fire Wrap," report date May 4, 1995. (NUDOCS Accession No. 9507140090)

Test Report Provided by Omega Point Laboratories, "ASTM E136-94, Behavior of Materials in a Vertical Tube Furnace at 750°C, 3M E-50 Interam Series Mat," report date January 17, 1995. (NUDOCS Accession No. 9705050067)

Letter from R.W. Brown (Peak Seals, Inc.) to L.B. Marsh (U.S. NRC), "Inform that Peak Seals has become Master Distributor of 3M Interam Fire Wrap Sys for Commercial Nuclear Power Plants, Response to Specific Questions," dated October 3, 1997. (NUDOCS Accession No. 9802040354)

Letter from J.K. Wood (Centerior Energy) to U.S. NRC, "Combustibility Testing of 3M Interam Material," dated February 7, 1997. (NUDOCS Accession No. 9702200311)

Letter from R. Licht (3M Fire Protection Products) to C.E. McCracken (U.S. NRC), "Acknowledges Receipt of 930504 Letter Requesting Info on 3M Fire Barrier Systems for Protection of Electrical Raceways. Informs that 3M Would Like to Supply Requested Info in Three Parts, Covering Flexible Wrap Systems, Rigid Panel Systems, and FS195 Systems," dated May 18, 1993. (NUDOCS Accession No. 9308310099)

Letter from D.R. Coy (3M Ceramic Materials Department), "Discusses NUMARC Meeting 931201-02 re Performance of Certain Fire Barrier materials. 3M Additional Test SVC Program Will Provide Technical Support, Supply of 3M Fire Protection Products, and Fire Testing at Cottage Grove Facility," dated January 14, 1994. (NUDOCS Accession No. 9401310171)

Letter from K.W. Howell (Underwriters Laboratories, Inc.) to B.J. Youngblood (U.S. NRC), "Qualification of 3M Fire Wrap," dated October 22, 1984. (NUDOCS Accession No. 8410240232)

Letter from D.R. Coy (3M Ceramic Materials Department), "Informs that 3M Will be Providing Complete Document Package for 3M Interam E-50 Series Material, 1-and-3 Hour Systems, Including Fire Test Reports, per 10CFR50, App R," dated August 3, 1993. (NUDOCS Accession No. 9308120140)

Letter from D.R. Coy (3M Ceramic Materials Department), "Advises that 3M Will Continue to Supply Interam E-50 Series Materials and all Peripheral 3M Fire Protection Products Used in Installation Process Under Original Nuclear Product Designations," dated March 1, 1994. (NUDOCS Accession No. 9403170125)

Test Report Provided by 3M Fire Protection Products, "3M Fire Test Reports 94-27 and 94-42, re Upgrading TSI Material for 3-Hour Conduit Systems using 3M's Interam E-50 Series Mats," dated March 17, 1994. (NUDOCS Accession No. 9403280162)

Letter from Dwight E. Nunn (Tennessee Valley Authority) to U.S. NRC, "Watts Bar Nuclear Plant (WBN) – 3M Fire Barrier Material, Cable Compressive Load Testing," dated June 17, 1994. (NUDOCS Accession No. 9406240089)

Letter from D.R. Coy (3M Fire Protection Products) to Nuclear Power Utility Customers, "Letter Advising that Effective 950628 3M Fire Protection Products Certified Nuclear Installer Program Will End and that Peak Seals, Inc Will Become Exclusive Supplier of 3M Fire Protection Products, Effective 950629," dated April 28, 1995. (NUDOCS Accession No. 9505230123)

Test Report Provided by 3M Fire Protection Products, "Results of Fire Test Conducted on Latest NRC Criteria With 3M Interam E-50 Series Material," dated May 25, 1995. (NUDOCS Accession No. 9506220297)

Test Report Provided by U.S. NRC, "Trip to Omega Point Laboratories – Peak Seals 3M Interam Raceway Fire Barrier Fire Endurance Test Program (April 20, 1995) (TAC No. M82809)," dated May 1, 1995. (NUDOCS Accession No. 9507140050)

3M Fire Protection Products, "3M Advanced Training Program: 3M Interam E-50 Series Fire Protection Systems for the Nuclear Industry," dated June 15, 1993. (NUDOCS Accession No. 9308040286)

Letter from C.E. McCracken (U.S. NRC) to R.R. Licht (3M Fire Protection Products), "NRC Intent to Review Fire Barrier Systems Used by Licensees Re Compliance with NRC Fire Protection Requirements," dated May 4, 1993. (NUDOCS Accession No. 9403170119)

SwRI Test Report, "Ampacity Derating of Fire-Protected Cables in Conduit / Cable Trays Using 3M Incorporated's Passive Fire Protection Systems Identified as 3M Interam E-50A, E-50D, E-53A, and E-50D/E-53A," dated September, 30 1986 (NUDOCS Accession No. 9308040280)

3M Fire Protection Products, "3M Interam E-50 Series Fire Protection Mat, 1-Hour Flexible Wrap System for Electrical Raceways, Installation Booklet Including Quality Assurance Guidelines and Typical Drawings," dated June 19, 1987. (NUDOCS Accession No. 9308040230)

Test Report Provided by Twin City Testing Corporation, "Qualification Fire Tests of the 3M Interam E-50D Fire Protection Mat for 3-Hour Rated Electrical Raceways," dated March 1986. (NUDOCS Accession No. 9308040267)

SwRI Test Report No. 01-7912, "Qualification Fire Test of a Protective Envelope System," dated June 1984. (NUDOCS Accession No. 9308040243)

3M Fire Protection Products, "3M Interam E-54A Fire Protection Mat, 3-Hour Flexible Wrap System for Electrical Raceways, Installation Booklet Including Quality Assurance Guidelines and Typical Drawings," dated October 27, 1987. (NUDOCS Accession No. 9308040228)

3M Fire Protection Products, "3M Interam E-50 Series 1-hour and 3-hour Flexible Wrap Fire Protection Systems." (NUDOCS Accession No. 9308040224)

SwRI Project Report No. 01-7912a(1), "Qualification Fire Test of a Protective Envelope System," dated June 1985. (NUDOCS Accession No. 9308040263)

Test Report Provided by Twin City Testing Corporation, "Qualification Fire Tests of the 3M Interam E-50 Series Fire Protection Mat for 1-Hour Rated Electrical Raceways," dated September 1986. (NUDOCS Accession No. 9308040275)

SwRI Project Report No. 01-7912(2), "Qualification Fire Test of a Protective Envelope System," dated June 1985. (NUDOCS Accession No. 9308040254)

Letter from R. Licht (3M Fire Protection Products) to C.E. McCracken (U.S. NRC), "Response to Questions Noted in 930504 Letter and Provides Info Intended to Validate Use of 3M Interam E-50 Series 1-Hour and 3 Hour Fire Protection Systems," dated June 30, 1993. (NUDOCS Accession No. 9308040054)

Letter from R.W. Brown (Peak Seals) to P.M. Madden (U.S. NRC), "Response: 3M Interam Fire Wrap Systems," dated August 7, 1995. (NUDOCS Accession No. 9509050173)

Test Report Provided by Central Laboratories Services, "Testing to Determine Ampacity Derating Factors for 3M Fire Barrier Wrapped Conduits and Air Drops, Job Number 94-0357, Revision 0," dated February 3, 1994. (NUDOCS Accession No. 9403030227)

SwRI Test Report No. 01-8818-208/-209d, "Ampacity Derating of Fire-Protected Cables in Conduit/Cable Trays Using 3M Incorporated's Passive Fire Protection Systems Identified as 3M Interam E-50A (Verification Tests)," dated October 6, 1986. (NUDOCS Accession No. 9308040249)

SwRI Test Report No. 1208-001, "Nuclear Component Qualification Test Report for the Generic Seismic Qualification of 3M Interam E-50D 3-Hour Fire Protection System," dated July 1986. (NUDOCS Accession No. 9308040186)

Test Report Provided by 3M Fire Protection Products, "3M Fire Test #92-115: 3-hour Fire Protection on Conduits with the 3M Interam E-50 Series Mats," dated August, 6, 1992. (NUDOCS Accession No. 9308040119)

Test Report Provided by 3M Fire Protection Products, "3M Fire Test #87-79: 3-hour Fire Protection on a Cable Tray With the 3M Interam E-50 Series Mats," dated July 24, 1992. (NUDOCS Accession No. 9308040106)

Test Report Provided by 3M Fire Protection Products, "3M Fire Test #92-167: 1-hour Fire Protection on 1" Sch. 40 Steel Conduits Using the 3M Interam E-53A Mats," dated June 25, 1993. (NUDOCS Accession No. 9308040101)

Test Report Provided by 3M Fire Protection Products, "3M Fire Test #92-141: 1-hour and 3-Hour Fire Protection on 1" Sch. 40 Steel Conduits Using the 3M Interam E-54A Mats," dated August 27, 1992. (NUDOCS Accession No. 9308040099)

Test Report Provided by 3M Fire Protection Products, "3M Fire Test #87-40: 1-hour Fire Protection on a Cable Tray With the 3M Interam E-50 Series Materials," dated April 3, 1992. (NUDOCS Accession No. 9308040097)

Test Report Provided by 3M Fire Protection Products, "3M Fire Test Report #87-57: 3M Chemolite Building 66, Large Scale Furnace," dated May 27, 1987. (NUDOCS Accession No. 9308040094)

Test Report Provided by 3M Fire Protection Products, "3M Fire Test #87-76: 60 Minute Fire Protection on Conduits with Interam E-53A Mat and FireDam 150 Caulk," dated June 25, 1987. (NUDOCS Accession No. 9308040090)

Letter from R. Licht (3M Ceramic Materials) to C.E. McCracken (U.S. NRC), "Response to Questions Noted in 930504 Letter and Provides Info Intended to Validate Use of 3M Interam E-50 Series 1-Hour and 3-Hour Fire Protection Systems," dated June 30, 1993. (NUDOCS Accession No. 9308040054)

# Appendix E  Fire Protection Regulations Cited from 10 CFR 50

Disclaimer: The following has been reproduced from 10 CFR Part 50 and may contain errors. This reproduction is only meant to provide a quick reference. An official copy of 10 CFR Part 50 should be consulted for regulatory matters.

## § 50.48 Fire protection.

(a)(1) Each holder of an operating license issued under this part or a combined license issued under part 52 of this chapter must have a fire protection plan that satisfies Criterion 3 of appendix A to this part. This fire protection plan must:

(i) Describe the overall fire protection program for the facility;

(ii) Identify the various positions within the licensee's organization that are responsible for the program;

(iii) State the authorities that are delegated to each of these positions to implement those responsibilities; and

(iv) Outline the plans for fire protection, fire detection and suppression capability, and limitation of fire damage.

(2) The plan must also describe specific features necessary to implement the program described in paragraph (a)(1) of this section such as--

(i) Administrative controls and personnel requirements for fire prevention and manual fire suppression activities;

(ii) Automatic and manually operated fire detection and suppression systems; and

(iii) The means to limit fire damage to structures, systems, or components important to safety so that the capability to shut down the plant safely is ensured.

(3) The licensee shall retain the fire protection plan and each change to the plan as a record until the Commission terminates the reactor license. The licensee shall retain each superseded revision of the procedures for 3 years from the date it was superseded.

(4) Each applicant for a design approval, design certification, or manufacturing license under part 52 of this chapter must have a description and analysis of the fire protection design features for the standard plant necessary to demonstrate compliance with Criterion 3 of appendix A to this part.

(b) Appendix R to this part establishes fire protection features required to satisfy Criterion 3 of appendix A to this part with respect to certain generic issues for nuclear power plants licensed to operate before January 1, 1979.

(1) Except for the requirements of Sections III.G, III.J, and III.O, the provisions of Appendix R to this part do not apply to nuclear power plants licensed to operate before January 1, 1979, to the extent that--

(i) Fire protection features proposed or implemented by the licensee have been accepted by NRC staff as satisfying the provisions of Appendix A to Branch Technical Position (BTP) APCSB 9.5-1 reflected in NRC fire protection safety evaluation reports issued before the effective date of February 19, 1981; or

(ii) Fire protection features were accepted by NRC staff in comprehensive fire protection safety evaluation reports issued before Appendix A to Branch Technical Position (BTP) APCSB 9.5-1 was published in August 1976.

(2) With respect to all other fire protection features covered by Appendix R, all nuclear power plants licensed to operate before January 1, 1979, must satisfy the applicable requirements of Appendix R to this part, including specifically the requirements of Sections III.G, III.J, and III.O.

(c) *National Fire Protection Association Standard NFPA 805. - (1) Approval of incorporation by reference.* National Fire Protection Association (NFPA) Standard 805, "Performance-Based Standard for Fire Protection for Light Water Reactor Electric Generating Plants, 2001 Edition" (NFPA 805), which is referenced in this section, was approved for incorporation by reference by the Director of the Federal Register pursuant to 5 U.S.C. 552(a) and 1 CFR part 51. Copies of NFPA 805 may be purchased from the NFPA Customer Service Department, 1 Batterymarch Park, P.O. Box 9101, Quincy, MA 02269-9101 and in PDF format through the NFPA Online Catalog (http://www.nfpa.org) or by calling 1-800-344-3555 or (617) 770-3000. Copies are also available for inspection at NRC Library, Two White Flint North, 11545 Rockville Pike, Rockville, Maryland 20852-2738, and at NRC Public Document Room, Building One White Flint North, Room O1-F15, 11555 Rockville Pike, Rockville, Maryland 20852-2738. Copies are also available at the National Archives and Records Administration (NARA). For information on the availability of this material at NARA, call (202) 741-6030, or go to: http://www.archives.gov/federal_register/code_of_federal_regulations/ibr_locations.html.

(2) *Exceptions, modifications, and supplementation of NFPA 805*. As used in this section, references to NFPA 805 are to the 2001 Edition, with the following exceptions, modifications, and supplementation:

(i) *Life Safety Goal, Objectives, and Criteria*. The Life Safety Goal, Objectives, and Criteria of Chapter 1 are not endorsed.

(ii) *Plant Damage/Business Interruption Goal, Objectives, and Criteria*. The Plant Damage/Business Interruption Goal, Objectives, and Criteria of Chapter 1 are not endorsed.

(iii) *Use of feed-and-bleed*. In demonstrating compliance with the performance criteria of Sections 1.5.1(b) and (c), a high-pressure charging/injection pump coupled with the pressurizer power-operated relief valves (PORVs) as the sole fire-protected safe shutdown path for maintaining reactor coolant inventory, pressure control, and decay heat removal capability (i.e., feed-and-bleed) for pressurized-water reactors (PWRs) is not permitted.

(iv) *Uncertainty analysis*. An uncertainty analysis performed in accordance with Section 2.7.3.5 is not required to support deterministic approach calculations.

(v) *Existing cables*. In lieu of installing cables meeting flame propagation tests as required by Section 3.3.5.3, a flame-retardant coating may be applied to the electric cables, or an automatic fixed fire suppression system may be installed to provide an equivalent level of protection. In addition, the italicized exception to Section 3.3.5.3 is not endorsed.

(vi) *Water supply and distribution*. The italicized exception to Section 3.6.4 is not endorsed. Licensees who wish to use the exception to Section 3.6.4 must submit a request for a license amendment in accordance with paragraph (c)(2)(vii) of this section.

(vii) *Performance-based methods*. Notwithstanding the prohibition in Section 3.1 against the use of performance-based methods, the fire protection program elements and minimum design requirements of Chapter 3 may be subject to the performance-based methods permitted elsewhere in the standard. Licensees who wish to use performance-based methods for these fire protection program elements and minimum design requirements shall submit a request in the form of an application for license amendment under § 50.90. The Director of the Office of Nuclear Reactor Regulation, or a designee of the Director, may approve the application if the Director or designee determines that the performance-based approach;

(A) Satisfies the performance goals, performance objectives, and performance criteria specified in NFPA 805 related to nuclear safety and radiological release;

(B) Maintains safety margins; and

(C) Maintains fire protection defense-in-depth (fire prevention, fire detection, fire suppression, mitigation, and post-fire safe shutdown capability).

(3) *Compliance with NFPA 805*. (i) A licensee may maintain a fire protection program that complies with NFPA 805 as an alternative to complying with paragraph (b)

of this section for plants licensed to operate before January 1, 1979, or the fire protection license conditions for plants licensed to operate after January 1, 1979. The licensee shall submit a request to comply with NFPA 805 in the form of an application for license amendment under § 50.90. The application must identify any orders and license conditions that must be revised or superseded, and contain any necessary revisions to the plant's technical specifications and the bases thereof. The Director of the Office of Nuclear Reactor Regulation, or a designee of the Director, may approve the application if the Director or designee determines that the licensee has identified orders, license conditions, and the technical specifications that must be revised or superseded, and that any necessary revisions are adequate. Any approval by the Director or the designee must be in the form of a license amendment approving the use of NFPA 805 together with any necessary revisions to the technical specifications.

(ii) The licensee shall complete its implementation of the methodology in Chapter 2 of NFPA 805 (including all required evaluations and analyses) and, upon completion, modify the fire protection plan required by paragraph (a) of this section to reflect the licensee's decision to comply with NFPA 805, before changing its fire protection program or nuclear power plant as permitted by NFPA 805.

(4) *Risk-informed or performance-based alternatives to compliance with NFPA 805.* A licensee may submit a request to use risk-informed or performance-based alternatives to compliance with NFPA 805. The request must be in the form of an application for license amendment under § 50.90 of this chapter. The Director of the Office of Nuclear Reactor Regulation, or designee of the Director, may approve the application if the Director or designee determines that the proposed alternatives:

(i) Satisfy the performance goals, performance objectives, and performance criteria specified in NFPA 805 related to nuclear safety and radiological release;

(ii) Maintain safety margins; and

(iii) Maintain fire protection defense-in-depth (fire prevention, fire detection, fire suppression, mitigation, and post-fire safe shutdown capability).

(d)-(3) (Reserved).

(f) Licensees that have submitted the certifications required under § 50.82(a)(1) shall maintain a fire protection program to address the potential for fires that could cause the release or spread of radioactive materials (i.e., that could result in a radiological hazard). A fire protection program that complies with NFPA 805 shall be deemed to be acceptable for complying with the requirements of this paragraph.

(1) The objectives of the fire protection program are to--

(i) Reasonably prevent these fires from occurring;

(ii) Rapidly detect, control, and extinguish those fires that do occur and that could result in a radiological hazard; and

(iii) Ensure that the risk of fire-induced radiological hazards to the public, environment and plant personnel is minimized.

(2) The licensee shall assess the fire protection program on a regular basis. The licensee shall revise the plan as appropriate throughout the various stages of facility decommissioning.

(3) The licensee may make changes to the fire protection program without NRC approval if these changes do not reduce the effectiveness of fire protection for facilities, systems, and equipment that could result in a radiological hazard, taking into account the decommissioning plant conditions and activities.

(65 FR 38190, June 20, 2000; 69 FR 33550, June 16, 2004; 72 FR 49495, Aug. 28, 2007)

## Appendix A to Part 50 – General Design Criteria 3 – Fire Protection

Structures, systems, and components important to safety shall be designed and located to minimize, consistent with other safety requirements, the probability and effect of fires and explosions. Noncombustible and heat resistant materials shall be used wherever practical throughout the unit, particularly in locations such as the containment and control room. Fire detection and fighting systems of appropriate capacity and capability shall be provided and designed to minimize the adverse effects of fires on structures, systems, and components important to safety. Firefighting systems shall be designed to assure that their rupture or inadvertent operation does not significantly impair the safety capability of these structures, systems, and components.

## Appendix R to Part 50 – Fire Protection Program for Nuclear Power Facilities Operating Prior to January 1, 1979

I. Introduction and Scope

This appendix applies to licensed nuclear power electric generating stations that were operating prior to January 1, 1979, except to the extent set forth in § 50.48(b) of this part. With respect to certain generic issues for such facilities it sets forth fire protection features required to satisfy Criterion 3 of Appendix A to this part.

Criterion 3 of Appendix A to this part specifies that "Structures, systems, and components important to safety shall be designed and located to minimize, consistent with other safety requirements, the probability and effect of fires and explosions."

When considering the effects of fire, those systems associated with achieving and maintaining safe shutdown conditions assume major importance to safety because damage to them can lead to core damage resulting from loss of coolant through boil off.

The phrases "important to safety," or "safety-related," will be used throughout this Appendix R as applying to all safety functions. The phrase "safe shutdown" will be used throughout this appendix as applying to both hot and cold shutdown functions.

Because fire may affect safe shutdown systems and because the loss of function of systems used to mitigate the consequences of design basis accidents under post fire conditions does not per se impact public safety, the need to limit fire damage to systems required to achieve and maintain safe shutdown conditions is greater than the need to limit fire damage to those systems required to mitigate the consequences of design basis accidents. Three levels of fire damage limits are established according to the safety functions of the structure, system, or component:

| Safety Function | Fire damage limits |
| --- | --- |
| Hot Shutdown... | One train of equipment necessary to achieve hot shutdown from either the control room or emergency control station(s) must be maintained free of fire damage by a single fire, including an exposure fire.1 |
| Cold Shutdown | Both trains of equipment necessary to achieve cold shutdown may be damaged by a single fire, including an exposure fire, but damage must be limited so that at least one train can be repaired or made operable within 72 hours using onsite capability. |
| Design Basis Accidents | Design Basis Accidents Both trains of equipment necessary |

for mitigation of consequences following design basis accidents may be damaged by a single exposure fire.

---
[1] *Exposure Fire.* An exposure fire is a fire in a given area that involves either in situ or transient combustibles and is external to any structures, systems, or components located in or adjacent to that same area. The effects of such fire (e.g., smoke, heat, or ignition) can adversely affect those structures, systems, or components important to safety. Thus, a fire involving one train of safe shutdown equipment may constitute an exposure fire for the redundant train located in the same area, and a fire involving combustibles other than either redundant train may constitute an exposure fire to both redundant trains located in the same area.

The most stringent fire damage limit shall apply for those systems that fall into more than one category. Redundant systems used to mitigate the consequences of other design basis accidents but not necessary for safe shutdown may be lost to a single exposure fire. However, protection shall be provided so that a fire within only one such system will not damage the redundant system.

II. General Requirements

A. Fire protection program. A fire protection program shall be established at each nuclear power plant. The program shall establish the fire protection policy for the protection of structures, systems, and components important to safety at each plant and the procedures, equipment, and personnel required to implement the program at the plant site.

The fire protection program shall be under the direction of an individual who has been delegated authority commensurate with the responsibilities of the position and who has available staff personnel knowledgeable in both fire protection and nuclear safety.

The fire protection program shall extend the concept of defense-in-depth to fire protection in fire areas important to safety, with the following objectives:

To prevent fires from starting;

To detect rapidly, control, and extinguish promptly those fires that do occur;

To provide protection for structures, systems, and components important to safety so that a fire that is not promptly extinguished by the fire suppression activities will not prevent the safe shutdown of the plant.

B. Fire hazards analysis. A fire hazards analysis shall be performed by qualified fire protection and reactor systems engineers to (1) consider potential in situ and transient fire hazards; (2) determine the consequences of fire in any location in the plant on the ability to safely shut down the reactor or on the ability to minimize and control the release of radioactivity to the environment; and (3) specify measures for fire prevention, fire detection, fire suppression, and fire containment and alternative shutdown capability as required for each fire area containing structures, systems, and components important to safety in accordance with NRC guidelines and regulations.

C. Fire prevention features. Fire protection features shall meet the following general requirements for all fire areas that contain or present a fire hazard to structures, systems, or components important to safety.

1. In situ fire hazards shall be identified and suitable protection provided.

2. Transient fire hazards associated with normal operation, maintenance, repair, or modification activities shall be identified and eliminated where possible. Those transient fire hazards that can not be eliminated shall be controlled and suitable protection provided.

3. Fire detection systems, portable extinguishers, and standpipe and hose stations shall be installed.

4. Fire barriers or automatic suppression

systems or both shall be installed as necessary to protect redundant systems or components necessary for safe shutdown.

5. A site fire brigade shall be established, trained, and equipped and shall be on site at all times.

6. Fire detection and suppression systems shall be designed, installed, maintained, and tested by personnel properly qualified by experience and training in fire protection systems.

7. Surveillance procedures shall be established to ensure that fire barriers are in place and that fire suppression systems and components are operable.

D. Alternative or dedicated shutdown capability. In areas where the fire protection features cannot ensure safe shutdown capability in the event of a fire in that area, alternative or dedicated safe shutdown capability shall be provided.

III. Specific Requirements

A. Water supplies for fire suppression systems. Two separate water supplies shall be provided to furnish necessary water volume and pressure to the fire main loop.

Each supply shall consist of a storage tank, pump, piping, and appropriate isolation and control valves. Two separate redundant suctions in one or more intake structures from a large body of water (river, lake, etc.) will satisfy the requirement for two separated water storage tanks. These supplies shall be separated so that a failure of one supply will not result in a failure of the other supply.

Each supply of the fire water distribution system shall be capable of providing for a period of 2 hours the maximum expected water demands as determined by the fire hazards analysis for safety-related areas or other areas that present a fire exposure hazard to safety-related areas.

When storage tanks are used for combined service-water/fire-water uses the minimum volume for fire uses shall be ensured by means of dedicated tanks or by some physical means such as a vertical standpipe for other water service. Administrative controls, including locks for tank outlet valves, are unacceptable as the only means to ensure minimum water volume.

Other water systems used as one of the two fire water supplies shall be permanently connected to the fire main system and shall be capable of automatic alignment to the fire main system. Pumps, controls, and power supplies in these systems shall satisfy the requirements for the main fire pumps. The use of other water systems for fire protection shall not be incompatible with their functions required for safe plant shutdown. Failure of the other system shall not degrade the fire main system.

B. Sectional isolation valves. Sectional isolation valves such as post indicator valves or key operated valves shall be installed in the fire main loop to permit isolation of portions of the fire main loop for maintenance or repair without interrupting the entire water supply.

C. Hydrant isolation valves. Valves shall be installed to permit isolation of outside hydrants from the fire main for maintenance or repair without interrupting the water supply to automatic or manual fire suppression systems in any area containing or presenting a fire hazard to safety-related or safe shutdown equipment.

D. Manual fire suppression. Standpipe and hose systems shall be installed so that at least one effective hose stream will be able to reach any location that contains or presents an exposure fire hazard to structures, systems, or components important to safety.

Access to permit effective functioning of

the fire brigade shall be provided to all areas that contain or present an exposure fire hazard to structures, systems, or components important to safety.

Standpipe and hose stations shall be inside PWR containments and BWR containments that are not inerted. Standpipe and hose stations inside containment may be connected to a high quality water supply of sufficient quantity and pressure other than the fire main loop if plant-specific features prevent extending the fire main supply inside containment. For BWR drywells, standpipe and hose stations shall be placed outside the dry well with adequate lengths of hose to reach any location inside the dry well with an effective hose stream.

E. Hydrostatic hose tests. Fire hose shall be hydrostatically tested at a pressure of 150 psi or 50 psi above maximum fire main operating pressure, whichever is greater. Hose stored in outside hose houses shall be tested annually. Interior standpipe hose shall be tested every three years.

F. Automatic fire detection. Automatic fire detection systems shall be installed in all areas of the plant that contain or present an exposure fire hazard to safe shutdown or safety-related systems or components. These fire detection systems shall be capable of operating with or without offsite power.

G. Fire protection of safe shutdown capability. 1. Fire protection features shall be provided for structures, systems, and components important to safe shutdown. These features shall be capable of limiting fire damage so that:

a. One train of systems necessary to achieve and maintain hot shutdown conditions from either the control room or emergency control station(s) is free of fire damage; and

b. Systems necessary to achieve and maintain cold shutdown from either the control room or emergency control station(s) can be repaired within 72 hours.

2. Except as provided for in paragraph G.3 of this section, where cables or equipment, including associated non-safety circuits that could prevent operation or cause maloperation due to hot shorts, open circuits, or shorts to ground, of redundant trains of systems necessary to achieve and maintain hot shutdown conditions are located within the same fire area outside of primary containment, one of the following means of ensuring that one of the redundant trains is free of fire damage shall be provided:

a. Separation of cables and equipment and associated non-safety circuits of redundant trains by a fire barrier having a 3-hour rating. Structural steel forming a part of or supporting such fire barriers shall be protected to provide fire resistance equivalent to that required of the barrier;

b. Separation of cables and equipment and associated non-safety circuits of redundant trains by a horizontal distance of more than 20 feet with no intervening combustible or fire hazards. In addition, fire detectors and an automatic fire suppression system shall be installed in the fire area; or

c. Enclosure of cable and equipment and associated non-safety circuits of one redundant train in a fire barrier having a 1-hour rating, In addition, fire detectors and an automatic fire suppression system shall be installed in the fire area;

Inside noninerted containments one of the fire protection means specified above or one of the following fire protection means shall be provided:

d. Separation of cables and equipment and associated non-safety circuits of redundant trains by a horizontal distance of more than 20 feet with no intervening combustibles or fire hazards;

e. Installation of fire detectors and an automatic fire suppression system in the fire area; or

f. Separation of cables and equipment and associated non-safety circuits of redundant trains by a noncombustible radiant energy shield.

3. Alternative of dedicated shutdown capability and its associated circuits,1 independent of cables, systems or components in the area, room, zone under consideration should be provided:

a. Where the protection of systems whose function is required for hot shutdown does not satisfy the requirement of paragraph G.2 of this section; or

b. Where redundant trains of systems required for hot shutdown located in the same fire area may be subject to damage from fire suppression activities or from the rupture or inadvertent operation of fire suppression systems.

In addition, fire detection and a fixed fire suppression system shall be installed in the area, room, or zone under consideration.

H. Fire brigade. A site fire brigade trained and equipped for fire fighting shall be established to ensure adequate manual fire fighting capability for all areas of the plant containing structures, systems, or components important to safety. The fire brigade shall be at least five members on each shift. The brigade leader and at least two brigade members shall have sufficient training in or knowledge of plant safety-related systems to understand the effects of fire and fire suppressants on safe shutdown capability. The qualification of fire brigade members shall include an annual physical examination to determine their ability to perform strenuous fire fighting activities. The shift supervisor shall not be a member of the fire brigade. The brigade leader shall be competent to assess the potential safety consequences of a fire and advise control room personnel. Such competence by the brigade leader may be evidenced by possession of an operator's license or equivalent knowledge of plant safety-related systems.

The minimum equipment provided for the brigade shall consist of personal protective equipment such as turnout coats, boots, gloves, hard hats, emergency communications equipment, portable lights, portable ventilation equipment, and portable extinguishers. Self-contained breathing apparatus using full-face positive-pressure masks approved by NIOSH (National Institute for Occupational Safety and Health--approval formerly given by the U.S. Bureau of Mines) shall be provided for fire brigade, damage control, and control room personnel. At least 10 masks shall be available for fire brigade personnel. Control room personnel may be furnished breathing air by a manifold system piped from a storage reservoir if practical. Service or rated operating life shall be a minimum of one-half hour for the self-contained units.

At least a 1-hour supply of breathing air in extra bottles shall be located on the plant site for each unit of self-contained breathing appratus. In addition, an onsite 6-hour supply of reserve air shall be provided and arranged to permit quick and complete replenishment of exhausted air supply bottles as they are returned. If compressors are used as a source of breathing air, only units approved for breathing air shall be used and the compressors shall be operable assuming a loss of offsite power. Special care must be taken to locate the compressor in areas free of dust and contaminants.

I. Fire brigade training. The fire brigade training program shall ensure that the capability to fight potential fires is established and maintained. The program shall consist of an initial classroom instruction program followed by periodic classroom instruction, fire fighting practice, and fire drills:

1. Instruction

a. The initial classroom instruction shall include:

(1) Indoctrination of the plant fire fighting plan with specific identification of each individual's responsibilities.

(2) Identification of the type and location of fire hazards and associated types of fires that could occur in the plant.

(3) The toxic and corrosive characteristics of expected products of combustion.

(4) Identification of the location of fire fighting equipment for each fire area and familiarization with the layout of the plant, including access and egress routes to each area.

(5) The proper use of available fire fighting equipment and the correct method of fighting each type of fire. The types of fires covered should include fires in energized electrical equipment, fires in cables and cable trays, hydrogen fires, fires involving flammable and combustible liquids or hazardous process chemicals, fires resulting from construction or modifications (welding), and record file fires.

(6) The proper use of communication, lighting, ventilation, and emergency breathing equipment.

(7) The proper method for fighting fires inside buildings and confined spaces.

(8) The direction and coordination of the fire fighting activities (fire brigade leaders only).

(9) Detailed review of fire fighting strategies and procedures.

(10) Review of the latest plant modifications and corresponding changes in fire fighting plans.

Note: Items (9) and (10) may be deleted from the training of no more than two of the non-operations personnel who may be assigned to the fire brigade.

b. The instruction shall be provided by qualified individuals who are knowledgeable, experienced, and suitably trained in fighting the types of fires that could occur in the plant and in using the types of equipment available in the nuclear power plant.

c. Instruction shall be provided to all fire brigade members and fire brigade leaders.

d. Regular planned meetings shall be held at least every 3 months for all brigade members to review changes in the fire protection program and other subjects as necessary.

e. Periodic refresher training sessions shall be held to repeat the classroom instruction program for all brigade members over a two-year period. These sessions may be concurrent with the regular planned meetings.

2. Practice

Practice sessions shall be held for each shift fire brigade on the proper method of fighting the various types of fires that could occur in a nuclear power plant. These sessions shall provide brigade members with experience in actual fire extinguishment and the use of emergency breathing apparatus under strenuous conditions encountered in fire fighting. These practice sessions shall be provided at least once per year for each fire brigade member.

3. Drills

a. Fire brigade drills shall be performed in the plant so that the fire brigade can practice as a team.

b. Drills shall be performed at regular intervals not to exceed 3 months for each shift fire brigade. Each fire brigade member should participate in each drill, but must participate in at least two drills per year.

A sufficient number of these drills, but not less than one for each shift fire brigade per year, shall be unannounced to determine the fire fighting readiness of the plant fire brigade, brigade leader, and fire protection systems and equipment. Persons planning and authorizing an unannounced drill shall ensure that the responding shift fire brigade members are not aware that a drill is being planned until it is begun. Unannounced drills shall not be scheduled closer than four weeks.

At least one drill per year shall be performed on a "back shift" for each shift fire brigade.

c. The drills shall be preplanned to establish the training objectives of the drill and shall be critiqued to determine how well the training objectives have been met. Unannounced drills shall be planned and critiqued by members of the management staff responsible for plant safety and fire protection. Performance deficiencies of a fire brigade or of individual fire brigade members shall be remedied by scheduling additional training for the brigade or members. Unsatisfactory drill performance shall be followed by a repeat drill within 30 days.

d. At 3-year intervals, a randomly selected unannounced drill must be critiqued by qualified individuals independent of the licensee's staff. A copy of the written report from these individuals must be available for NRC review and shall be retained as a record as specified in section III.I.4 of this appendix.

e. Drills shall as a minimum include the following:

(1) Assessment of fire alarm effectiveness, time required to notify and assemble fire brigade, and selection, placement and use of equipment, and fire fighting strategies.

(2) Assessment of each brigade member's knowledge of his or her role in the fire fighting strategy for the area assumed to contain the fire. Assessment of the brigade member's conformance with established plant fire fighting procedures and use of fire fighting equipment, including self-contained emergency breathing apparatus, communication equipment, and ventilation equipment, to the extent practicable.

(3) The simulated use of fire fighting equipment required to cope with the situation and type of fire selected for the drill. The area and type of fire chosen for the drill should differ from those used in the previous drill so that brigade members are trained in fighting fires in various plant areas. The situation selected should simulate the size and arrangement of a fire that could reasonably occur in the area selected, allowing for fire development due to the time required to respond, to obtain equipment, and organize for the fire, assuming loss of automatic suppression capability.

(4) Assessment of brigade leader's direction of the fire fighting effort as to thoroughness, accuracy, and effectiveness.

4. Records

Individual records of training provided to each fire brigade member, including drill critiques, shall be maintained for at least 3 years to ensure that each member receives training in all parts of the training program. These records of training shall be available for NRC review. Retraining or broadened training for fire fighting within buildings shall be scheduled for all those brigade members whose performance records show deficiencies.

J. Emergency lighting. Emergency lighting units with at least an 8-hour battery power supply shall be provided in all areas needed for operation of safe shutdown equipment and in access and egress routes thereto.

K. Administrative controls. Administrative controls shall be established to minimize fire hazards in areas containing structures, systems, and components important to safety. These controls shall establish procedures to:

1. Govern the handling and limitation of the use of ordinary combustible materials, combustible and flammable gases and liquids, high efficiency particulate air and charcoal filters, dry ion exchange resins, or other combustible supplies in safety-related areas.

2. Prohibit the storage of combustibles in safety-related areas or establish designated storage areas with appropriate fire protection.

3. Govern the handling of and limit transient fire loads such as combustible and flammable liquids, wood and plastic products, or other combustible materials in buildings containing safety-related systems or equipment during all phases of operating, and especially during maintenance, modification, or refueling operations.

4. Designate the onsite staff member responsible for the inplant fire protection review of proposed work activities to identify potential transient fire hazards and specify required additional fire protection in the work activity procedure.

5. Govern the use of ignition sources by use of a flame permit system to control welding, flame cutting, brazing, or soldering operations. A separate permit shall be issued for each area where work is to be done. If work continues over more than one shift, the permit shall be valid for not more than 24 hours when the plant is operating or for the duration of a particular job during plant shutdown.

6. Control the removal from the area of all waste, debris, scrap, oil spills, or other combustibles resulting from the work activity immediately following completion of the activity, or at the end of each work shift, whichever comes first.

7. Maintain the periodic housekeeping inspections to ensure continued compliance with these administrative controls.

8. Control the use of specific combustibles in safety-related areas. All wood used in safety-related areas during maintenance, modification, or refueling operations (such as lay-down blocks or scaffolding) shall be treated with a flame retardant. Equipment or supplies (such as new fuel) shipped in untreated combustible packing containers may be unpacked in safety-related areas if required for valid operating reasons. However, all combustible materials shall be removed from the area immediately following the unpacking. Such transient combustible material, unless stored in approved containers, shall not be left unattended during lunch breaks, shift changes, or other similar periods. Loose combustible packing material such as wood or paper excelsior, or polyethylene sheeting shall be placed in metal containers with tight-fitting self-closing metal covers.

9. Control actions to be taken by an individual discovering a fire, for example, notification of control room, attempt to extinguish fire, and actuation of local fire suppression systems.

10. Control actions to be taken by the control room operator to determine the need for brigade assistance upon report of a fire or receipt of alarm on control room annunciator panel, for example, announcing location of fire over PA system, sounding

fire alarms, and notifying the shift supervisor and the fire brigade leader of the type, size, and location of the fire.

11. Control actions to be taken by the fire brigade after notification by the control room operator of a fire, for example, assembling in a designated location, receiving directions from the fire brigade leader, and discharging specific fire fighting responsibilities including selection and transportation of fire fighting equipment to fire location, selection of protective equipment, operating instructions for use of fire suppression systems, and use of preplanned strategies for fighting fires in specific areas.

12. Define the strategies for fighting fires in all safety-related areas and areas presenting a hazard to safety-related equipment. These strategies shall designate:

a. Fire hazards in each area covered by the specific prefire plans.

b. Fire extinguishants best suited for controlling the fires associated with the fire hazards in that area and the nearest location of these extinguishants.

c. Most favorable direction from which to attack a fire in each area in view of the ventilation direction, access hallways, stairs, and doors that are most likely to be free of fire, and the best station or elevation for fighting the fire. All access and egress routes that involve locked doors should be specifically identified in the procedure with the appropriate precautions and methods for access specified.

d. Plant systems that should be managed to reduce the damage potential during a local fire and the location of local and remote controls for such management (e.g., any hydraulic or electrical systems in the zone covered by the specific fire fighting procedure that could increase the hazards in the area because of overpressurization or electrical hazards).

e. Vital heat-sensitive system components that need to be kept cool while fighting a local fire. Particularly hazardous combustibles that need cooling should be designated.

f. Organization of fire fighting brigades and the assignment of special duties according to job title so that all fire fighting functions are covered by any complete shift personnel complement. These duties include command control of the brigade, transporting fire suppression and support equipment to the fire scenes, applying the extinguishant to the fire, communication with the control room, and coordination with outside fire departments.

g. Potential radiological and toxic hazards in fire zones.

h. Ventilation system operation that ensures desired plant air distribution when the ventilation flow is modified for fire containment or smoke clearing operations.

i. Operations requiring control room and shift engineer coordination or authorization.

j. Instructions for plant operators and general plant personnel during fire.

L. Alternative and dedicated shutdown capability. 1. Alternative or dedicated shutdown capability provided for a specific fire area shall be able to (a) achieve and maintain subcritical reactivity conditions in the reactor; (b) maintain reactor coolant inventory; (c) achieve and maintain hot standby2 conditions for a PWR (hot shutdown2 for a BWR); (d) achieve cold shutdown conditions within 72 hours; and (e) maintain cold shutdown conditions thereafter. During the postfire shutdown, the reactor coolant system process variables shall be maintained within those predicted for a loss of normal a.c. power, and the fission product boundary integrity shall not be affected; i.e., there shall be no fuel clad damage, rupture of any primary coolant boundary, of rupture of the containment boundary.

2. The performance goals for the shutdown functions shall be:

a. The reactivity control function shall be capable of achieving and maintaining cold shutdown reactivity conditions.

b. The reactor coolant makeup function shall be capable of maintaining the reactor coolant level above the top of the core for BWRs and be within the level indication in the pressurizer for PWRs.

c. The reactor heat removal function shall be capable of achieving and maintaining decay heat removal.

d. The process monitoring function shall be capable of providing direct readings of the process variables necessary to perform and control the above functions.

e. The supporting functions shall be capable of providing the process cooling, lubrication, etc., necessary to permit the operation of the equipment used for safe shutdown functions.

3. The shutdown capability for specific fire areas may be unique for each such area, or it may be one unique combination of systems for all such areas. In either case, the alternative shutdown capability shall be independent of the specific fire area(s) and shall accommodate postfire conditions where offsite power is available and where offsite power is not available for 72 hours. Procedures shall be in effect to implement this capability.

4. If the capability to achieve and maintain cold shutdown will not be available because of fire damage, the equipment and systems comprising the means to achieve and maintain the hot standby or hot shutdown condition shall be capable of maintaining such conditions until cold shutdown can be achieved. If such equipment and systems will not be capable of being powered by both onsite and offsite electric power systems because of fire damage, an independent onsite power system shall be provided. The number of operating shift personnel, exclusive of fire brigade members, required to operate such equipment and systems shall be on site at all times.

5. Equipment and systems comprising the means to achieve and maintain cold shutdown conditions shall not be damaged by fire; or the fire damage to such equipment and systems shall be limited so that the systems can be made operable and cold shutdown can be achieved within 72 hours. Materials for such repairs shall be readily available on site and procedures shall be in effect to implement such repairs. If such equipment and systems used prior to 72 hours after the fire will not be capable of being powered by both onsite and offsite electric power systems because of fire damage, an independent onsite power system shall be provided. Equipment and systems used after 72 hours may be powered by offsite power only.

6. Shutdown systems installed to ensure postfire shutdown capability need not be designed to meet seismic Category I criteria, single failure criteria, or other design basis accident criteria, except where required for other reasons, e.g., because of interface with or impact on existing safety systems, or because of adverse valve actions due to fire damage.

7. The safe shutdown equipment and systems for each fire area shall be known to be isolated from associated non-safety circuits in the fire area so that hot shorts, open circuits, or shorts to ground in the associated circuits will not prevent operation of the safe shutdown equipment. The separation and barriers between trays and conduits containing associated circuits of one safe shutdown division and trays and conduits containing associated circuits or

safe shutdown cables from the redundant division, or the isolation of these associated circuits from the safe shutdown equipment, shall be such that a postulated fire involving associated circuits will not prevent safe shutdown.3

M. Fire barrier cable penetration seal qualification. Penetration seal designs must be qualified by tests that are comparable to tests used to rate fire barriers. The acceptance criteria for the test must include the following:

1. The cable fire barrier penetration seal has withstood the fire endurance test without passage of flame or ignition of cables on the unexposed side for a period of time equivalent to the fire resistance rating required of the barrier;

2. The temperature levels recorded for the unexposed side are analyzed and demonstrate that the maximum temperature is sufficiently below the cable insulation ignition temperature; and

3. The fire barrier penetration seal remains intact and does not allow projection of water beyond the unexposed surface during the hose stream test.

N. Fire doors. Fire doors shall be self-closing or provided with closing mechanisms and shall be inspected semiannually to verify that automatic hold-open, release, and closing mechanisms and latches are operable.

One of the following measures shall be provided to ensure they will protect the opening as required in case of fire:

1. Fire doors shall be kept closed and electrically supervised at a continuously manned location;

2. Fire doors shall be locked closed and inspected weekly to verify that the doors are in the closed position;

3. Fire doors shall be provided with automatic hold-open and release mechanisms and inspected daily to verify that doorways are free of obstructions; or

4. Fire doors shall be kept closed and inspected daily to verify that they are in the closed position.

The fire brigade leader shall have ready access to keys for any locked fire doors.

Areas protected by automatic total flooding gas suppression systems shall have electrically supervised self-closing fire doors or shall satisfy option 1 above.

O. Oil collection system for reactor coolant pump. The reactor coolant pump shall be equipped with an oil collection system if the containment is not inerted during normal operation. The oil collection system shall be so designed, engineered, and installed that failure will not lead to fire during normal or design basis accident conditions and that there is reasonable assurance that the system will withstand the Safe Shutdown Earthquake.4

Such collection systems shall be capable of collecting lube oil from all potential pressurized and unpressurized leakage sites in the reactor coolant pump lube oil systems. Leakage shall be collected and drained to a vented closed container that can hold the entire lube oil system inventory. A flame arrester is required in the vent if the flash point characteristics of the oil present the hazard of fire flashback. Leakage points to be protected shall include lift pump and piping, overflow lines, lube oil cooler, oil fill and drain lines and plugs, flanged connections on oil lines, and lube oil reservoirs where such features exist on the reactor coolant pumps. The drain line shall be large enough to accommodate the largest potential oil leak.

(45 FR 76611, Nov. 19, 1980; 46 FR 44735, Sept. 8, 1981, as amended at 53 FR 19251, May 27, 1988; 65 FR 38191, June

20, 2000)

1 Alternative shutdown capability is provided by rerouting, relocating, or modifying existing systems; dedicated shutdown capability is provided by installing new structures and systems for the function of post-fire shutdown.

2 As defined in the Standard Technical Specifications.

3 An acceptable method of complying with this alternative would be to meet Regulatory Guide 1.75 position 4 related to associated circuits and IEEE Std 384-1974 (Section 4.5) where trays from redundant safety divisions are so protected that postulated fires affect trays from only one safety division.

4 See Regulatory Guide 1.29--"Seismic Design Classification" paragraph C.2.

# Appendix F  Summary of GL 06-03 Responses

## Compiled Listing of ERFBS used in NPP as of December 21, 2007

The following list has been compiled from feedback obtained through GL 06-03. This list differs from the table presented in Table 1.1 beginning on page 1-2 of this report. Table 1.1. is based on docketed information for licensees and feedback received during public comment. Table F.1 has been generated from docketed responses to GL 06-03. Both tables should be used for reference purposes only and do not constitute licensing bases.

### Table F.1. Summary of GL 06-03 Responses

| Plant Name | ADAMS Accession No. | | Type(s) of Barrier |
|---|---|---|---|
| | Licensee Response | NRC Closeout | |
| Arkansas Nuclear One Units, 1 and 2 | ML061720459 | ML062620115 | Hemyc, Thermo-Lag, Versa Wrap |
| Beaver Valley Power Station, Units 1 and 2 | ML061710429 ML070370315 | ML070680131 | Thermo-Lag, 3M Interam, Darmatt |
| Braidwood Station, Units, 1 and 2 | ML061640343 ML071520085 | ML071700766 | 3M Interam |
| Browns Ferry Nuclear Plant Units, 1, 2, and 3 | ML061600208 | ML070250411 | Thermo-Lag |
| Brunswick Steam Electric Plant Units, 1 and 2 | ML061640386 | ML071580106 | 3M Interam |
| Byron Station, Units, 1 and 2 | ML061640343 | ML071000347 | Darmatt |
| Callaway Plant, Unit 1 | ML061570382 ML062060383 | ML062680005 | Darmatt |
| Calvert Cliffs Nuclear Power Plant Units, 1 and 2 | ML061650026 | ML070880103 | None |
| Catawba Nuclear Station, Units, 1 and 2 | ML061640310 | ML071430127 | Hemyc |

Table F.1. Summary of GL 06-03 Responses (Continued)

| Plant Name | ADAMS Accession No. | | Type(s) of Barrier |
|---|---|---|---|
| | Licensee Response | NRC Closeout | |
| Clinton Power Station, Unit 1 | ML061640343 ML071520085 | ML071700766 | Thermo-Lag, 3M Interam |
| Columbia Generating Station | ML061710470 | ML062850088 | 3M Interam, Darmatt |
| Comanche Peak Steam Electric Station, Units, 1 and 2 | ML061660092 | ML071230006 | Hemyc, Thermo-Lag |
| Cooper Nuclear Station | ML061530275 | ML061650200 | None |
| Crystal River Unit 3 Nuclear Generating Plant | ML061570390 | ML071580594 | Thermo-Lag, Mecatiss |
| Davis Besse Nuclear Power Station, Unit 1 | ML061710429 ML070370315 | ML070680131 | 3M Interam |
| Diablo Canyon Nuclear Power Plant, Units, 1 and 2 | ML061720079 | ML063390066 | 3M Interam, Pyrocrete |
| Donald C. Cook Nuclear Power Plant Units, 1 and 2 | ML061600213 | ML070180221 | Thermo-Lag, Darmatt, Mecatiss |
| Dresden Nuclear Power Station, Units, 2 and 3 | ML061640343 | ML063000065 ML071360223 | 3M Interam |
| Duane Arnold Energy Center | ML061640269 | ML070860462 | Darmatt |
| Edwin I. Hatch Nuclear Plant | ML061600376 ML063330230 | ML071280045 | 3M Interam, Promat |
| Fermi-2 | ML061660087 ML070580135 | ML070860419 | 3M Interam |
| James A. Fitzpatrick Nuclear Power Plant | ML061650025 | ML062960164 | Hemyc, FP-60 |

Table F.1. Summary of GL 06-03 Responses (Continued)

| Plant Name | ADAMS Accession No. | | Type(s) of Barrier |
|---|---|---|---|
| | Licensee Response | NRC Closeout | |
| Fort Calhoun Station, Unit 1 | ML061530476 ML070850193 | ML071060295 | 3M Interam, Pyrocrete, Pabco |
| R. E Ginna Nuclear Power Plant | ML061650026 | ML070940337 | Hemyc, MT |
| Grand Gulf Nuclear Station | ML061570135 | ML061650383 | Thermo-Lag, 3M Interam |
| Shearon Harris Nuclear Power Plant, Unit 1 | ML061240052 ML061710062 | ML062900541 | Hemyc, MT, Thermo-Lag, 3M Interam |
| Edwin I. Hatch Nuclear Plant | ML061600376 ML072060088 | ML072180188 | Promat |
| Hope Creek Generating Station | ML061660080 | ML061810011 | None |
| Indian Point Nuclear Generating, Units, 2 and 3 | ML061720091 | ML073320029 | Hemyc, 3M Interam |
| Kewaunee Power Station | ML061590505 ML071520515 | ML072500079 | 3M Interam |
| LaSalle County Station, Unit 1 | ML061640343 | ML062300114 ML071360223 | Darmatt, Kaowool |
| LaSalle County Station, Unit 2 | ML061640343 | ML062300114 ML071360223 | Darmatt, Kaowool |
| Limerick Generating Station, Units, 1 and 2 | ML061640343 | ML071000347 | Thermo-Lag, Darmatt |
| McGuire Nuclear Station, Units, 1 and 2 | ML061640310 | ML071430162 | Hemyc |
| Millstone Power Station, Units, 2 and 3 | ML061590505 ML071520515 | ML073060163 | None |
| Monticello Nuclear Generating Plant, Unit 1 | ML061600209 | ML061810437 | None |

Table F.1. Summary of GL 06-03 Responses (Continued)

| Plant Name | ADAMS Accession No. | | Type(s) of Barrier |
| --- | --- | --- | --- |
| | Licensee Response | NRC Closeout | |
| Nine Mile Point Nuclear Station, Units, 1 and 2 | ML061650026 | ML070880123 | None |
| North Anna Power Station, Units, 1 and 2 | ML061590505 ML071520515 | ML071910366 | 3M Interam |
| Oconee Nuclear Station, Units, 1, 2, and 3 | ML061640310 | ML061650421 | None |
| Oyster Creek Nuclear Generating Station | ML061640343 | ML071000347 | Thermo-Lag, Mecatiss |
| Palisades Nuclear Plant | ML061600209 | ML070660064 | Concrete |
| Palo Verde Nuclear Generating Station, Units, 1, 2, and 3 | ML061650261 | ML063540027 | Thermo-Lag |
| Peach Bottom Atomic Power Station, Units, 2 and 3 | ML061640343 | ML071000347 | Thermo-Lag, Darmatt |
| Perry Nuclear Power Plant, Unit 1 | ML061710429 ML070370315 | ML070680131 | 3M Interam |
| Pilgrim Nuclear Power Station | ML061640132 | ML063620110 | 3M Interam, Mecatiss |
| Point Beach Nuclear Plant, Units, 1 and 2 | ML061600209 ML062550167 | ML061640009 | 3M Interam |
| Prairie Island Nuclear Generating Plant, Units, 1 and 2 | ML061600209 | ML062050077 | 3M Interam, Darmatt |
| Quad Cities Nuclear Power Station, Units, 1 and 2 | ML061640343 ML071630310 | ML071700766 | 3M Interam, Darmatt, Versa Wrap |
| River Bend Station, Unit 1 | ML061570394 ML061670210 | ML061650386 | Thermo-Lag |
| H. B. Robinson Steam Electric Plant, Unit 2 | ML061640136 ML072250063 | ML071070583 | 3M Interam |

Table F.1. Summary of GL 06-03 Responses (Continued)

| Plant Name | ADAMS Accession No. Licensee Response | NRC Closeout | Type(s) of Barrier |
|---|---|---|---|
| St. Lucie Plant, Units, 1 and 2 | ML061640269 ML062680162 | ML063070029 | Hemyc, Thermo-Lag, Mecatiss |
| Salem Nuclear Generating Station, Units, 1 and 2 | ML061660091 | ML061810077 | 3M Interam |
| San Onofre Nuclear Generating Station, Units, 2 and 3 | ML061590310 ML071710548 | ML071920538 ML072770906 | 3M Interam, Cerablanket |
| Seabrook Station, Unit 1 | ML061640269 ML071990101 | ML072010149 | 3M Interam |
| Sequoyah Nuclear Plant, Units, 1 and 2 | ML061600208 | ML070250184 | Thermo-Lag |
| South Texas Project, Units, 1 and 2 | ML061510352 | ML071130024 | Thermo-Lag |
| Virgil C. Summer Nuclear Station, Unit 1 | ML061590311 ML062220348 | ML061660200 | 3M Interam, Kaowool |
| Surry Power Station, Units, 1 and 2 | ML061590505 ML071520515 | ML071910366 | Pyrocrete |
| Susquehanna Steam Electric Station, Units 1 and 2 | ML061660076 | ML062160010 | Thermo-Lag, Darmatt |
| Three Mile Island Nuclear Station, Unit 1 | ML061640343 | ML061810093 | Thermo-Lag, Mecatiss |
| Turkey Point Nuclear Generating, Units, 3 and 4 | ML061640269 | ML062910197 | Thermo-Lag |
| Vermont Yankee Nuclear Power Station | ML061630231 | ML063620129 | 3M Interam |
| Vogtle Electric Generating Plant, Units, 1 and 2 | ML061600376 | ML063490324 | 3M Interam, Cementitious material |
| Waterford Steam Electric Station, Unit 3 | ML061600210 | ML062300315 | Hemyc, 3M Interam |
| Watts Bar Nuclear Plant, Unit 1 | ML061600208 | ML070250345 | Thermo-Lag |

Table F.1. Summary of GL 06-03 Responses (Continued)

| Plant Name | ADAMS Accession No. | | Type(s) of Barrier |
| --- | --- | --- | --- |
| | Licensee Response | NRC Closeout | |
| Wolf Creek Generating Station, Unit 1 | ML061570375 | ML061650179 | Thermo-Lag, Darmatt |

# Appendix G    Additional Information on ERFBS Acceptance Criteria

## G.1 UL Subject 1724

UL Subject 1724, "Outline of Investigations for Fire Tests for Electrical Circuit Protective Systems," is an acceptable method of qualifying ERFBS provided the cable qualification testing of UL 1724 Appendix B and Generic Letter 86-10 Supplement 1 is performed.

Appendix B to UL Subject 1724 provides a method acceptable to NRC to determine circuit integrity of insulated electrical cables protected with ERFBS. This method evaluates the circuit integrity independent of use of an ERFBS. The method consists of exposing unprotected cable samples to elevated temperatures in a circulating air oven. The exposure temperatures are based on fire endurance test temperature data collected on a bare # 8 American Wire Gauge (AWG) conductor protected in a raceway by an ERFBS (data from separate test).

The cables under evaluation are arranged in a cable raceway (i.e., conduit or ladder-backed or solid cable tray) along with a bare #8 AWG conductor that is used to monitor and control the air oven temperature. All conductors are energized and monitored for electrical circuit faults (1) between individual conductors in a multiconductor cable, (2) between adjacent individual conductors (cables), and (3) between the electrical conductors and ground or raceway. The air oven exposes the cables to the thermal environment experienced within an ERFBS. The testing is conducted until the air oven temperature reaches the maximum interior ERFBS endurance test temperature or when a circuit fault occurs.

During the test, the cables are under constant compression loading to simulate the maximum allowable fill of insulated electrical cable. In addition, the test assembly is subjected to an impact test representative of the impact force and frequency of impacts that could be encountered by the raceway from falling material (e.g., ceiling) during a fire. Circuit integrity is monitored during these impact tests.

Appendix B is typically used when the ERFBS fire endurance testing temperature rise acceptance criteria were not met. UL Subject 1724 provides one method to demonstrate the functionality of the electrical cables protected with an ERFBS exposed to elevated temperatures.

## G.2 NRC Acceptance Criteria

Supplement 1 to GL 86-10, "Fire Endurance Test Acceptance Criteria for Fire Barrier Systems Used to Separate Redundant Safety Shutdown Trains within the Same Fire Area," and RG 1.189, "Fire Protection for NPPs," provide guidance related to the criteria found acceptable to NRC for qualifying ERFBSs. It should be understood that these guidance documents only provide one particular method that is acceptable to NRC; however, other acceptable methods exist such as those used by TVA to license Watts Bar Unit 1 in 1995.

Based on past reviews, NRC staff acceptance is based on the barriers performance in the following areas:

- Fire Endurance
  - Test Specimen Construction
  - Hose Stream Test
  - Cable Functionality
- Combustibility
- Ampacity Derating
- Seismic Qualification

*G.2.1 Acceptance Criteria – ERFBS Fire Endurance Test*

NRC considers the fire endurance qualification test for fire barrier materials applied directly to a raceway or component to be successful when exposed to a standard time-temperature fire endurance exposure if the following conditions are met:

- The average unexposed side temperature of the fire barrier system, as measured on the exterior surface of the raceway or component, did not exceed 139°C (250°F) above its initial temperature.

- Irrespective of the unexposed side temperature rise during the fire test, a visual inspection should be performed if cables or components are included in the fire barrier test specimen. Cables should not show signs of degraded conditions resulting from the thermal affects of the fire exposure.

- The cable tray, raceway, or component fire barrier system remained intact during the fire exposure and water hose stream test without developing any openings through which the cable tray, raceway, or component (e.g., cables) is visible.

NFPA 251 and ASTM E-119 allow the temperature criteria to be determined by averaging thermocouple temperature readings. For the purposes of the first criterion, thermocouple averaging can be used provided similar series of thermocouples (e.g., cable tray side rail) are averaged together to determine temperature performance of the raceway fire barrier system. In addition, conditions of acceptance are placed on the temperatures measured by a single thermocouple. If any single thermocouple exceeds 30 percent of the maximum allowable temperature rise (i.e., 139°C + 42°C = 181°C (250°F + 75°F = 325°F)), the test exceeded the temperature criteria limit. This is a single point failure and often indicates a joint failure in an otherwise acceptable ERFBS.

Because of the poor thermal conductivity of cable jacket and insulation material, measuring cable temperatures is not considered a reliable means for determining excessive temperature conditions that may occur at any point along the length of the cable during the fire test. In lieu of measuring the unexposed surface temperature of the fire barrier test specimen, methods that will adequately measure the surface temperature of the raceway (e.g., exterior of the conduit, side rails of cable trays, bottom and top of cable tray surfaces, junction box external surfaces) can be considered as equivalent if the raceway components used to construct the fire test specimen represent plant-specific components and configuration. The metal surfaces of the raceway, under fire test conditions, exhibit good thermal conductivity properties. Temperatures measured on these surfaces provide a reliable indication of the actual temperature rise within the fire barrier system.

The basic premise of NRC fire resistance criteria is that fire barriers that do not exceed the maximum allowable temperature rise and pass the hose stream test to provide adequate assurance that the shutdown capability is protected without further analyses. If the temperature criteria are exceeded, sufficient additional information is needed to perform an engineering evaluation to demonstrate that the shutdown capability is protected.

The following are acceptable placement of thermocouple for determining the thermal performance of raceway or cable tray fire barrier systems that contain cables during the fire exposure:

- Conduits – Unexposed surface of the fire barrier system should be measured by placing thermocouples every 152 mm (6-in) on the exterior conduit surface underneath the fire barrier material. The thermocouples should be attached to the exterior conduit surface located opposite the test deck and closest to the furnace fire source. Thermocouples should also be placed immediately adjacent to all structural members, supports, and barrier penetrations.

- Cable Trays – The temperature rise on the unexposed surface of a fire barrier system installed on a cable tray should be measured by placing the thermocouples on the exterior surface of the tray side rails between the cable tray side rail and the fire barrier material. In addition to placing thermocouples on the side rails, thermocouples should be attached to two AWG 8 stranded bare copper conductors. The first copper conductor should be installed on the bottom of the cable tray rungs along the entire length and down the longitudinal center of the cable tray run. The second conductor should be installed along the outer top surface of the cables closest to the top and toward the center of the fire barrier. The bare copper wire is more responsive than cable jackets to temperature rise within the fire barrier enclosure. The temperature changes measured along the bare copper conductors provide indication of joint failure or material burn through conditions. Thermocouples should be placed every 152 mm (6-in) down the longitudinal center along the outside surface of the cable tray side rails and along the bare copper conductors. Thermocouples also should be placed immediately adjacent to all structural members, supports, and barrier penetrations.

- Junction Boxes – The temperature rise on the unexposed surface of a fire barrier system installed on junction boxes should be measured by placing thermocouples on either the inside or the outside of each junction box (JB) surface. Each JB surface or face should have a minimum of one thermocouple located at its geometric center. In addition, one thermocouple should be installed for every 1 square foot of JB surface area. These thermocouples should be located at the geometric centers of the 1-square-foot areas. At least one thermocouple should also be placed within 25 mm (1-in) of each penetration connector/interface.

- Airdrops – The internal airdrop temperatures should be measured by thermocouples placed every 305 mm (12-in) on the cables routed within the air drop and by a stranded AWG 8 bare copper conductor routed inside and along the entire length of the airdrop system with thermocouples installed every 152 mm (6-in) along the length of the copper conductor. The copper conductor should be in close proximity with the unexposed surface of the fire barrier material. Thermocouples also should be placed immediately adjacent to all supports and barrier penetrations.

With the exception of airdrops, the installation of thermocouples on cables is optional and is left to the discretion of the licensee, test sponsor, or test laboratory. Cable thermocouples are to be used for engineering purposes only. Cable thermocouples alone are not acceptable for the demonstration of fire barrier performance. However, cable thermocouples may support fire barrier deviation conditions.

Regarding the second criteria, examples of thermal cable degradation include:

- Jacket swelling,
- Splitting,
- Cracking,
- Blistering,
- Melting, or discoloration,
- Exposed shield,
- Jacket hardening;
- Exposed conductor insulation, degraded, or discolored; and
- Exposed bare copper conductor.

For those cases where signs of thermal degradation are present, the fire barrier did not perform its intended fire-resistive function. For these barriers, a deviation based on demonstrating that the functionality of thermally degraded cables or component was maintained and that the cables or component would have adequately performed their intended function during and after a postulated fire exposure may be acceptable. Refer to Section G.2.4 for more information on cable functionality testing.

When evaluating the test results to configurations installed in the plant, the installed ERFBS configuration can be considered to be bounded by a tested ERFBS configuration only if the physical configuration (dimensions of the raceway, number of layers, interfering items, protection of supports, etc.) are the same as the tested configuration and the weight of the raceway, including cables, equals or exceeds the weight of the tested configuration. For example, a 4-inch conduit with a raceway total mass (including conduit and cables) of 16.5 pounds (7.5 kilograms) would bound a 4-inch conduit with a raceway total mass of 9.1 kilograms (20 pounds). However, 4-inch conduit with a total raceway of 6.8 kilograms (15 pounds) would be considered indeterminate when using this test data. Explained differently, for a plant configuration to be bounded by a qualified test configuration, the plant configuration must be of the same physical construction and have a thermal mass equivalent or greater than the tested configuration.

*G.2.2 Acceptance Criteria – Test Specimen Construction*

In addition to the above criteria, GL 86-10 Supplement 1 provides guidance on acceptable methods for conducting the fire endurance test, including:

- Raceway fire barrier system construction should be representative of the end use
  - (i.e., if raceway supports are not protected in actual plant applications, then they should not be protected in test),
- Test program should encompass or bound raceway sizes and the various configurations for those fire barrier systems installed in the plant, and
- Tests should be conducted without cables.

Supplement 1 to GL 86-10 also provides guidance on acceptable placement of thermocouples, which includes placing a bare single conductor 8 AWG cable, instrumented every 152 mm (6-in) with thermocouples. The bare copper wire is more responsive than cable jackets to temperature rise within the fire barrier enclosure. Where exact replication of a tested configuration cannot be achieved, the field installation should meet all of the following criteria:

- The continuity of the fire barrier material is maintained.
- The thickness of the barrier is maintained.
- The nature of the support assemblies is unchanged from the tested configuration.
- The application or "end use" of the fire barrier is unchanged from the tested configuration. For example, the use of cable tray barriers to protect a cable tray which differs in configuration from those that were tested would be acceptable. However, the use of structural steel fire proofing to protect a cable tray assembly may not be acceptable.
- The configuration has been reviewed by a qualified fire protection engineer and found to provide an equivalent level of protection.

Although cables may be placed within the raceway, NRC has determined that measuring cable temperature is not a reliable means for determining excessive temperature conditions that may occur at any point along the length of the cable during the fire test. Monitoring cable temperature as the primary method of determining cable tray or raceway fire barrier performance is a nonconservative approach. The additional thermal mass added by the cables may cause the internal fire barrier temperature rise conditions to be masked. As stated in the acceptance criteria above, temperature monitored on the exterior surface of the raceway provides a more representative indication of fire barrier performance. The metal surfaces of the raceway, under fire test conditions, exhibit good thermal conductivity properties. As such, temperatures measured on these surfaces provide an indication of the actual temperature rise within the fire barrier system. The following provides a technical basis for the effects of cable mass on ERFBS thermal performance, based on an evaluation of Thermo-Lag fire test results.

G.2.2.1 Effects of Cable Mass in the Thermal Performance of ERFBS

As a part of the TVA Thermo-Lag 330-1 ERFBS program, Salley and Brown investigated the effects of cable fill on the thermal performance of the ERFBS. The following is their description of the cable tray testing and resultant relationships.

The first three Phase II fire tests of the joint TVA/TSI program were dedicated to cable tray configurations. TVA test 6.1.7 "Fire Endurance Test of Thermo-Lag 330-1 Fire Protective Envelopes (Three 18 in. Cable Trays and a 3 in. Conduit)" (41) consisted of three 18-inch wide, ladder back, steel cable trays with identical upgraded ERFBS and varying cable fill. The left tray in the test deck represented a maximum filled tray (i.e., 289 4/C #16 AWG (69.36 lbs. cable/linear ft)). The center tray in the test deck was filled with a single layer of cables (i.e., 26 4/C #16 AWG (6.24 Lbs. cable/linear ft)). The protected cable trays were constructed by the same installers and subjected to the same test fire to reduce as many variables as possible. The right tray in the test deck represented an empty tray i.e., no cables). The results of the fire test are shown in Figure G-1.

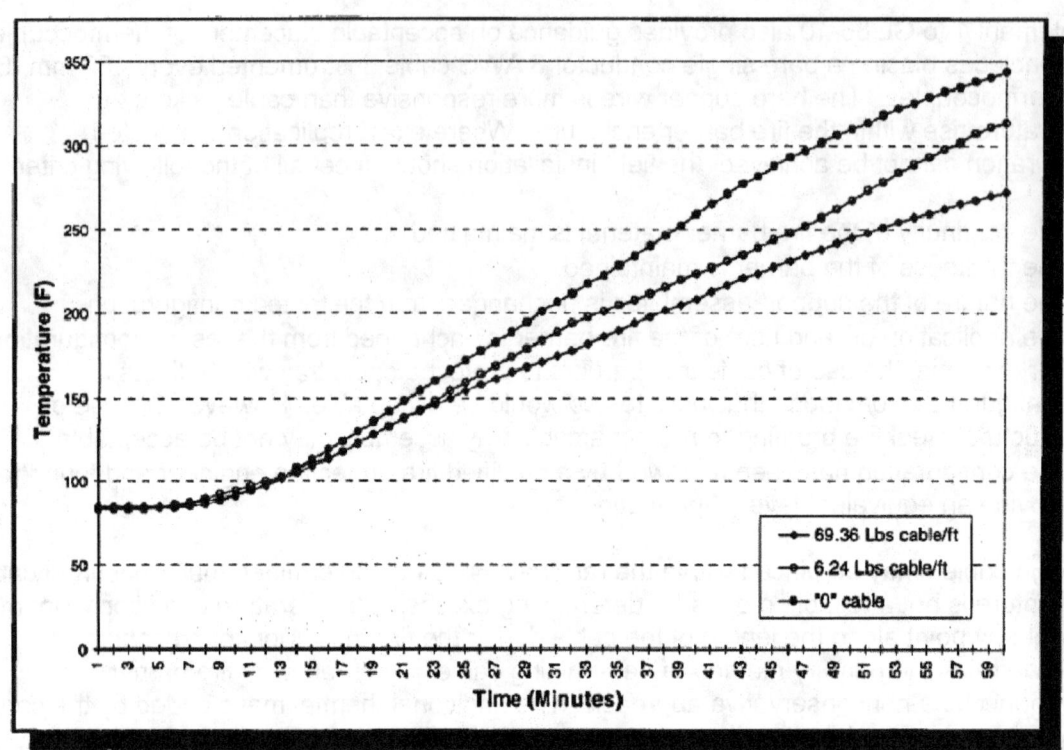

Figure G-1. Effects of Cable Mass on Cable Tray ERFBS Performance
(M.H. Salley Thesis UMd 2000)

At the end of the test, the only thermocouples to exceed the acceptance temperature were those on the instrumented bare #8 AWG copper cable inside the empty tray. This occurred at 56 minutes into the 1-hour test. The ambient temperature at the start of the test was 28 °C (83 °F) which dictated a maximum average temperature of 167 °C (333 °F) at 60 minutes (ambient temperature plus 121 °C (250 °F) allowable rise). By plotting the weight of each cable tray system (i.e., the weight of the tray and cables and not including the weight of the Thermo-Lag 330-1 ERFBS that was about approximately the same for each tray) versus its temperature at 60 minutes, an expression for the effects of cable mass can be developed (lumped heat formation).

$$\dot{q}''A = mC_p \partial T / \partial t \quad \text{(Equation G-1)}$$

$$\partial T / \partial t = \dot{q}''A / mC_p$$

$$T - T_\infty = \dot{q}''A / mC_p (t)$$

$$T = T_\infty + \dot{q}''A / mC_p (t) \quad \text{(Equation G-2)}$$

where:

m = mass of raceway ($m_r$) + mass of cable ($m_c$)
$\dot{q}''$ = rate of heat transfer
A = area

$C_p$ = specific heat

The correlation of the curves could be further defined as follows:

$$\dot{q}''A = (m_r C_{p,r} + m_c C_{p,c}) \partial T / \partial t \qquad \text{(Equation G-3)}$$

$$T = T_\infty + \dot{q}''A / (m_r C_{p,r} + m_c C_{p,c})t \qquad \text{(Equation G-4)}$$

The test laboratory, OPL, developed an exact equation using a computer model based on a "best fit" curve approach with a logarithmic relationship from the data shown in Figure G-2 (i.e., results of linear regression, method of least squares).[20]

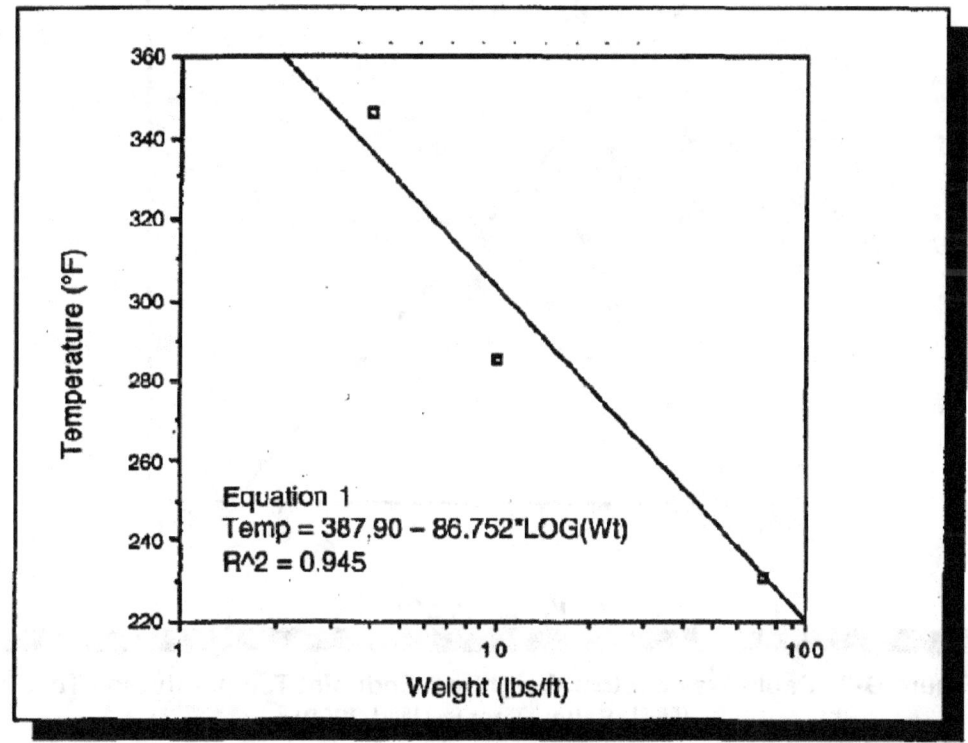

Figure G-2. Cable Tray System Weight vs. Endpoint Temperatures - Test 1
(M.H. Salley Thesis UMd 2000)

The equation is:

$$\text{Final Temp} = 387.9 - 86.75 * \text{Log (Weight)} \qquad \text{(Equation G-5)}$$

where:
Final Temp. = Degrees Fahrenheit[21]

---

[20] The constants developed for these equations were originally based on the British Units and no International System of Units (SI) conversions were performed.

[21] The ambient starting temperature of 83 °F must be used for the equation to be valid. The figure has

Weight = Lbs/ft of cable tray and cables

This equation is valid for 18-inch-wide cable trays protected with the TVA-designed Thermo-Lag 330-1 ERFBS having cable fills ranging from 6.24 lbs/ft up thru 69.36 lbs/ft. Further review was performed on the results of the single-layer-filled cable tray (6.24 lbs/ft of cable) and the empty cable tray (0.0 lbs/ft of cable). This was determined to be necessary because the effects of adding cables over the first layer becomes less important because of the cable insulation slowing the heat transfer to the copper conductors. Figure G-3 shows this data.

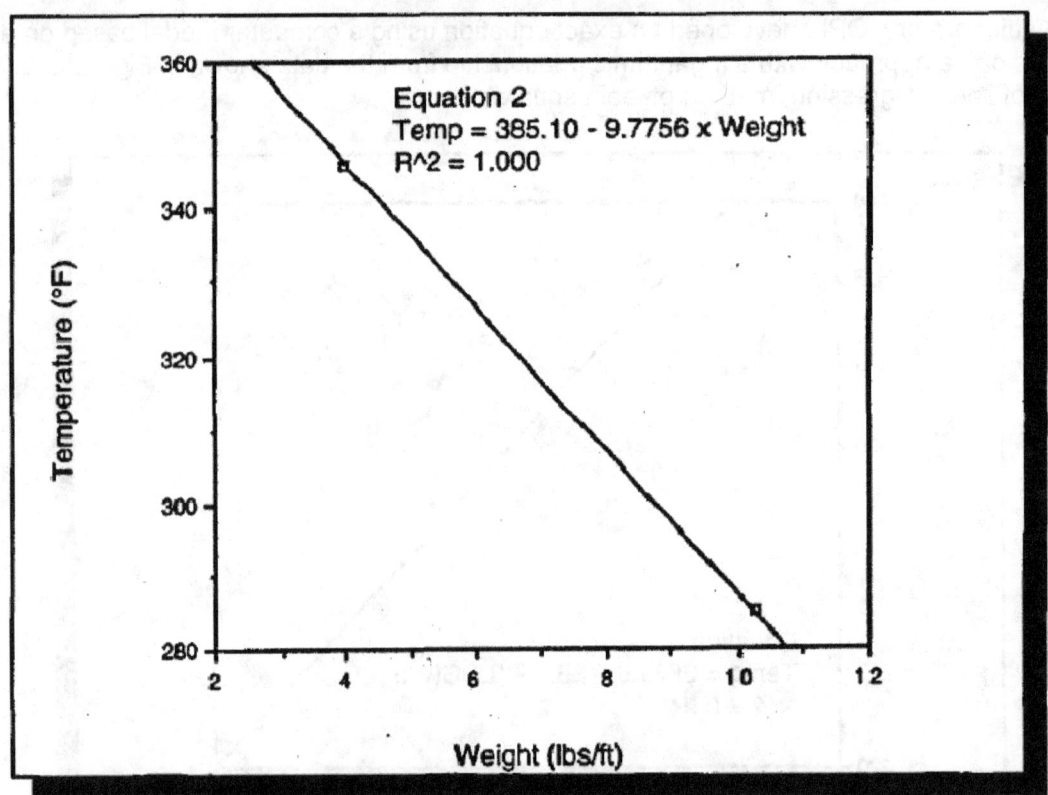

Figure G-3. Cable Tray System Weight vs. Endpoint Temperatures - Test 2
(M.H. Salley Thesis UMd 2000)

Conservatively, a plot was constructed of the temperatures for the empty cable tray (0.0 lbs/ft of cable) and the single-layer cable tray (6.24 lbs/ft of cable). The resulting linear equation given below conservatively predicts the system's thermal response at low cable fills (i.e., less than 6.24 lbs/ft of cable).

$$\text{Final Temp.} = 385.10 - 9.7756 * (w) \qquad \text{(Equation G-6)}[22]$$

where:

Final Temp. = Degree Fahrenheit[23]

---

been simplified (i.e., the ambient temperature subtracted) to graphically show the allowable temperature rise (i.e., $\Delta T = 250°F$).

[22] The constants developed for these equations were originally based on the British Units and no International System of Units (SI) conversions were performed.

w = total weight of cable tray and cables (Lbs/ft)

Solving this linear equation in the range of acceptable temperatures indicates that a cable tray system with a weight of 5.33 lbs/ft would maintain acceptable temperatures for 60 minutes. Subtracting the weight of the cable tray (4.00 lbs/ft) from the system yields a cable loading of 1.33 lbs/ft. Based on the cables used in the test (4C #16 AWG = 0.24 lbs/ft), a minimum of six cables are needed to produce acceptable temperatures (i.e., $\Delta T < 250$ °F).

The TVA has also performed similar tests on Thermo-Lag 330-1 ERFBS for aluminum conduits as installed at SQN (42). The testing consisted of three, 76.2 mm (3 in.) diameter aluminum conduits with identical minimum 12.7 mm (1/2 in.) thick Thermo-Lag 330-1 ERFBS. Conduit "B" had no cable fill, conduit "C" had five 7/C #16 AWG (0.85 Lb/linear ft) cables installed, and "D" had 16 7/C #16 AWG (2.70 Lbs./linear ft) cables installed. Figure G-4 shows the plot of the system weight vs. endpoint data for the surface of the conduit while Figure G-5 shows the plot of the system weight vs. endpoint data for the bare copper conductor located inside the conduit.

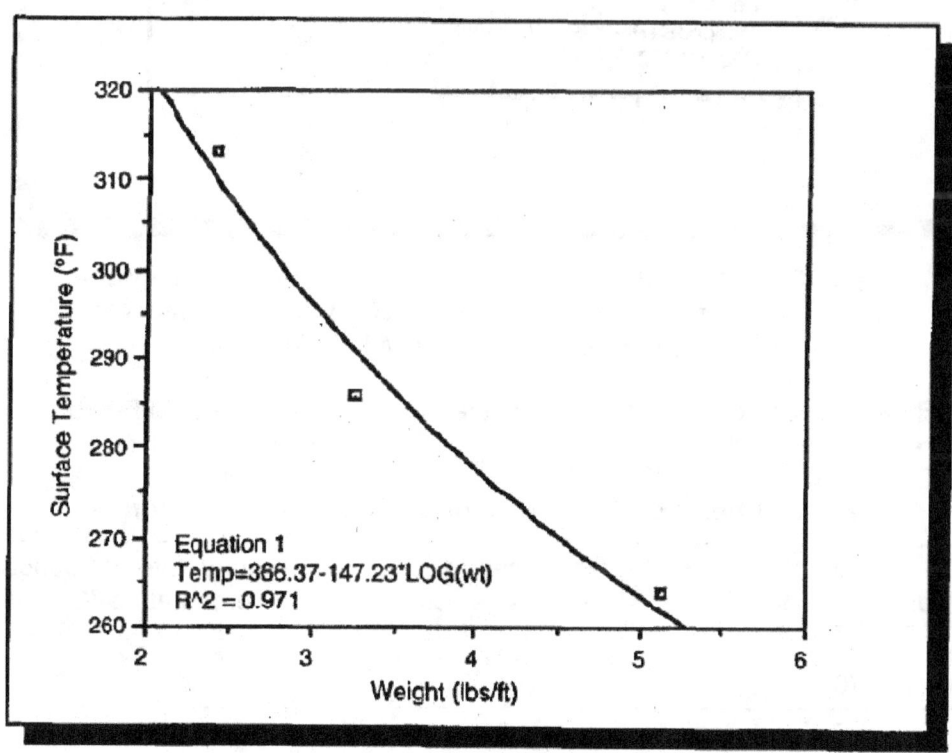

Figure G-4. Conduit System Weight vs. Endpoint Temperatures - Measured on the Conduit (M.H. Salley Thesis UMd 2000)

---

[23] The ambient starting temperature of 83 °F must be used for the equation to be valid. The figure has been simplified (i.e., the ambient temperature subtracted) to graphically show the allowable temperature rise (i.e., $\Delta T = 250$ °F).

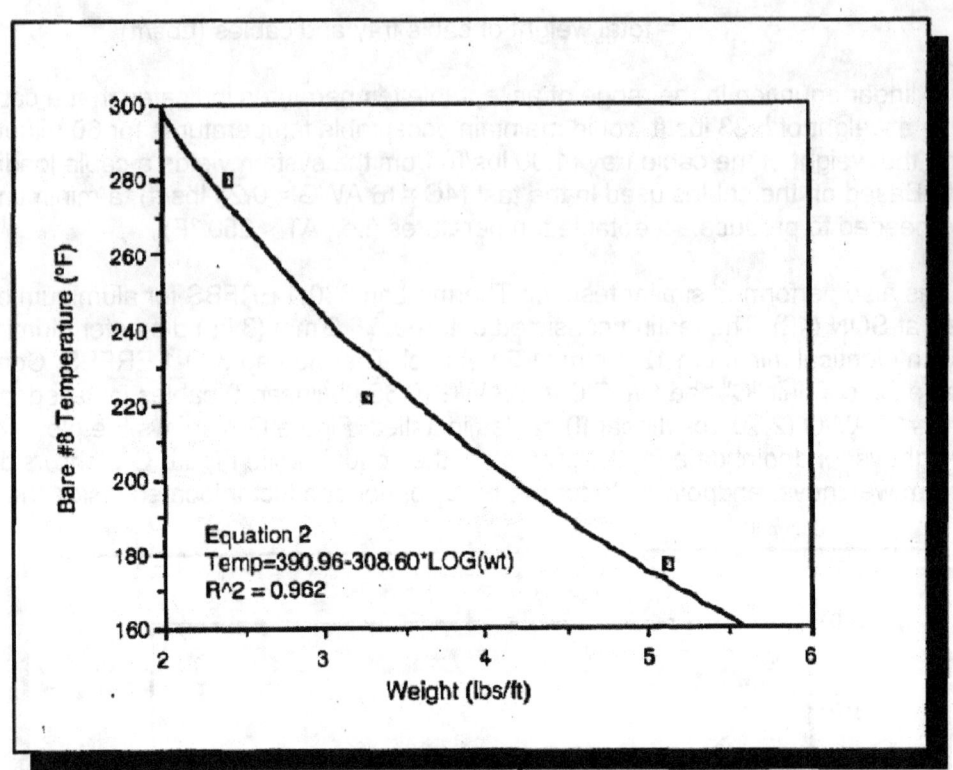

Figure G-5. Conduit System Weight vs. Endpoint Temperatures Measured on the Bare Copper Conductor Inside the Raceway (M.H. Salley Thesis UMd 2000)

Table G-1 shows the temperatures recorded at the end of the 1-hour ASTM E 119 fire exposure.

Table G-1. Effects of Cable Mass on ERFBS Thermal Performance

| Conduit (Number of Cables) | Average External conduit Temperature °C (°F) | Average Internal Conduit Temperature °C (°F) |
| --- | --- | --- |
| B (0) | 138 (280) | 156 (313) |
| C (5) | 105 (221) | 141 (286) |
| D (16) | 81 (177) | 129 (264) |

Reviewing the temperature profiles, the plots demonstrated that the rate of temperature rise was inversely proportional to the cable mass. Performing a linear regression, method of least squares, OPL developed the following relationships to predict the end point temperature based on cable mass:

External Conduit Temperature (°F) = 366.37 − 147.23*Log(w+2.41)     (Equation G-7)
Internal Conduit Temperature (°F) = 390.96 − 308.60*Log(w+2.41)     (Equation G-8)

where:
Temperature = final temperature in °F at 60 minutes of exposure with an assumed starting temperature of 65°°F
w = weight in pounds of cable per linear foot (Lbs/ft)

Another noteworthy observation is the average external surface temperature of the aluminum conduit compared to the average internal temperature as shown in Table G-2. The cable fill not only reduces the external temperature rise of the raceway as shown by the values in Table G-2, but an even greater temperature reduction occurs inside the raceway.

Table G-2. ERFBS Protected Conduit - External Raceway Surface vs. Internal Area Temperature Differential as a Function of Cable Mass (M.H. Salley Thesis UMd 2000)

| Conduit (Number of Cables) | External vs. Internal Temperature Difference as a Function of Cable Fill °C (°F) |
|---|---|
| B (0) | 18 (64) |
| C (5) | 36 (97) |
| D (16) | 48 (118) |

In summary, the TVA research demonstrated that properly designed and installed Thermo-Lag 330-1 provides an effective ERFBS. The TVA research also provides insight to the effects of thermal mass inside a protected electrical raceway, and the temperature gradients across the assembly.

*G.2.3 Acceptance Criteria – Hose Stream Test*

The purpose of the hose stream test is to evaluate the cooling, impact and erosion aspects of the ERFBS. NRC identified that the hose stream application specified in NFPA 251 an acceptable method of application for safe-shutdown-related fire barrier systems that have been exposed to the fire endurance exposure. NFPA 251 required that as a condition of acceptance a hose stream test be conducted on a duplicate specimen subjected to a fire exposure test of one-half the duration of the resistance period but no more than 1 hour, followed by a hose stream impact test. The standard provided specifics on hose stream equipment and water pressures varied by resistance.

As an alternative to the procedure specified in NFPA 251, NRC also found it acceptable to perform the hose stream immediately after the completion of the full fire endurance test period on a single specimen, provided that one of the following applications methods are used.

- The stream applied at random to all exposed surfaces of the test specimen through a 6.4 cm (2.5 inch) national standard playpipe with a 2.9 cm (1.1 inch) orifice at a pressure of 207 kiloPascals (30 lbs per in$^2$) at a distance of 6.1 m (20 feet) from the specimen. (Duration of the hose stream application - 1 minute for a 1-hour barrier and 2 minutes for a 3-hour barrier); or

- The stream applied at random to all exposed surfaces of the test specimen through a 3.8-cm (1.5-inch) fog nozzle set at a discharge angle of 30 degrees with a nozzle pressure of 517 kiloPascals (75 lbs per in$^2$) and a minimum discharge of 284 liters per minute (75 gallons per minute) with the tip of the nozzle at a maximum of 1.5 m (5 ft) from the test specimen. (Duration of the hose stream application 5 minutes for both 1-hour and 3-hour barriers); or

- The stream applied at random to all exposed surfaces of the test specimen through 3.8-cm (1.5-inch) fog nozzle set at a discharge angle of 15 degrees with a nozzle pressure of 517 kiloPascals (75 lbs per in$^2$) and a minimum discharge of 284 liters per minute (75 gallons per minute) with the tip of the nozzle at a maximum of 3 m (10 ft) from the test specimen. (Duration of the hose stream application - 5 minutes for both 1-hour and 3-hour barriers.)

To perform this during actual testing, the test specimen (assembly deck) is typically removed from the test furnace, raised 1.2 to 2.4 m (4.0 to 8.0 ft) off of the ground and slowly turned (nominally 6 revolutions to 8 revolutions per minute) while being exposed to the impact erosion and cooling effects of a hose stream directed perpendicularly at the exposed surface of the test specimen. Figure G-6 shows the performance of a hose stream test following the fire endurance portion of testing.

**Figure G-6. Post-Fire Exposure – Hose Stream Test (NRC Hemyc Tests 2005)**

*G.2.4 Acceptance Criteria – Cable Functionality*

When the ERFBS doesn't pass the endurance testing criteria specified above, a deviation based on demonstrating that the functionality of thermally degraded cables or component was maintained and that the cables or component would have adequately performed their intended function during and after a postulated fire exposure may be acceptable. Enclosure 2 to GL 86-10, Supplement 1, provides NRC staff guidance on acceptable methods to perform the cable functionality testing and engineering determination (often called an "engineering evaluation"). These methods include evaluation of equipment qualification tests, cable insulation tests, and air oven tests. The remainder of this section will describe these methods. However, it should be noted that Enclosure 2 also stated that circuit integrity monitoring, as specified in the ANI standard, is not acceptable to NRC staff for determination of cable integrity.

The American Nuclear Insurers (ANI) testing standard provided the first method available to perform cable integrity testing (for insurance purposes only). As was mentioned above, this ANI method used low voltages that would not typically be used in nuclear power plants for safety-related equipment important to safe shutdown. NRC and licensees felt that the use of this low-voltage testing would only indicate a dead short or open circuit fault. Realistic circuits' loss-of-cable insulation conditions can exist during the fire tests without a dead short occurring. As a result of this determination, the use of circuit-integrity monitoring during the fire-endurance test is not a valid method for demonstrating that the protected shutdown circuits are capable of performing their required function during and after the test fire exposure. Therefore, circuit-integrity monitoring is not required to satisfy NRC acceptance criteria for fire barrier qualification.

The purpose of the functionality test is to justify observed deviations in fire barrier performance. For those fire barrier test specimens that are tested without cables, an engineering analysis justifying internal fire barrier temperature conditions greater than allowed can be based on a comparison of the fire barrier internal temperature profile measured during the fire-endurance test to existing cable-specific performance data such as environmental qualification (EQ) tests.

Comparison of the fire barrier internal time-temperature profile measured during the fire-endurance test to existing cable performance data, such as data from the EQ tests, could be submitted to the NRC staff as a method for demonstrating cable functionality. The EQ testing was performed to rigorous conditions, including rated voltage and current. Correlating the EQ test time-temperature curve to that of the ERFBS fire endurance test would provide a viable mechanism to ensure cable functionality. In addition, the large volume of EQ test data presents a cost-effective approach to addressing cable functionality for fire testes for those cases where the 163°C (325°F) limit is exceeded.

Cable insulation testing can be used as an acceptable method for demonstrating cable functionality. Supplement 1 to GL 86-10, provides one method acceptable to NRC for demonstrating cable functionality through the use of cable insulation testing. Table G-3 summarizes the acceptable cable insulation testing approach. This approach uses a megom to test conductors for any insulation damage (often referred to as megger testing). The megom applies a direct current (dc) voltage between the two conductive planes being tested and outputs the resistance of the insulation between the two conductive planes. The Megger test provides a nondestructive method to evaluate cable performance.

**Table G-3: Summary of Acceptable Cable Insulation Testing Approach**

| Type | Operating Voltage | Megger Test Voltage | High Potential Test Voltage |
|---|---|---|---|
| Power | ≥ 1000 V ac | 2500 V dc | 60% x 80 V/mil (ac)<br>60% x 240 V/mil (dc) |
| Power | < 1000 V ac | 1500 V dc* | None |
| Instrument and Control | ≤ 250 V dc<br>≤ 120 V ac | 500 V dc | |

\* A Megger test voltage of 1000 V dc is acceptable provided a Hi-Pot test is performed after the Megger test for power cables rated at less than 1000 V ac.

To provide reasonable assurance that the cables would have functioned during and after the fire exposure, Megger tests need to be performed before the fire tests, at multiple time intervals during the test (for instrument cable only), and immediately after the fire test before the hose stream test. Megger testing should be done immediately after the fire test such that the cable insulation does not reset after the cables are removed from elevated temperatures. As a result of cable insulation tendency to reset, Megger tests of insulated cables after the fire endurance test and after the cables have cooled may not detect degradation in the insulation resistance. In addition, Megger testing should be done conductor-to-conductor for multiconductor cables and conductor-to-ground for all cables. The minimum acceptable insulation resistance (IR) value, using the test voltage shown above in Table G-3, is determined by using the following expression:

$$\text{IR (Mega-ohms)} \geq \frac{(K+1 \text{ Mega-ohm}) * 1000 \text{ (ft)}}{\text{Length (ft)}} \quad \text{(Equation G-9)}$$

When the ERFBS test specimen is tested without cables and doesn't pass the endurance testing on internal temperature rise, then an Air Oven test may be used to evaluate the functionality of cables. This testing method consists of exposing insulated wires and cables at rated voltage to elevated temperatures. The Air Oven temperature profile will be the temperature measured by the # 8 AWG bare copper conductor during the fire exposure of those ERFBS test specimen that were tested without cables. NRC staff determined that the test method described by UL Subject 1724, "Outline of Investigations of Fire Tests for Electrical Circuit Protective Systems," Issue Number 2, August 1991, Appendix B, "Qualification Test for Circuit Integrity of Insulated Electrical Wires and Cables in Electrical Circuit Protection Systems," was acceptable with the following modifications:

- During the air oven test, the cables should be energized at rated voltage. The cables should to be monitored for conductor-to-conductor faults in multi-conductor cables and conductor-to-ground in all cables.

- The cables being evaluated should be subjected to the Megger and high-potential tests.

- The impact force test, which simulated the force of impact imposed on the raceway by the solid stream test, described in UL 1724, Appendix B, Paragraph B3.16, is not required to be performed.

The last acceptable method for determining cable functionality, as presented in Supplement 1, involves the comparison of cable operating temperature within the ERFBS at the time of failure along with the cable thermal exposure threshold (TET). The difference between the cable TET and internal ERFBS cable operation temperature presents the maximum temperature rise allowed within the ERFBS. The cable TET limits in conjunction with a post test visual cable inspection and the Hi-Pot test should demonstrate the functionality of the cable circuit during and after a fire.

### G.2.5 *Acceptance Criteria – Combustibility of the ERFBS*

NRC's fire protection guidelines and requirements establish the need for each nuclear power plant to perform a plant-specific fire hazard analysis. The fire hazard analysis should consider the potential for in situ and transient fire hazards and combustibles. With respect to building materials (e.g., cable insulation and jackets, plastics, thermal insulation, fire barrier materials), the combustibility, ease of ignition, and flame spread over the surface of a material should be considered by the fire hazards analysis. This is especially important when licensee's have installed ERFBS inside containment as "Radiant Energy Shields" that are required to be noncombustible per Appendix R to 10 *CFR* Part 50, which states:

III.G.2

> Inside noninerted containments one of the fire protection means specified above or one of the following fire protection means shall be provided:
>
> d. Separation of cables and equipment and associated non-safety circuits of redundant trains by a horizontal distance of more than 20 feet with no intervening combustibles or fire hazards;
>
> e. Installation of fire detectors and an automatic fire suppression system in the fire area; or
>
> f. Separation of cables and equipment and associated non-safety circuits of redundant trains by a noncombustible radiant energy shield.

A radiant energy shield is a shield designed to provide protection from redundant essential raceways or fire safe shutdown equipment against the radiant energy from an exposure fire. Radiant energy shields are typically installed within containment. Numerous ERFBS have been used to construct these shields and, in some cases, licensees have had to replace or modify the shields because they were later determined to be combustible.

In Branch Technical Position (BTP) CMEB 9.5-1, "Guidelines for Fire Protection for NPPs," dated July 1981, noncombustible materials are defined as:

a. A material which in the form in which it is used and under the conditions anticipated, will not ignite, burn, support combustion, or release flammable vapors when subjected to fire or heat.

b. Material having a structural base of noncombustible material, as defined in a., above, with a surface not over 0.318 cm (0.125 inch) thick that has a flame spread rating not higher than 50 when measured using ASTM E-84.

Per the guidance in BTP CMEB 9.5-1, an acceptable method to test a materials combustibility/flame spread characteristic is to subject it to the ASTM E-84, "Standard Test Method for Surface Burning Characteristics of Building Materials." If the testing results in a flame spread not higher than 50, then the material is considered noncombustible. Ease of ignition can be determined by the flashover ignition temperature derived by ASTM-D1929, "Standard Test Method for Determining Ignition Temperature of Plastics."

In addition, Supplement 1 to Generic Letter 86-10 accepted ASTM E-136, "Standard Test Method for Behavior of Materials in a Vertical Tube Furnace at 750°C (1382°F)," as an acceptable method for determining the combustibility of a fire barrier material. The criteria for passing the ASTM E-136 test for classifying a material as noncombustible are that three of the four test specimens must meet the following conditions: (1) the increase in the recorded temperatures of internal and external thermocouples may not exceed 30°C (86°F), (2) no flaming occurs from the test specimen after the first 30 seconds, and (3) if the weight loss of the specimen exceeds 50 percent, then the increase in the recorded temperatures of the internal and external thermocouples may not exceed the furnace temperature at the beginning of the test and the specimen may not flame.

*G.2.6 Acceptance Criteria – Ampacity Derating*

Title 10 of the Code of Federal Regulations, Part 50, Appendix A, General Design Criterion (GDC) 17, "Electric power systems," if applicable, requires that onsite and offsite electric power systems be provided to permit the functioning of structures, systems, and components important to safety. The safety function of either electrical power system (assuming the other system is not functioning) is to provide sufficient capacity and capability to ensure that vital functions are maintained. Cables routed in electrical raceways are derated to ensure that systems have sufficient capacity and capability to perform their intended safety functions. Other factors that affect ampacity derating include cable fill, cable loading, cable type, raceway construction, and ambient temperature.

NRC requires that cable derating due to the use of fire retardant coatings be considered by utilities during plant design or when design changes are made to existing electrical system configurations. The utility is responsible for evaluating the ampacity derating effect of ERFBS and applying those factors when designing the current carrying capacity of individual cables.

Cable derating calculations that are based on inaccurate or non-conservative derating factors could result in the installation of undersized cables or overfilling of raceways. Either of these conditions could cause operating temperatures to exceed design limits within the raceways, thereby reducing the expected design life of the cables. The National Electrical Code, Insulated Cable Engineers Association publications, and other industry standards provide general ampacity derating factors for open-air installations but do not include derating factors for fire barrier systems. The Insulated Conductors Committee of the Institute of Electrical and Electronics Engineers (IEEE) Power Engineering Society, Task Force 12-45, has developed IEEE Standard Procedure 848, "Procedure for the Determination of the Ampacity Derating of Fire Protected Cables," for use as an industry standard.

Appendix A provides a detailed description of ampacity derating.

### G.2.7 *Acceptance Criteria – Seismic Qualification*

The regulations that address the need for fire protection at nuclear power plants (i.e., 10 CFR Section 50.48; Part 50, Appendix A, General Design Criterion 3; and Part 50 Appendix R) do not explicitly require fire barriers to be seismically qualified (i.e., to maintain their functionality after postulated seismic events). However, provision C.2 of Regulatory Guide (RG) 1.29, "Seismic Design Classification," addresses the issue of seismic Category II versus seismic Category I. In that context, fire barriers are considered a seismic Category II component. Based on these provisions, the fire barriers are allowed to undergo damage during the postulated seismic events. However, the fire barriers may not lose its position and potentially fall on Class 1E equipment. This requirement for position retention is a requirement where Category II commodities are located above safety related equipment.

### G.2.8 *Test Assembly*

ERFBS fire endurance testing often involves an elaborate test assembly consisting of several raceway configurations (e.g., conduit, cable tray, junction box, raceway supports, raceway bends, etc.). A typical test deck has a steel framework that provides the structural support for raceways and junction boxes. Either a sheet of steel with an insulation fiber blanket or a concrete slab is used to provide a continuous enclosure surface to the test deck. The test deck has predetermined holes that provide penetration points for the raceways. At the raceway penetration point the edges of the raceway are completely filled with a suitable fire stop material-usually cement grout, ceramic fiber packed tightly or silicon foam. The penetration seal is considered a part of the support system and is not in itself being evaluated by the test. Figure G-7 shows a 90° angled test assembly deck with protected raceways installed as test specimens.

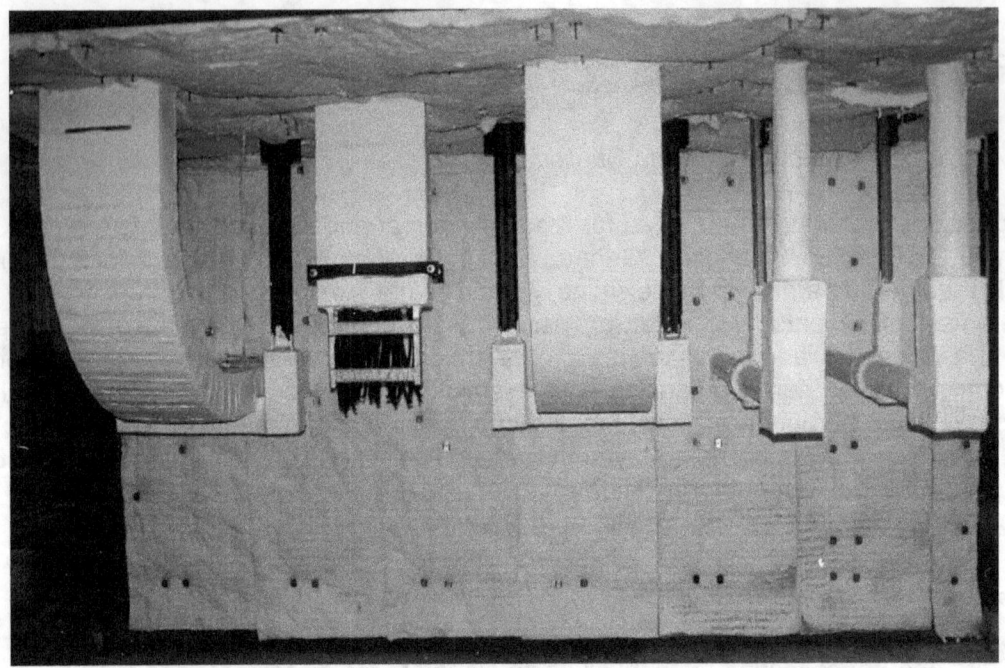

**Figure G-7: Thermo-Lag 90 Degree Test Assembly (TVA Thermo-Lag Test Report)**

Some utilities testing used the 90*-degree test assembly shown in Figure G-7, while others used a horizontal test assembly. The horizontal test configurations resulted in the raceways penetrating the ceiling of the test deck vertically, extending a few feet into the oven space before making a 90-degree bend, then traveling several m horizontally followed by another 90-degree bend back up through the test deck. Figure G-8 and G-9 depict a typical horizontal test deck and associated test oven, respectively. In either case, the test decks provided the mobility to allow the installers easy access to the raceways and immediately following the fire endurance portion of the test, the test assembly could be hoisted to a location suitable for hose stream application. Though the horizontal test configuration was commonly used, several other test assembly configurations were used.

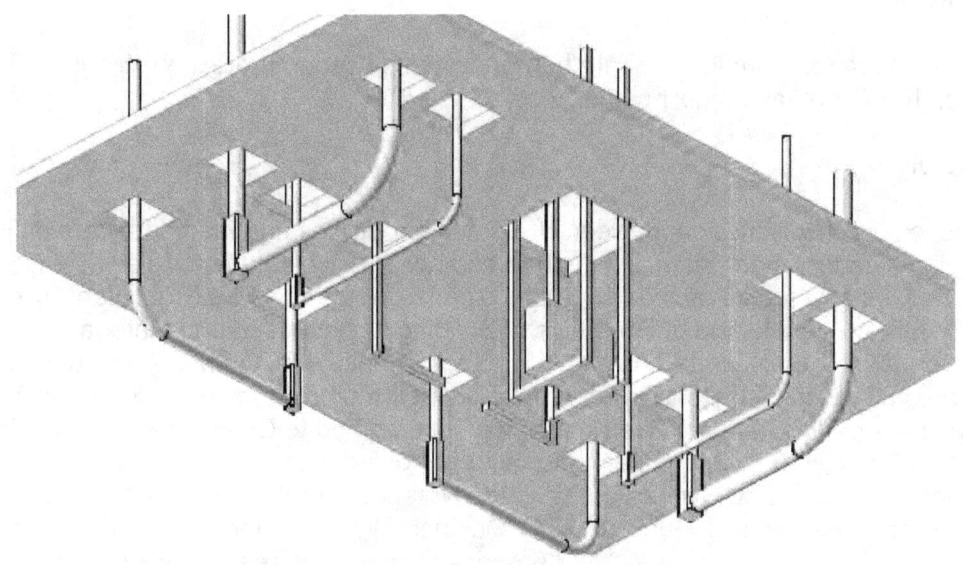

Figure G-8. Isometric View of Typical Base Horizontal Test Assembly Prior to ERFBS Installation (NRC Hemyc Report by OPL)

Figure G-9. 12' x 18' Horizontal Furnace (Top View) (NRC Hemyc Tests)

## G.3 TVA Position

TVA position of fire testing criteria for fire barrier systems used to protect electrical cabling required for 10 CFR 50 appendix r compliance

### BACKGROUND

There is considerable discussion between NRC, nuclear utilities and manufacturers of fire barrier systems on the appropriate test method and acceptance criteria for electrical fire barrier systems. NRC has based its methodology and criteria on National Fire Protection Association (NFPA) 251, "Standard Method of Fire Tests of Building Construction and Materials," Chapter 7, "Tests of Nonbearing Walls and Partitions."[1] Thermal Science, Inc. (TSI), the manufacturer of Thermo Lag, and most nuclear utilities, have based their methodology and criteria on American Nuclear Insurers (ANI) "Standard Fire Endurance Test Method to Qualify a Protective Envelope for Class 1E Electrical Circuits."[2] Other manufacturers of fire barrier systems, such as 3M and Thermal Ceramics, Inc., have typically used Underwriters Laboratory (UL) test methods and acceptance criteria such as UL Subject 1724, "Outline of Investigation for Fire Tests for Electrical Circuit Protective Systems."[3] The American Society for Testing and Materials (ASTM) has recognized the need to develop a unique test method and acceptance criteria for electrical fire barrier systems. They have been working for approximately the last five years on this issue but have not issued a standard.

### DISCUSSION

The Code of Federal Regulations (CFR), Title 10 Part 50 Domestic Licensing of Production and Utilization Facilities, Appendix R, Fire Protection Program for Nuclear Power Facilities Operating Prior to January 1, 1979, paragraph III.G.2 provides the requirements for fire protection and safe shutdown capability. If redundant trains are located in the same fire area and a licensee does not provide alternative or dedicated shutdown systems for the redundant equipment in that fire area, the three acceptable methods of ensuring that one of the trains is free from fire damage are:

a.) Separation of cables and equipment and associated non-safety circuits of redundant trains by a fire barrier having 3-hour rating. Structural steel forming a part of or supporting such fire barriers shall be protected to provide fire resistance equivalent to that required of the barrier;

b.) Separation of cables and equipment and associated non-safety circuits of redundant trains by a horizontal distance of more than 20 feet with no intervening combustible or fire hazards. In addition, fire detectors and an automatic fire suppression system shall be installed in the fire area; or

c.) Enclosure of cable and equipment and associated non-safety circuits of one redundant train in a fire barrier having 1-hour rating. In addition, fire detectors and an automatic fire suppression system shall be installed in the fire area.[4]

A fire wall design that has passed on appropriate test method (e.g., NFPA 251) is considered a "rated" barrier. Components which penetrate fire walls, such as mechanical and electrical penetrations, fire doors, and HVAC fire dampers, are presently not generally accepted test method and acceptance criteria specifically applicable to fire barrier enclosures applied to electrical cable systems. Existing methods intended for other purposes have been utilized to test such barrier systems, but none of these standards are fully appropriate to this unique application of fire barrier materials. In an attempt to define a test method for electrical circuit protection, American Nuclear Insurers (ANI) prepared "Guidelines for Fire Stop and Wrap Systems at Nuclear Facilities". However, this test method was intended to be used "for insurance purposes only".[2] The method and acceptance criteria in the ANI document are not definitive.

POSITION

The fire test methodology and acceptance criteria for electrical cable systems should be unique to these systems. Underwriters Laboratory currently has an appropriate test method (Subject 1724), which addresses the uniqueness of electrical cable fire barrier systems. This test method was developed by UL specifically to address issues such as Appendix R electrical fire barrier rating requirements. The scope of the test method is:

a.) Measurement of temperature changes within the electrical circuit protective system caused by the heat transfer through the electrical circuit protective system to the electrical conductor or raceway, or both, during the external fire exposure test.

b.) Determination of the integrity of the electrical circuit protective system during the external fire exposure and water hose stream test.

c.) Determination of the ability of insulated electrical conductors to maintain electrical circuit integrity at the temperature conditions present within the electrical circuit protective system during the external fire exposure test and during the water hose stream test.[3]

Details such as thermocouple types and placements are discussed in this test method. The test follows the standard time-temperature curve specified in ASTM E-119, as used in other fire endurance tests (e.g., NFPA 251). The test allows the use of the actual installed cables or a No. 8 AWG (3.38 mm$^2$) bare copper conductor to simulate the electrical circuits. With the bare conductor method the thermocouple measurements can be correlated to actual cable qualification tests as described in Appendix B of UL Subject 1724.

TVA considers that UL Subject 1724 is the most appropriate test method currently available for determining the fire resistance rating of electrical fire barrier systems. TVA will use UL Subject 1724 with the following clarifications to perform tests of Thermo-Lag 330 electrical circuit protective systems intended for use at Watts Bar:

a.) The exterior surface temperature of the electrical raceway will be recorded (cold side of the barrier). If the average temperature recorded by the exterior thermocouples is less than 250°F (121°C) above their initial temperature and no individual thermocouple is in excess of 325°F (163°C) above its initial temperature, the fire barrier will be considered acceptable for use with any type cable.[5]

b.) Section 6, Internal Fire Exposure Test, will not be used. TVA considers that this portion of the testing is not necessary, since an internally generated cable tray fire would be extremely unlikely. Circuits are protected with a fuse or breaker that will actuate prior to the jacket of a faulted cable reaching its auto-ignition temperature (for existing designs) or reaching its insulation damage temperature (for new designs) for all credible low impedance and bolted faults.[6] No other ignition sources exist within the protective barrier.

c.)       Section 5, Hose Stream Test. TVA will follow the criteria for hose stream testing described in NUREG-0800 using one and on-half inch fog nozzle set at a discharge angle of 15° with a nozzle pressure of 75 psi and a minimum discharge of 75 gpm.[7] TVA considers that this would accurately represent the mechanical impact, erosion and cooling effects that would exist in TVA's nuclear power plant environment. The hose stream test shall be performed within ten minutes of the completion of the fire test. The duration and application will follow the requirements of UL 1724 Table 5.1. The nozzle will be located a maximum of ten feet measured horizontally from the outside edge of the testing assembly. Acceptance shall be based on the fire barrier system remaining intact with minimal material flaking. (The alternative test called for by the UL document, involving a one and 1-eighth inch solid bore National Standard Playpipe operating at 30 psi, is not a realistic simulation of the challenge to barrier systems as installed in a nuclear power plant).

Appendix G References

M.H. Salley, "An Examination of the Methods and Data Used to Determine Functionality of Electrical Cables When Exposed to Elevated Temperatures as a Results of a Fire in a Nuclear Power Plant," University of Maryland Masters of Science Thesis, 2000, U.S. NRC (ADAMS Accession No. ML051450082).

OPL Report for NRC sponsored Hemyc Tests, 2005, ML051190026

TVA Thermo-Lag Test Report, 1994, (NUDOCS Accession No. ML9501120204)

Supplement 1 to GL 86-10, "Fire Endurance Test Acceptance Criteria for Fire Barrier Systems Used to Separate Redundant Safety Shutdown Trains within the Same Fire Area"

RG 1.189, "Fire Protection for NPPs,"

UL Subject 1724, "Outline of investigation for fire tests for electrical circuit protective systems"

IEEE Standard 848-1996, "Standard Procedure for the Determination of the Ampacity Derating of Fire Protected Cables," The Institute of Electrical and Electronics Engineers, Inc., 345 East 47th Street, New York, NY 10017-2394, USA.

| NRC FORM 335 (9-2004) NRCMD 3.7 | U.S. NUCLEAR REGULATORY COMMISSION<br>BIBLIOGRAPHIC DATA SHEET<br>(See instructions on the reverse) | 1. REPORT NUMBER<br>(Assigned by NRC, Add Vol., Supp., Rev., and Addendum Numbers, if any.)<br>**NUREG-1924** |
|---|---|---|
| 2. TITLE AND SUBTITLE<br>Electric Raceway Fire Barrier Systems in U.S. Nuclear Power Plants | | 3. DATE REPORT PUBLISHED<br>MONTH: May    YEAR: 2010 |
| | | 4. F N OR GRANT NUMBER |
| 5. AUTHOR(S)<br>Gabriel J. Taylor and Mark Henry Salley | | 6. TYPE OF REPORT<br>Technical |
| | | 7. PERIOD COVERED (Inclusive Dates)<br>10/1/2008 – 6/1/2009 |

8. PERFORMING ORGANIZATION – NAME AND ADDRESS (If NRC, provide Division, Office or Region, U.S. Nuclear Regulatory Commission, and mailing address: if contractor, provide name and mailing address.)

Division of Risk Analysis
Office of Nuclear Regulatory Research
U.S. Nuclear Regulatory Commission
Washington, DC 20555-0001

9. SPONSORING ORGANIZATION – NAME AND ADDRESS (If NRC, type "Same as above", if contractor, provide NRC Division, Office or Region, U.S. Nuclear Regulatory Commission, and mailing address.)

Division of Risk Analysis
Office of Nuclear Reactor Regulation
U.S. Nuclear Regulatory Commission
Washington, DC 20555-0001

10. SUPPLEMENTARY NOTES

11. ABSTRACT (200 words or less)

On March 22, 1975, the Tennessee Valley Authority (TVA) Browns Ferry Nuclear Plant (BFN) experienced a serious fire in their cable spreading room (CSR) and Unit 1 reactor building. The fire lasted over seven hours and damaged over 1,600 electrical cables, rendering all Unit 1 Emergency Core Cooling Systems inoperable. This near miss accident illustrated the vulnerability of essential electric cables to fire damage. In response to this fire, the United States Nuclear Regulatory Commission (NRC) issued Appendix R to Title 10 of the Code of Federal Regulations Part 50 (10 CFR 50) as a backfit to operating reactors and similar requirements implemented on reactors under construction. In order to comply with this new requirement, most operating United States (US) NPPs (NPPs) installed 1- or 3-hour Electrical Raceway Fire Barrier Systems (ERFBS) to protect electrical cables essential to Post-Fire Safe-Shutdown. At the time of initial installation, there were no definitive test standards for the ERFBS qualification. This resulted in most NPPs using the ERFBS to re-qualify/upgrade/change out their barriers. This report traces this history and how US NPPs returned their ERFBS to compliance.

| 10.5.1    12. KEY WORDS/DESCRIPTORS (List words or phrases that will assist researchers in locating the report.)<br>Electric Raceway, Fire Barrier, ERFBS, Thermo-Lag, Hemyc, MT, Mecatiss, Darmatt, Versawrap, 3M Interam, Kaowool, Cerablanket, FP-60, Pabco, Promat, cables, fire, cable failure, fire risk, fire PRA, post-fire safe shutdown analysis. | 13. AVA LABILITY STATEMENT<br>unlimited |
|---|---|
| | 14. SECURITY CLASSIFICATION<br>(this page)<br>unclassified<br>this report)<br>unclassified |
| | 15. NUMBER OF PAGES |
| | 16. PRICE |

NRC FORM 335 (9-2004)

www.ingramcontent.com/pod-product-compliance
Lightning Source LLC
Chambersburg PA
CBHW081105170526
45165CB00008B/2329